KJ Lee
Nov. 13, 2000

Topics in Applied Physics
Volume 78

Available Online

Topics in Applied Physics is part of the Springer LINK service. For all customers with standing orders for Topics in Applied Physics we offer the full text in electronic form via LINK free of charge. Please contact your librarian who can receive a password for free access to the full articles by registration at:

http://link.springer.de/orders/index.htm

If you do not have a standing order you can nevertheless browse through the table of contents of the volumes and the abstracts of each article at:

http://link.springer.de/series/tap/

There you will also find more information about the series.

Springer

Berlin
Heidelberg
New York
Barcelona
Hong Kong
London
Milan
Paris
Singapore
Tokyo

Physics and Astronomy

ONLINE LIBRARY

http://www.springer.de/phys/

Topics in Applied Physics

Topics in Applied Physics is a well-established series of review books, each of which presents a comprehensive survey of a selected topic within the broad area of applied physics. Edited and written by leading research scientists in the field concerned, each volume contains review contributions covering the various aspects of the topic. Together these provide an overview of the state of the art in the respective field, extending from an introduction to the subject right up to the frontiers of contemporary research.

Topics in Applied Physics is addressed to all scientists at universities and in industry who wish to obtain an overview and to keep abreast of advances in applied physics. The series also provides easy but comprehensive access to the fields for newcomers starting research.

Contributions are specially commissioned. The Managing Editors are open to any suggestions for topics coming from the community of applied physicists no matter what the field and encourage prospective editors to approach them with ideas.

See also: http://www.springer.de/phys/books/TAP

Managing Editors

Dr. Claus E. Ascheron

Springer-Verlag Heidelberg
Topics in Applied Physics
Tiergartenstr. 17
69121 Heidelberg
Germany
Email: ascheron@springer.de

Dr. Hans J. Kölsch

Springer-Verlag Heidelberg
Topics in Applied Physics
Tiergartenstr. 17
69121 Heidelberg
Germany
Email: koelsch@springer.de

Assistant Editor

Dr. Werner Skolaut

Springer-Verlag Heidelberg
Topics in Applied Physics
Tiergartenstr. 17
69121 Heidelberg
Germany
Email: skolaut@springer.de

Roland Diehl (Ed.)

High-Power Diode Lasers
Fundamentals, Technology, Applications

With Contributions by Numerous Experts

With 260 Figures and 20 Tables

 Springer

Dr. Roland Diehl (Ed.)
Fraunhofer-Institut für Angewandte Festkörperphysik
Tullastrasse 72
79108 Freiburg
Germany
Email: diehl@iaf.fhg.de

Library of Congress Cataloging-in-Publication Data

High-power diode lasers : fundamentals, technology, applications, with contributions by
numerous experts / Roland Diehl, (Ed.).
    p. cm. -- (Topics in applied physics ; v. 78)
    Includes bibliographical references and index.
    ISBN 3540666931 (alk. paper)
    1. Semiconductor lasers. 2. Diodes, Semiconductor. I. Diehl, R. D. (Renee D.) II.
Series.

TA1700 .H52 2000
621.36'6--dc21

00-028510

Physics and Astronomy Classification Scheme (PACS): 42.55.-f; 42.60.-v; 42.62.Cf

ISSN print edition: 0303-4216
ISSN electronic edition: 1437-0859
ISBN 3-540-66693-1 Springer-Verlag Berlin Heidelberg New York

Springer-Verlag Berlin Heidelberg New York
a member of BertelsmannSpringer Science+Business Media GmbH

© Springer-Verlag Berlin Heidelberg 2000
Printed in Germany

Data conversion by DA-TEX Blumenstein - Seidel GbR, Leipzig
Cover design: design & production GmbH, Heidelberg

Printed on acid-free paper   SPIN: 10702280   57/3144/mf   5 4 3 2 1 0

Preface

Indisputably, the laser is a key technology in highly industrialized economies. In many industries, application of the laser in materials processing has modified fabrication processes with speed, quality, reliability, and flexibility of manufacturing having been substantially increased. Nevertheless, with respect to its technical potential and market diffusion, laser processing is still far away from saturation. Among other things, this is due to process technologies not sufficiently adapted to laser employment and to the fact that today's laser systems are generally heavy and voluminous with relatively high costs for operation and maintenance.

This situation is going to change profoundly as the diode laser is on the move to revolutionize laser technology, as the transistor did with electrical engineering. Diode lasers have long been used as light emitters in fiber-optic telecommunications, as barcode readers, and for implementing the write–read functions of optical disks. Nowadays, diode lasers do not merely deliver bits but also optical power. They are increasingly found in applications such as materials processing (welding, cutting, drilling, derusting, surface hardening, etc.) as well as in printing and graphical arts, in displays, and medical applications.

In fact, since the advent of the high-power diode laser, laser technology is experiencing a fundamental structural change, as this semiconductor device has become the key element of a new breed of laser systems that are competing with gas lasers and lamp-pumped solid-state lasers. High-power diode lasers are continuously making inroads into industrial applications, as they are compact, easy to cool, yield a power efficiency beyond 50%, which is about five times higher than any other kind of laser has to offer, and their costs are becoming increasingly attractive. To exploit the tremendous application potential of high-power diode lasers, research and development (R & D) programs are performed in many industrial countries.

Such programs are operational in Germany as well. The idea to write this book arose when R & D teams from research institutes and industry combined their forces to cooperate in a network the objectives of which were to get access to high-power diode lasers with increasing power, brightness and lifetime, to reliable mounting technologies for fabricating high-brightness laser modules, and to demonstrate novel all-solid-state laser systems. As the general coordinator of this network the editor accepted the responsibility to motivate a team of authors who are actively researching the field of high-power diode lasers. Their competence, knowledge, experience, viewpoints, ideas and visions have been brought together to bring into existence this update monograph on the principles, technology, and innovative applications of high-power diode lasers, summarizing the state-of-the-art in a greatly dynamic field at the threshold to a new millennium.

Readers who want a fundamental introduction to the underlying physical principles, design considerations, and the basics of device technology will find the first contribution by *Peter Unger* a compact stand-alone monograph on diode lasers for high-power operation. Those who want to probe further will find an extensive theoretical treatment of the microscopic spatio-temporal dynamics of large-active-area high-power diode lasers and their impact on the macroscopic laser characteristics in the contribution by *Edeltraud Gehrig* and *Ortwin Hess*. From a practical point of view, the analysis and prediction of the temporal and spatial dynamics of broad-area lasers are extremely important when designing high-power diode-laser systems.

Ideal material systems to fabricate high-power diode lasers are the III–V compound semiconductors. Through solid solutions among various III–V materials, a large spectrum of laser wavelengths has become accessible. It ranges from the blue to the mid-infrared, being extended to the far-infrared through the unipolar quantum cascade laser. Power performance is so far restricted to wavelengths ranging from 630 nm to 1600 nm. One of the key steps in the fabrication of high-power diode lasers is the epitaxial process. Sophisticated stacks of III–V epilayers comprising up to 100 single layers have to be deposited with high crystalline quality. *Markus Weyers* and coauthors describe in detail the technologies and peculiarities of epitaxy for high-power diode-laser layer structures.

Despite the tremendous progress III–V epitaxy has made in the last decade, nothing would have been achieved if low-dislocation-density substrate wafers were not to hand. As gallium arsenide (GaAs) is by far the most important substrate material for high-power diode lasers, the contribution by *Georg Müller* et al. has been solely and comprehensively devoted to the growth of GaAs-substrate crystals of high perfection, emphasizing the importance of a sound materials base for long-lifetime devices.

Epiwafers of high crystalline perfection are patterned to laser structures by applying semiconductor process technologies rather similar to processes applied in microelectronics. High-power diode lasers usually come in the form of bars (approximately 10 mm × 0.6 mm) comprising about 25 monolithic groups of up to 20 parallel single-laser stripes. *Götz Erbert* and coauthors present detailed design considerations, technologies and device characteristics of single-diode lasers and bars with large optical cavities supported by sections on the simulation of thermal behavior and on reliability considerations.

A major drawback of high-power broad-area diode lasers is their unsatisfactory beam quality. This allows only limited focusing of the total beam as it is the addition of many single beams. Hence, power density at the workpiece is limited as well, leaving high-power diode lasers with restricted application opportunities. But this is changing as they are going to pick up brightness to approach diffraction-limited performance by introducing special designs such as, e.g., the tapered resonator/amplifier. Two contributions address this important subject. *Klaus-Jochen Boller* et al. report on properties of high-brightness diode-laser systems with their potential for frequency conversion, whereas *Michael Mikulla* addresses the design and performance of tapered diode lasers exhibiting both high power and high brightness.

Output power per bar is approaching the 100 W level, which substantially reduces the cost per watt of delivered optical power. Crucial for reliability and lifetime of bars is proper heat sinking. Although power efficiency is extremely high, one half of the absorbed pump power has to be removed as waste heat. Mounting high-power diode-laser bars on cooling elements requires high precision and the complete mastering of the electrical, thermal, and mechanical junction process. *Peter Loosen* describes practical aspects of heat sinking applying active coolers.

By stacking many such mounted bars, optical output powers in the kilowatt range can be achieved. This is the fundamental concept for direct-diode applications. Stacked arrays integrating sophisticated micro-optical beam shaping will push power densities to the 10^5 W$/$cm^2 realm with the prospects to achieve even the MW$/$cm^2 level – sufficient for the welding of sheet steel. For this purpose a large number of individual oscillators have to be coupled to a single powerful beam. Following *Hans Opower's* introduction, *Peter Loosen* deals with currently applied methods of incoherent beam combining whereas *Uwe Brauch* treats coherent beam combining, both to achieve direct-diode-laser systems with power levels of some kilowatts.

Another approach besides coherent beam combining to achieve high brightness is the use of high-power diode lasers to pump solid-state lasers – one of the earliest applications of these semiconductor devices and still their predominant one – the solid-state gain medium acting as a brightness transducer. Diode-Pumped Solid-State Lasers (DPSSLs) are extending their presence at the expense of their lamp-pumped counterparts due to better beam quality, reduced maintenance, and an improving cost-of-ownership situation. Classic DPSSLs will not be addressed in this book but rather novel concepts that became feasible only through diode pumping. Such a novel concept is the fiber laser. This type of laser promises innovative applications, as it is particularly attractive to couple the laser output to an optical fiber so that the optical power becomes available at the free end and can be directed wherever the flexible fiber is placed. This confers a great deal of flexibility on the application of laser power previously unavailable to traditional lasers. Moreover, using double-clad fibers, multimode input is transformed to single-mode diffraction-limited output. Also, a new star in the laser sky is the disk laser – a thin diode-pumped crystal plate (approximately 0.2 mm thick and up to 12 mm in diameter) with axial cooling. It provides an optimum combination of high beam quality, high efficiency, and high optical output power. After their introduction to diode-laser pumping, *Andreas Tünnermann* et al. treat in detail the fiber laser; *Adolf Giesen* and *Karsten Contag* summarize the still short history and current status of the disk laser.

Finally, the editor wants to express his sincerest thanks to all his colleagues who accepted the challenge and the burden to spend much of their free time and effort to author the various chapters, responding so constructively to the editor's criticism, suggestions, and recommendations. Special thanks are due to the German Federal Ministry of Education and Research (BMBF) and to the VDI Technology Center for consistent support of the joint effort which yielded most of the results reported in this book. Furthermore the team of authors owes thanks to Werner Skolaut and Claus Ascheron of Springer-Verlag Heidelberg for their patience and continuous support. Last but not least the editor thanks his wife Christel for forgiving her husband's virtual absence during countless weekends, and for gallons of good coffee.

Freiburg, June 2000   *Roland Diehl*

Contents

GaAs Substrates for High-Power Diode Lasers
Georg Müller, Patrick Berwian, Eberhard Buhrig
and Berndt Weinert

High-Power Broad-Area Diode Lasers and Laser Bars
Götz Erbert, Arthur Bärwolff, Jürgen Sebastian and Jens Tomm

Introduction to Power Diode Lasers

Peter Unger

Department of Optoelectronics, University of Ulm,
89069 Ulm, Germany
peter.unger@e-technik.uni-ulm.de

Abstract. An introduction to the physics, design, and fabrication of semiconductor-diode lasers is presented with emphasis on high-power operation. Beginning with a general section about fundamental aspects and elementary physics of these optoelectronic devices, topics like optical gain, quantum-well structures, optical resonators, mirror coatings, optical waveguides, mode patterns, beam profiles, laser rate equations, device properties, high-power design, epitaxy, and process technology are discussed in more detail.

1 Fundamental Aspects of Diode Lasers

This section provides a basic understanding of the physical phenomena in diode lasers. A more comprehensive treatment can be found in books on solid-state physics [1], semiconductor physics [2,3], optoelectronics [4,5,6,7], and semiconductor-diode lasers [8,9,10,11,12].

1.1 Emission and Absorption in Semiconductors

Gas and solid-state lasers have electronic energy levels which are nearly as sharp as the energy levels of isolated atoms. In semiconductors, these energy levels are broadened into energy bands due to the overlap of the atomic orbitals. In an undoped semiconductor with no external excitation at a temperature of $T = 0\,\mathrm{K}$, the uppermost energy band, called the conduction band, is completely empty and the energy band below the conduction band, called the valence band, is completely filled with electrons. Conduction and valence bands are separated by an energy gap, which has a value of $E_{\mathrm{g}} = 0.5\text{--}2.5\,\mathrm{eV}$ for semiconductor materials which power diode lasers are made of.

Two types of carriers contribute to electronic conduction, these are electrons in the conduction band and holes (missing electrons) in the valence band. A free electron has a kinetic energy of $E = p^2/(2m_0)$ where $m_0 = 9.109\,534 \times 10^{-31}\,\mathrm{kg}$ is the free-electron rest mass and p the mechanical momentum. When treated as a quantum-mechanical particle, the momentum $p = \hbar k$ is proportional to the wavenumber $k = 2\pi/\lambda$ with the reduced Planck constant $\hbar = h/(2\pi) = 6.582\,173 \times 10^{-16}\,\mathrm{eVs}$ and the wavelength λ. Thus, for a free electron, the dependency of energy versus wavenumber is $E(k) = (\hbar^2 k^2)/(2m_0)$. In semiconductors, the electron energies in the

R. Diehl (Ed.): High-Power Diode Lasers, Topics Appl. Phys. **78**, 1–53 (2000)
© Springer-Verlag Berlin Heidelberg 2000

conduction band $E_c(k)$ and in the valence band $E_v(k)$ behave similarly for small wavenumbers k.

$$E_c(k) = E_g + \frac{\hbar^2 k^2}{2m_e}, \qquad\qquad E_v(k) = -\frac{\hbar^2 k^2}{2m_h}. \qquad (1)$$

Figure 1 shows this behavior which is called the nearly-free carrier approximation. The interaction of the carriers with the solid-state lattice is taken into account by the introduction of effective masses for the electrons m_e and for the holes m_h which are in general different from the free-electron rest mass m_0. Since the $E(k)$ dependence in the valence band is a negative parabolic curve, holes can be regarded as particles with positive charge.

Radiative band-to-band transitions are generation and recombination of electron–hole pairs associated with absorption or emission of photons. For these transitions, conservation of energy E and momentum $\hbar k$ must be fulfilled. Due to the high value for the speed of light $c = 2.997\,925 \times 10^{10}\,\mathrm{cm/s}$, the momentum of the photon $\hbar k = \hbar\omega/c = E_{ph}/c$ for photon energies E_{ph} in the 0.5–2.5 eV range can be neglected in comparison to the momentum of the electronic carriers. Thus, a radiative transition between an electron in the conduction band with energy $E_2(k_2)$ and a hole in the valence band with

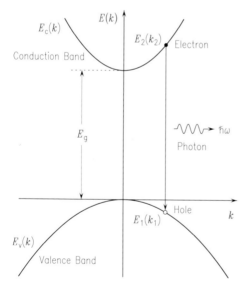

Fig. 1. Parabolic band structure $E(k)$ for electrons in a direct semiconductor. The conduction band is separated from the valence band by an energy gap E_g. A recombination of an electron at $E_2(k_2)$ in the conduction band and a hole at $E_1(k_1)$ in the valence band generates a photon with energy $\hbar\omega$. Since the momentum of the photon $\hbar k$ is negligibly small, radiative electronic transitions between conduction and valence bands only occur at the same wavenumber k

energy $E_1(k_1)$ under emission or absorption of a photon can only occur at the same wavenumber k.

$$E_{\mathrm{ph}} = \hbar\omega = E_2 - E_1 , \qquad\qquad k_2 = k_1 . \qquad\qquad (2)$$

As shown in Fig. 1, these transitions can be illustrated by vertical arrows with the length of the photon energy $\hbar\omega$ pointing upwards for generation and downwards for recombination of an electron–hole pair. In thermal equilibrium, the carriers tend to occupy the states with lowest energy. For electrons, these are states at the minimum of the conduction band. On the ordinate axis of a band diagram, the electron energy is plotted; therefore, the minimum energy of the positively charged holes is in the maximum of the valence band. The valence-band maximum and the conduction-band minimum of direct semiconductors are both located at the point $k = 0$. In indirect semiconductors like silicon and germanium, minimum and maximum have different k-values; therefore, band-to-band recombinations can only occur with the contribution of phonons or traps. These transitions are unsuitable for laser activity, because the spatial density of phonons and traps is very low. Furthermore, they are mostly nonradiative and rather unlikely since more partners are involved.

In thermal equilibrium at a temperature T, the probability whether a state with the energy level E is occupied by an electron is expressed by the Fermi function $f(E, T)$.

$$f(E, T) = \frac{1}{\exp\left(\frac{E - E_{\mathrm{F}}}{k_{\mathrm{B}} T}\right) + 1} . \qquad\qquad (3)$$

At $T = 0\,\mathrm{K}$, the Fermi function is a step function which has a value of 1 (all electronic states filled) below the Fermi-level energy E_{F} and a value of 0 (all states empty) for higher energies. In undoped semiconductors, the Fermi level is located in the middle between conduction and valence band edges. For higher temperatures T, the Fermi function is smeared out in the range $E_{\mathrm{F}} \pm 2k_{\mathrm{B}}T$, with $k_{\mathrm{B}} = 8.617\,347 \times 10^{-5}\,\mathrm{eV/K}$ being the Boltzmann constant.

For a fixed photon energy $\hbar\omega$, it is entirely correct to consider only two discrete energy levels $E_1(k)$ and $E_2(k)$ since the transition can only occur at the same wavenumber k as illustrated in Fig. 1. Three types of radiative band-to-band transitions can be found in semiconductors, which are sketched in Fig. 2. The first process is called spontaneous emission, where a recombination of an electron–hole pair leads to the emission of a photon. This is the predominant process in Light-Emitting Diodes (LEDs). The emission of the photon is random in direction, phase, and time resulting in incoherent radiation. Since this process depends on the existence of an electron at E_2 and a hole at E_1 simultaneously, the transition rate for spontaneous emission $\mathcal{R}_{\mathrm{sp}}$ is proportional to the product of the electron density at E_2 and the hole density at E_1. The electron density at the energy E_2 is the product

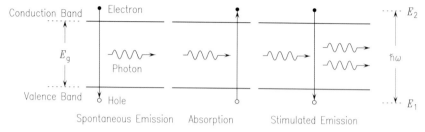

Fig. 2. Radiative band-to-band transitions in semiconductors

of the density of electronic states $D(E_2)$ and the probability that they are occupied by electrons given by the Fermi function $f(E_2, T)$. The hole density at the energy E_1 is the density of electronic states $D(E_1)$ multiplied with the probability of not being occupied by electrons $[1 - f(E_1, T)]$. So, the transition rate per volume for spontaneous emission of photons with fixed energy $\hbar\omega = E_2 - E_1$ can be written as

$$\mathcal{R}_{sp} = A\, D(E_2) f(E_2, T)\, D(E_1)\, [1 - f(E_1, T)] \,, \tag{4}$$

with A being the proportionality constant for spontaneous emission.

Absorption, also called stimulated absorption, is the second process illustrated in Fig. 2. A photon is absorbed and an electron–hole pair is generated. This is a three-particle process and the transition rate $\mathcal{R}_{12}$ therefore is proportional to the product of three particle densities: first, the density of nonoccupied states $D(E_2)\,[1 - f(E_2, T)]$ in the conduction band at the energy E_2, second, the density of states occupied by electrons $D(E_1)\, f(E_1, T)$ in the valence band at E_1, and third, the density of the photons $\rho(\hbar\omega)$ with energy $\hbar\omega = E_2 - E_1$.

$$\mathcal{R}_{12} = B_{12}\, \rho(\hbar\omega)\, D(E_1) f(E_1, T)\, D(E_2)\, [1 - f(E_2, T)] \,. \tag{5}$$

B_{12} is a proportionality constant for stimulated absorption.

The third process is stimulated emission. A recombination of an electron–hole pair is stimulated by a photon and a second photon is generated simultaneously which has the same direction and phase as the first photon. This process can be used to amplify optical radiation, since the photons are emitted into the optical mode of the stimulating photon resulting in coherent radiation. Light sources based on this emission process are called lasers, which is an abbreviation of light amplification by stimulated emission of radiation. Analogous to the stimulated absorption (5), the transition rate $\mathcal{R}_{21}$ for stimulated emission can be described as

$$\mathcal{R}_{21} = B_{21}\, \rho(\hbar\omega)\, D(E_2) f(E_2, T)\, D(E_1)\, [1 - f(E_1, T)] \,, \tag{6}$$

with B_{21} being the proportionality constant for stimulated emission.

When the semiconductor is in thermal equilibrium with the photons, no energy is transferred from the semiconductor to the optical radiation field; thus, absorption and emission must be balanced:

$$\mathcal{R}_{12} = \mathcal{R}_{21} + \mathcal{R}_{\text{sp}} . \tag{7}$$

Using (4), (5), and (6) for the rates $\mathcal{R}_{\text{sp}}$, $\mathcal{R}_{12}$, and $\mathcal{R}_{21}$, respectively, gives

$$\frac{B_{21}\,\rho(\hbar\omega) + A}{B_{12}\,\rho(\hbar\omega)} = \frac{f(E_1,T)\,[1 - f(E_2,T)]}{f(E_2,T)\,[1 - f(E_1,T)]} , \tag{8}$$

$$\frac{B_{21}}{B_{12}} + \frac{A}{B_{12}}\frac{1}{\rho(\hbar\omega)} = \frac{f(E_1,T) - f(E_1,T)\,f(E_2,T)}{f(E_2,T) - f(E_1,T)\,f(E_2,T)}$$

$$= \frac{\frac{1}{f(E_2,T)} - 1}{\frac{1}{f(E_1,T)} - 1} . \tag{9}$$

The spectral photon density $\rho(\hbar\omega)$ in thermal equilibrium does not depend on the specific function for the density of states $D(E)$. Inserting the Fermi function $f(E,T)$ from (3) and the relation $\hbar\omega = E_2 - E_1$ (2), gives

$$\frac{B_{21}}{B_{12}} + \frac{A}{B_{12}}\frac{1}{\rho(\hbar\omega)} = \frac{\exp\left(\frac{E_2 - E_\text{F}}{k_\text{B}T}\right)}{\exp\left(\frac{E_1 - E_\text{F}}{k_\text{B}T}\right)}$$

$$= \exp\left(\frac{\hbar\omega}{k_\text{B}T}\right) , \tag{10}$$

$$\rho(\hbar\omega) = \frac{A}{B_{12}\exp\left(\frac{\hbar\omega}{k_\text{B}T}\right) - B_{21}} . \tag{11}$$

The spectral energy density $u(\nu)\,d\nu$ at the frequency ν in a medium with refractive index n for radiation in thermal equilibrium is given by Planck's well-known formula for blackbody radiation.

$$u(\nu)\,d\nu = \frac{8\pi h n^3 \nu^3}{c^3}\frac{1}{\exp\left(\frac{h\nu}{k_\text{B}T}\right) - 1}\,d\nu . \tag{12}$$

Dividing the energy density $u(\nu)$ by the photon energy $\hbar\omega$ yields the photon density $\rho(\hbar\omega)$. Additionally, the relations $\omega = 2\pi\nu$, $h = 2\pi\hbar$, and $d(\hbar\omega) = d(h\nu) = h\,d\nu$ have been used.

$$\rho(\hbar\omega)\,d(\hbar\omega) = \frac{n^3\,(\hbar\omega)^2}{\pi^2\,\hbar^3 c^3}\frac{1}{\exp\left(\frac{\hbar\omega}{k_\text{B}T}\right) - 1}\,d(\hbar\omega) . \tag{13}$$

In thermal equilibrium of the semiconductor material with the radiation field, the spectral photon density described in (11) must be identical with the photon density of the blackbody radiation described by (13). Comparing these equations gives

$$B_{12} = B_{21} = B \,, \qquad A = \frac{n^3}{\pi^2 \hbar^3 c^3} \, (\hbar\omega)^2 \, B \,. \tag{14}$$

With the knowledge of these relations between the proportionality constants, the considerations can be extended to nonequilibrium conditions. When a p-n junction is forward biased, electrons and holes are injected into the depletion region where they can either recombine or travel further to the other side of the junction and recombine there with the majority carriers. In the transition zone, where electrons and holes coexist, the carrier distribution cannot be described by a single equilibrium Fermi function (3). Instead, separate quasi-Fermi functions are used for the electrons in the conduction band $f_c(E, T)$ and for the holes in the valence band $f_v(E, T)$.

$$f_c(E, T) = \frac{1}{\exp\left(\frac{E - E_{Fc}}{k_B T}\right) + 1} \,, \qquad f_v(E, T) = \frac{1}{\exp\left(\frac{E - E_{Fv}}{k_B T}\right) + 1} \,. \tag{15}$$

The equations are identical to the equilibrium Fermi function but different Fermi-level energies E_{Fc} and E_{Fv} are employed for the carrier distributions in the conduction and valence bands, respectively. The nonequilibrium situation can be described by replacing $f(E_1, T) \rightarrow f_v(E_1, T)$ and $f(E_2, T) \rightarrow f_c(E_2, T)$.

To determine whether an optical wave with quantum energy $\hbar\omega$ is absorbed or amplified by stimulated emission, the quotient of the corresponding rates $\mathcal{R}_{12}$ and $\mathcal{R}_{21}$ is calculated.

$$\begin{aligned}
\frac{\mathcal{R}_{12}}{\mathcal{R}_{21}} &= \frac{f_v(E_1, T) \left[1 - f_c(E_2, T)\right]}{f_c(E_2, T) \left[1 - f_v(E_1, T)\right]} = \frac{f_v(E_1, T) - f_v(E_1, T) \, f_c(E_2, T)}{f_c(E_2, T) - f_v(E_1, T) \, f_c(E_2, T)} \\[2mm]
&= \frac{\frac{1}{f_c(E_2, T)} - 1}{\frac{1}{f_v(E_1, T)} - 1} = \frac{\exp\left(\frac{E_2 - E_{Fc}}{k_B T}\right)}{\exp\left(\frac{E_1 - E_{Fv}}{k_B T}\right)} \\[2mm]
&= \exp\left[\frac{\hbar\omega - (E_{Fc} - E_{Fv})}{k_B T}\right] \,.
\end{aligned} \tag{16}$$

Again, the result does not depend on the specific density of states $D(E)$. In thermal equilibrium $E_{Fc} = E_{Fv} = E_F$, the exponent $[\hbar\omega/(k_B T)]$ is positive, the exponential function is larger than 1, and therefore the absorption rate $\mathcal{R}_{12}$ is always larger than the rate $\mathcal{R}_{21}$ of the stimulated emission. Laser operation in semiconductors can only be achieved if the condition

$$E_{Fc} - E_{Fv} > \hbar\omega > E_g \tag{17}$$

is fulfilled. In this status, which is called inversion, the exponential function is smaller than 1 and the rate of stimulated emission is larger than the absorption rate. Laser operation requires a process called pumping which builds up and maintains a nonequilibrium carrier distribution in the semiconductor material. From (16) it can be deduced that a laser-active transition always shows absorption in thermal equilibrium. Although pumping can also be provided by optical excitation of electron–hole pairs, one main advantage of semiconductor lasers over other types of lasers is the fact that they can be easily pumped with electrical currents as a forward-biased semiconductor diode as shown in Fig. 3. For this reason, electrically pumped semiconductor lasers are called diode lasers.

All state-of-the-art semiconductor-diode lasers use forward-biased double-hetero p-i-n structures to achieve carrier inversion. In this type of structure, an undoped semiconductor layer with a direct band gap is sandwiched between p-doped and n-doped material with a higher band gap. When the junction is forward biased, the quasi-Fermi levels E_{Fc} and E_{Fv} in the intrinsic layer are located inside the conduction and valence bands as illustrated in Fig. 3. Thus, this region acts as a laser-active layer which amplifies optical radiation by stimulated emission. Furthermore, the double heterostructure has two additional advantages. First, the carriers (electrons and holes) are confined between the double heterobarriers in the conduction and the valence bands and are therefore forced to recombine inside the intrinsic layer of direct semiconductor material. Second, this layer sequence works like an optical waveguide since for most semiconductor-material systems, the low-band-gap layer in the middle of the structure has a higher refractive index.

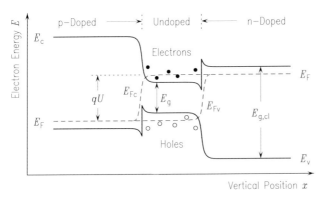

Fig. 3. Forward-biased double-heterostructure p-i-n junction. Conduction E_c and valence band edges E_v are plotted as solid lines. The Fermi-level energy E_F, represented by dashed lines, splits into quasi-Fermi levels E_{Fc} and E_{Fv} in the undoped transition region, where holes and electrons coexist. In this region, inversion is achieved since the quasi-Fermi levels are inside the bands

To assess which proportion of the carriers recombines by stimulated and spontaneous emission, the ratio of $\mathcal{R}_{21}$ and $\mathcal{R}_{\mathrm{sp}}$ is determined using (4), (6), and (14).

$$\frac{\mathcal{R}_{21}}{\mathcal{R}_{\mathrm{sp}}} = \frac{B}{A}\,\rho(\hbar\omega) = \frac{\pi^2\hbar^3 c^3}{n^3(\hbar\omega)^2}\,\rho(\hbar\omega)\,. \tag{18}$$

This equation shows that a high photon density $\rho(\hbar\omega)$ is necessary to suppress spontaneous emission. Since the term $(\hbar\omega)^2$ is in the denominator of (18), a higher value of the photon density $\rho(\hbar\omega)$ is required for lasers with higher photon energy $(\hbar\omega)$ to achieve the same suppression of spontaneous emission. To obtain a high photon density in semiconductor lasers, optical waveguides are implemented to confine the photons in the laser-active region of the device. Furthermore, an optical resonator, mostly a Fabry–Perot resonator, is used to increase the photon density in the resonator cavity. A semiconductor laser can be regarded as an optical oscillator consisting of an optically amplifying medium and a resonator which provides optical feedback to the amplifier. Waveguides and resonators for high-power semiconductor lasers are discussed in more detail in the next sections.

1.2   Basic Elements of Semiconductor-Diode Lasers

Several basic elements are necessary to realize a semiconductor-diode laser:

- a medium providing optical gain by stimulated emission,
- an optical waveguide which confines the photons in the active region of the device,
- a resonator creating optical feedback, and
- a lateral confinement of injected current, carriers, and photons which is required for operation in a fundamental lateral mode.

The optical-gain medium consists of an active undoped layer of direct semiconductor material embedded between high-band-gap p- and n-doped regions. When this p-i-n junction is forward biased, electrons and holes are injected into the active region and optical gain by stimulated emission becomes possible. Furthermore, the double heterobarriers confine the carriers in the active region. The active layer may consist of bulk material with a typical thickness of 100 nm or of one or more quantum wells having typical thicknesses of 10 nm. Quantum-well structures are discussed in Sect. 1.6.

A dielectric optical waveguide consists of a core film with high refractive index embedded in cladding material with lower refractive index. Figure 4 illustrates the optical waveguide for a double-heterostructure laser. The active film with band gap E_g, refractive index n_f, and thickness d is sandwiched between cladding layers with band gap $E_{g,\mathrm{cl}}$ and refractive index n_{cl}. If the index step $\Delta n = n_f - n_{\mathrm{cl}}$ and the core thickness d of the waveguide are small enough, only the fundamental mode with nearly Gaussian field distribution

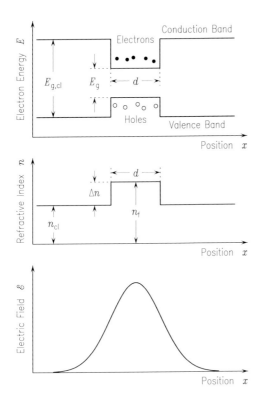

Fig. 4. Confinement of the electronic carriers (electrons and holes) and the electric field (photons) using a double heterostructure in the vertical direction x of an edge-emitting diode laser. Plotted are the energy-band diagram $E(x)$ with conduction and valence bands (*top*), the refractive-index profile $n(x)$ of the dielectric waveguide (*center*), and the electric-field distribution $\mathcal{E}(x)$ of the fundamental optical mode traveling in the z direction (*bottom*)

can propagate in the waveguide. The optical wave traveling in the direction of the waveguide experiences an effective refractive index n_{eff} which is different from the refractive indices of core and cladding ($n_{\mathrm{cl}} \leq n_{\mathrm{eff}} \leq n_{\mathrm{f}}$). Figure 4 shows a structure where the same layer provides the confinement of the carriers and the optical wave. In quantum-well lasers, so-called separate confinement structures are necessary, where the carriers are confined in quantum wells and the optical wave is confined in a separate dielectric-waveguide structure.

For high-power diode lasers, Fabry–Perot resonators are used. Figure 5 shows this type of resonator consisting of two mirrors with distance L around a laser-active material having an optical waveguide with effective refractive index n_{eff} in a propagation direction normal to the mirror surfaces. The

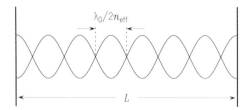

Fig. 5. A standing wave having $m = 7$ nodes in a Fabry–Perot resonator with a cavity length L. The wave propagates in a waveguide with an effective refractive index n_{eff}. The distance between two nodes is $\lambda_0/(2n_{\mathrm{eff}})$ with λ_0 being the vacuum wavelength

resonator provides feedback, when a standing wave develops between the mirrors.

$$L = m \, \frac{\lambda_0}{2n_{\mathrm{eff}}}, \qquad m = 1, 2, 3, \ldots \; . \tag{19}$$

m is the number of nodes of the standing wave, the order number of the longitudinal mode, and λ_0 is the vacuum wavelength. Lasers for optical communication systems have other types of optical resonators like Distributed FeedBack (DFB) or Distributed Bragg Reflector (DBR) resonators [13,14].

Up to this point, carrier and optical confinement have been discussed for a direction which is perpendicular to the active-layer plane. This direction is called the vertical direction. To obtain single-mode operation in both transversal directions, an additional lateral confinement is required. As discussed in Sect. 1.5, three types of lateral-confinement mechanisms are possible: current confinement leading to a gain-guided lateral waveguide, optical confinement providing index guiding, and buried heterostructures providing an additional carrier confinement.

1.3   Optical Gain and Threshold Condition

When passing through an absorbing material in the z direction, the intensity J of a planar optical wave exponentially decreases.

$$J(z) = J_0 \, \exp(-\alpha z), \tag{20}$$

with J_0 being the initial intensity and α the intensity-absorption coefficient. In laser-active semiconductor material, an amplification of the optical wave is achieved. In this case, the exponential increase in intensity can be described by a negative value of α which is referred to as optical gain $g = -\alpha$. In an optical waveguide, only a part of the intensity pattern of the optical mode overlaps with the active region, which usually is located in the core of the waveguide. One has to distinguish between the gain of the active material itself, called the material gain g, and the significantly lower gain of the optical mode, called the modal gain g_{modal}.

In Fig. 6, the material gain of GaAs at room temperature is plotted for different carrier densities N. The maximum gain is observed at photon energies which are slightly higher than the band gap energy. Figure 7 illustrates the optical-intensity pattern $J(x)$ of the fundamental optical mode in a double-heterostructure edge-emitting laser having an active-layer thickness d. The relation between modal gain g_{modal} and material gain g is expressed by defining a confinement factor Γ which depends on the overlap of the optical-mode pattern with the gain region of the laser.

$$g_{\text{modal}} = \Gamma\, g\,, \qquad \Gamma = \frac{\displaystyle\int_{-d/2}^{+d/2} J(x)\,dx}{\displaystyle\int_{-\infty}^{+\infty} J(x)\,dx}\,. \tag{21}$$

In double heterostructures with active-layer thicknesses of 50–300 nm, the confinement factor Γ has values in the range 10–70%. If the active layer consists of a quantum well with a typical thickness around 10 nm, confinement factors of a few percent are obtained.

For a mode traveling along the optical waveguide, the intensity-absorption coefficient α is usually split into two parts, one describing the intrinsic modal

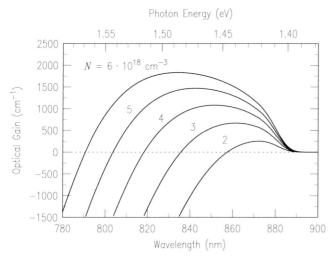

Fig. 6. Optical-gain spectrum $g(\lambda_0)$ of GaAs bulk material at carrier densities of $N = 2$–6×10^{18} cm^{-3}. For photon energies below the band-gap energy of GaAs of $E_{\text{g}} = 1.42$ eV, the material is transparent. Optical gain occurs for energies near the band gap. The maximum of the optical-gain curve shifts towards shorter wavelengths for higher carrier densities N due to the band-filling effect. If the photon energy is further increased, absorption takes place. The gain data have been calculated using the model described in [15]

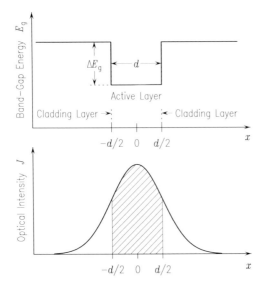

Fig. 7. The confinement factor Γ is defined by the overlap of the intensity pattern J of the optical mode with the active region. The illustration shows the intensity distribution $J(x)$ in the vertical direction x of a fundamental optical mode with nearly Gaussian shape for an edge-emitting laser with an active-layer thickness d

absorption α_i and the other describing the modal gain $g_{\mathrm{modal}} = \Gamma g$ which depends on the density of the injected carriers.

$$\alpha = \alpha_i - \Gamma g \,. \tag{22}$$

The intrinsic modal absorption is caused by scattering of the optical mode at defects or rough interfaces and by free-carrier absorption. Whereas scattering is extremely low for semiconductor-diode lasers with good crystalline quality, free-carrier absorption cannot be avoided since part of the optical-mode pattern overlaps with the p- and n-doped cladding regions. When the modal gain Γg is larger than the modal loss α_i, the propagating optical mode is amplified.

In a laser device, the optical waveguide is combined with a Fabry–Perot resonator having mirror reflectivities R_1 and R_2. Some optical intensity leaves the cavity at these mirrors and contributes to the laser output beam. As illustrated in Fig. 8, the intensity J_{rt} of the optical mode after a roundtrip in the cavity is given by

$$J_{\mathrm{rt}} = J_0\, R_1 R_2 \, \exp\left[2(\Gamma g - \alpha_i)L\right] \,. \tag{23}$$

Lasing occurs when the gain provided to the optical mode compensates the intrinsic absorption and the mirror losses for a roundtrip. The minimum gain g where the device starts lasing operation is called the threshold gain g_{th}.

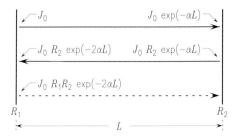

Fig. 8. Intensity of an optical wave during a roundtrip in a Fabry–Perot resonator with cavity length L and mirror reflectivities R_1 and R_2

In this case, the intensity J_{rt} after a roundtrip in the cavity again has its initial value J_0.

$$J_{rt} = J_0 ,$$
$$1 = R_1 R_2 \exp\left[2(\Gamma g_{th} - \alpha_i)L\right] ,$$
$$\Gamma g_{th} = \alpha_i + \frac{1}{2L} \ln\left(\frac{1}{R_1 R_2}\right) = \alpha_i + \alpha_{mirror} . \tag{24}$$

At laser threshold, the modal gain Γg_{th} is the sum of the two terms in (24), the intrinsic absorption α_i and the mirror losses α_{mirror}. The mirror losses depend on the cavity length L and the mirror reflectivities R_1 and R_2.

1.4   Edge- and Surface-Emitting Lasers

Figures 9 and 10 schematically show two basic implementations of semiconductor-diode lasers, the edge-emitting laser and the Vertical-Cavity Surface-Emitting Laser (VCSEL). The mirror facets of edge-emitting lasers are obtained by cleaving the wafer along crystal planes. The mirror reflectivities R_1 and R_2 are approximately 30% if the facets are uncoated. Mirror coatings can be applied to change these reflectivities and to passivate the surfaces. The propagation direction of the optical mode in the resonator is in plane with the substrate surface and is referred to as the axial direction. A planar optical waveguide and the laser-active region are formed by depositing a layer sequence onto the substrate surface using a growth technique called epitaxy where the deposited single-crystal layers are lattice-matched to the substrate. The growth direction which is perpendicular to the substrate surface is called the transverse or vertical direction. The lateral direction is in the substrate plane normal to the axial direction. The active region has a lateral width W, a vertical height d given by the thickness of the epitaxially grown active layer, and an axial length L which is identical to the cavity length. In Fig. 9, a stripe laser is shown, where the width of the active layer is defined by the top ohmic contact resulting in a gain-guided optical waveguide without carrier confinement. The reference coordinate system used to describe edge-emitting

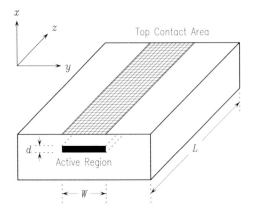

Fig. 9. Schematic drawing of an edge-emitting laser with the coordinate system. The laser cavity in the axial direction z is located in the substrate plane having a cavity length L in the range 300–2000 μm. The mirror facets are generated by cleaving the substrate along crystal planes. The direction perpendicular to the substrate plane is called the transverse or vertical direction x; the lateral direction y is in the substrate plane normal to the axial direction. The active region below the top contact area has a lateral width W and a vertical height d

lasers has its x axis in the vertical, the y axis in the lateral, and the z axis in the axial direction. Since edge-emitting lasers have typical cavity lengths L in the range 300–2000 μm, the order number m of longitudinal optical modes is very large ($m = 1000$–$20\,000$), the spectral density of the longitudinal modes is very high, and a lot of possible modes can exist within the bandwidth of the spectral gain.

In vertical-cavity lasers, the optical propagation (axial) direction is normal to the substrate surface and the effective cavity length is very short (typically 1–3 μm) allowing the existence of only a single longitudinal mode within the spectral-gain range. To avoid extremely high mirror losses in the short cavity, the reflectivity of the Fabry–Perot mirrors must be close to 100% according to (24). This can be achieved by using Bragg reflectors which consist of typically 20 pairs of epitaxially grown GaAs-AlAs layers having alternating high and low refractive index and a thickness of a quarter wavelength. Between the mirrors, a set of quantum wells is sandwiched, providing the optical gain in the active region. In the case of strained InGaAs quantum wells, all semiconductor material including the substrate is transparent for light in the wavelength range $\lambda_0 = 870$–1100 nm generated in these quantum wells. The choice of the appropriate mirror reflectivities allows the VCSEL to be operated as a top or a bottom emitter. In the schematic illustration of a VCSEL shown in Fig. 10, the electrical current is supplied through the p- and n-doped mirrors. The emitting lateral aperture normally has a circular geometry with a diameter of a few microns allowing single-lateral-mode oper-

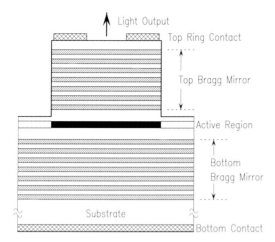

Fig. 10. Schematic drawing of the Vertical-Cavity Surface-Emitting Laser (VC-SEL). The laser cavity of the top-emitting VCSEL is perpendicular to the substrate plane. The mirrors of the Fabry–Perot resonator consist of Bragg reflectors with reflectivities close to 100%. The total length of the device in the vertical direction is about 7 μm and the effective cavity length is in the range 1–3 μm

ation and a highly effective coupling to optical fibers. Additional advantages are the ultra-low threshold current (< 1 mA), excellent dynamic properties, a high electrical-to-optical power-conversion efficiency, insensitivity to optical feedback, and the absence of sudden device failures attributed to mirror damage. Although the epitaxial growth of VCSEL structures is rather sophisticated, the remaining fabrication process is similar to the manufacturing of LEDs which allows wafer-scale processing and on-wafer device testing. VC-SELs can be easily arranged in two-dimensional arrays and coupled to parallel optical-fiber bundles. The single-mode output power of a VCSEL is in the mW range. Higher power levels can be achieved by enlarging the diameter of the emitting aperture or by a densely packed arrangement. Since VCSELs are rather novel devices, their full potential has not yet been exploited. Certainly, there will be an increasing range of applications for VCSELs in the high-power regime. More comprehensive information on the properties and applications of VCSELs can be found in [15,16,17].

1.5   Lateral Confinement

As already mentioned in Sect. 1.2, different implementations are utilized to achieve the lateral confinement of the current, the photons, and the carriers in edge emitting lasers. As illustrated at the top of Fig. 11, current confinement is provided by a current aperture which is mostly realized by a dielectric isolator as shown for a stripe laser at the top of Fig. 12. In another implementation, the current aperture is formed by ion implantation. Diode lasers

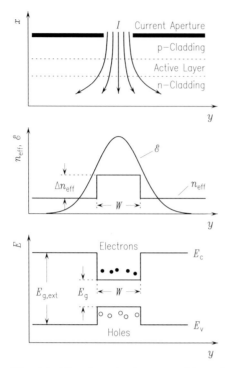

Fig. 11. The three basic types of lateral confinement. Current confinement (*top*): the current is injected through an aperture. Optical confinement (*center*): a step in the effective refractive index n_{eff} builds up a dielectric lateral waveguide for the optical mode. Carrier confinement (*bottom*): a double heterostructure barrier prevents the lateral diffusion of electrons and holes

which exclusively have current confinement are called gain-guided lasers. Only those modes are amplified, which are propagating under the stripe, since optical amplification only occurs in areas which are pumped by the electrical current. Outside the stripe, an optical wave experiences high optical losses. Fundamental-lateral-mode operation can be obtained in these devices for small stripe widths. A single-mode stripe laser is rather easy to fabricate but has some disadvantages. Compared to index-guided lasers, the threshold current is larger because the waveguide is rather lossy. Since the mode partially propagates in absorbing material, the phasefront of the mode is curved, leading to a significant astigmatism in the output beam.

The principle of index-guided lasers is illustrated in the center of Fig. 11. A lateral effective refractive-index step Δn_{eff} provides the waveguiding. Depending on this index step and the width W of the waveguide, single-lateral-mode operation can be obtained. A typical example of an index-guided laser is the ridge-waveguide laser shown in the center of Fig. 12. The index step is formed by a step in the thickness of the upper cladding layer. Since the

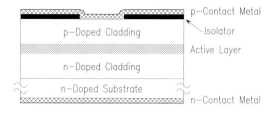

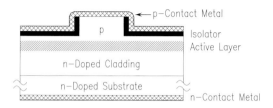

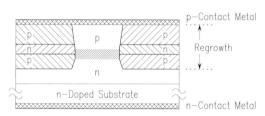

Fig. 12. Examples of laser structures with different lateral confinement. A stripe laser (*top*) only has a current confinement, the lateral waveguide is formed by gain guiding. The ridge-waveguide laser (*center*) has current confinement and optical confinement with index guiding. The buried-heterostructure laser (*bottom*) has current, photon, and carrier confinement

current is injected at the top of the ridge, current confinement is also implemented. The beam quality of ridge-waveguide lasers is very sensitive to the width and the height of the ridge. Therefore, a precise and reproducible control of ridge dimensions is necessary during device fabrication.

All three types of lateral confinement are combined in a diode laser having a buried heterostructure which is schematically shown at the bottom of Fig. 12. The lateral heterostructure is formed by an epitaxial regrowth technique. This heterostructure provides index guiding and carrier confinement since the barriers prevent the lateral diffusion of electrons and holes as illustrated at the bottom of Fig. 11. The p-n-p structure in the vertical direction works as a current-blocking layer providing carrier confinement. Buried-heterostructure lasers are mostly used in communication systems where an extremely low threshold current is required to allow good dynamic properties and a low power consumption.

1.6   Quantum-Well Structures

In double heterostructures, the typical thickness of the active layer is $d = 50$–$300\,\mathrm{nm}$, resulting in a confinement factor Γ in the range of 10–70%. The density of electronic states $D(E)$ increases with the square root of the energy at the band edge ($D(E) \propto \sqrt{E - E_\mathrm{g}}$). If the thickness of the active layer is shrunk to values of 5–10 nm, the electronic wave functions in this quantum well show quantization in the vertical direction x resulting in discrete energy levels. In this case, the density of electronic states $D(E)$ increases in steps which are located at the electronic energy levels of the quantum well. Thus, the density of states close to the lowest-energy level in a quantum well is much higher than the density of states at the band edge in bulk material. The density of electronic carriers at a given energy is the product of the density of states $D(E)$ and the probability of being occupied by electons $f_\mathrm{c}(E, T)$ or holes $[1 - f_\mathrm{v}(E, T)]$, which are exponentially decreasing functions (according to (15)). Thus, the carrier distribution for a quantum-well laser structure has a higher maximum value and a smaller energetic width.

Due to the small active volume of a quantum-well laser, low threshold currents can be obtained. Additionally, the material gain is higher and the spectral shift of the gain curve due to the band-filling effect is much smaller, because of the higher carrier density and its narrower energetic distribution. The main advantage of quantum-well structures, however, is the possibility

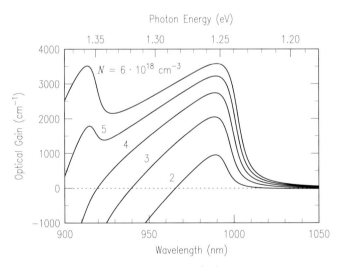

Fig. 13. Material gain spectrum $g(\lambda_0)$ of a single compressively strained 8 nm-thick $\mathrm{Ga}_{0.8}\mathrm{In}_{0.2}\mathrm{As}$ quantum well sandwiched in GaAs bulk material at carrier densities $N = 2$–$6 \times 10^{18}\,\mathrm{cm}^{-3}$. Due to the high density of states $D(E)$ in quantum wells, the maximum of the gain curve shows nearly no shift in wavelength. At higher carrier densities, transitions to the second subband of the quantum well contribute to the gain. The data have been calculated using the model described in [15]

to introduce compressive or tensile mechanical biaxial strain. In this way, the useable wavelength range of a particular material system can be extended, e.g. the incorporation of In instead of Ga into a thin GaAs quantum well-layer results in a compressively strained quantum well and the accessible wavelength now ranges from 870 nm for bulk GaAs into the long-wavelength region up to approximately 1100 nm. This is illustrated by the spectral gain of an 8 nm-thick $Ga_{0.8}In_{0.2}As$ quantum well plotted in Fig. 13 in comparison to the gain of GaAs bulk material shown in Fig. 6. The product of quantum-well film thickness and strain must be below a critical value. Above this value, the film experiences relaxation, which is associated with a high number of traps leading to nonradiative recombination. In compressively strained quantum wells, the light and heavy hole bands are split and the effective mass of the holes in the valence band is reduced. The effective masses of electrons and holes are now comparable $(m_e \approx m_h)$ resulting in a more-efficient population inversion in the quantum well. In the long-wavelength range, any kind of strain is beneficial due to the reduced inter-valence-band absorption and Auger recombination. Especially for compressively strained quantum-well lasers, a significantly improved reliability has been observed. For these reasons, strained quantum wells are the rule rather than the exception in state-of-the-art diode lasers.

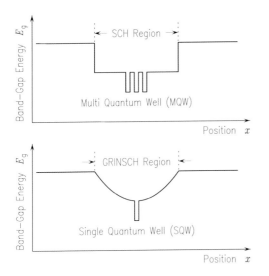

Fig. 14. Different vertical structures for separate confinement of electronic carriers and the optical mode. Plotted are the band-gap energies E_g versus vertical position x. In the *upper* diagram, a Multi-Quantum-Well Separate-Confinement Heterostructure (MQW-SCH) with three quantum wells is sketched. In the *lower* part of the figure, a Single-Quantum-Well GRaded-INdex Separate-Confinement Heterostructure (SQW-GRINSCH) is shown

Since quantum wells are very thin, the confinement of the optical mode is poor. This can be overcome by a Separate-Confinement Heterostructure (SCH) where the confinement of the optical mode is provided by a separate waveguide structure. Two examples of such vertical structures are shown in Fig. 14. If the waveguide includes a graded refractive-index profile, the structure is called a GRaded-INdex Separate-Confinement Heterostructure (GRINSCH).

2   Fabrication Technology

The multilayer structures of diode lasers are fabricated using epitaxial growth techniques. In these processes, single-crystal lattice-matched layers with precisely controlled thickness, material composition, and doping profiles are deposited onto GaAs and InP substrate wafers. The layer sequence consists of a buffer layer, p- and n-doped cladding layers, a layer for the ohmic contact, and the active region, which may be simple bulk material or a sophisticated structure containing one or more quantum wells with separate optical confinement. Figure 15 shows suitable III–V semiconductor compounds which can be epitaxially grown on GaAs and InP substrates.

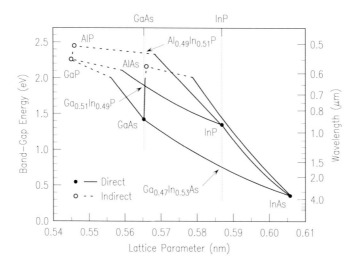

Fig. 15. Band-gap energy versus lattice parameter of III–V semiconductor alloys used for high-power laser diodes. The binary compounds are represented by dots, ternary alloys are drawn as lines. Direct semiconductors are plotted as full lines and dots whereas for indirect material dotted lines and open dots are used. Indicated with arrows are ternary compounds growing lattice-matched on the common substrate materials GaAs and InP. The data are derived from [18]

The following list gives an overview of material systems commonly used for high-power diode lasers. Not included are II–VI and GaN-based semiconductors for blue and green light emitters [19] since their output power is still rather low.

- $Al_xGa_{1-x}As$ grown on GaAs is the classical material for diode lasers. Since the radii of gallium and aluminum ions are nearly equal, $Al_xGa_{1-x}As$ can be grown lattice-matched for any composition x, and laser emission in the wavelength range 700–870 nm can be realized.
- Using an InGaAs quantum well, the emission range of the AlGaAs/GaAs system can be extended to longer wavelengths. Since an indium ion has a larger radius than a gallium ion, the lattice parameter of GaInAs is higher than the lattice parameter of GaAs. Therefore, only thin GaInAs quantum-well layers containing compressive mechanical strain can be grown keeping lattice-matching. The emission wavelength can be adjusted in the range 800–1100 nm by varying the thickness and indium content of the strained quantum well.
- The same wavelength range can be covered using the strained GaInAs quantum well in combination with GaInAsP separate-confinement layers and $Ga_{0.51}In_{0.49}P$ cladding layers. Like GaInAs-AlGaAs/GaAs, this system is also lattice-matched to GaAs but is completely aluminum-free [20,21].
- The visible-red short-wavelength range 600–700 nm can be accessed using $(Al_xGa_{1-x})_{0.5}In_{0.5}P$. Again, this material is lattice-matched to GaAs for any aluminum concentration x since the ion radii of aluminum and gallium are approximately equal.
- $Ga_xIn_{1-x}As_yP_{1-y}$ grows lattice-matched on InP substrates if the equation $x = 0.4\,y + 0.067\,y^2$ is fulfilled. The material has a direct band gap ranging from $E_g = 0.75\,eV$ for $Ga_{0.47}In_{0.53}As$ to $E_g = 1.35\,eV$ for InP. Lasers of this type are implemented in quartz-fiber communication systems at the wavelengths 1.3 µm and 1.55 µm.

The fabrication sequence following the epitaxial growth is illustrated in Fig. 16 for a ridge-waveguide laser. In the first lithographic step, the pattern of the lateral waveguide is defined and then transferred into the underlying semiconductor material using an etching process. For a ridge-waveguide laser, the width and the depth of the waveguide have to be controlled precisely to ensure a proper lateral-mode behavior of the device. This can be obtained using epitaxially grown etch-stop layers and anisotropic dry-etching processes. A dielectric isolator is then deposited at the bottom and the sidewalls of the ridge leaving uncovered only the highly doped layer for the ohmic contact on top of the ridge. For other types of lasers, the lateral waveguide can be produced by techniques like ion implantation or a sophisticated epitaxial regrowth technique. A second lithographic step followed by metal evaporation and a lift-off process in an organic solvent is used to define the structure of the top ohmic contact. Some diode lasers take advantage of a thick gold layer

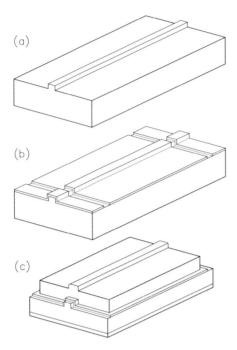

Fig. 16. Process sequence for the fabrication of ridge-waveguide diode lasers: **(a)** etching of the ridge, **(b)** deposition of top contact metallization, **(c)** electroplating of a gold heat spreader, substrate thinning, evaporation of the back-side contact, and mirror cleaving

which has been electroplated atop the ridge waveguide using a patterned photoresist as a mold for the galvanic deposition. The gold layer provides spreading of the heat generated in the laser leading to a significantly lower thermal resistance when the laser is mounted junction-side up. In addition, this layer protects the ridge against damage during processing, testing, and packaging and makes junction-side-down mounting more reliable.

After completion of the front-side processing, the wafer is thinned down to a thickness of around 100 μm to allow proper cleaving of the laser mirrors and to reduce the thermal resistance of junction-side-up-mounted devices. An ohmic contact is evaporated onto the back side and a baking process at high temperature provides alloying of the contact with the highly doped substrate material. To create the laser mirrors, the wafer is cleaved along crystallographic planes into bars having typical lengths of 1 cm. The cavity length L of the lasers is determined by the width of these bars. Figure 17 shows scanning electron micrographs of a laser chip on a bar with uncoated facets. After cleaving, the facets must be immediately coated to protect the mirrors against corrosion and to modify the reflectivity of the facets. Device testing is performed on the bar level. Single devices are cleaved from the bars

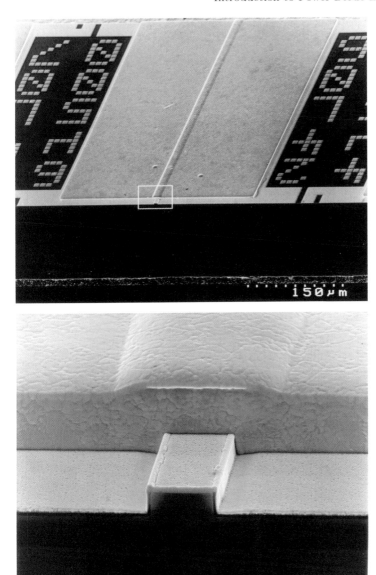

Fig. 17. Scanning electron micrographs of a (Al)GaInP/GaAs ridge-waveguide diode laser with a cavity length $L = 500\,\mu$m. A gold heat spreader is deposited atop the ohmic contact. The picture at the *top* features a single diode laser on a bar, the image at the *bottom* shows a closer view of the facet region ([25], © 1993 IEEE)

and then soldered into packages together with silicon p-i-n monitor diodes. For higher output powers, the chips have to be mounted junction-side down and submounts with low thermal conductivity have to be used to spread the heat. The packaged devices are again tested and often have to pass a burn-in procedure before being shipped to the customer. An alternative approach uses mirrors created by a highly anisotropic dry-etching process. Mirror coating and the device testing are performed on the full wafer. This process technology allows the monolithic integration of the devices and has the potential to make laser production easier and cheaper [22,23,24].

3   Optical Waveguides and Resonators

3.1   Effective Refractive Index

When a light wave propagates from a dielectric medium with refractive index n_f to a medium with refractive index $n_{cl} < n_f$ as illustrated in Fig. 18, refraction occurs at the interface according to Snell's refraction law

$$n_f \sin \varphi = n_{cl} \sin \varphi^\star , \tag{25}$$

with φ being the angle of incidence. The refraction angle $\varphi^\star > \varphi$ of the transmitted beam cannot be larger than $90°$ ($\sin \varphi^\star = 1$) corresponding to a critical angle of incidence φ_{crit} given by the equation

$$\sin \varphi_{crit} = \frac{n_{cl}}{n_f} . \tag{26}$$

If the angle of incidence φ is equal to or larger than the critical angle φ_{crit}, total reflection of the incoming wave occurs at the interface. The phase shift of the reflected wave depends on the incidence angle φ and the polarization of the wave. The incidence plane is defined as the plane that contains the wave vector k orientated in the propagation direction and the normal vector to the interface plane. For Transverse Electric (TE) polarization, where the

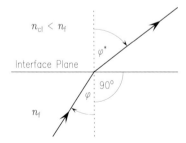

Fig. 18. Illustration of Snell's refraction law for a light ray coming from a medium with refractive index n_f and being transmitted into an optically thinner medium with refractive index $n_{cl} < n_f$

electrical field vector $\mathcal{E}$ is perpendicular to the incidence plane, the half phase-shift angle Φ_{TE} of the totally reflected wave is given by

$$\tan \Phi_{\text{TE}} = \frac{\sqrt{n_{\text{f}}^2 \sin^2 \varphi - n_{\text{cl}}^2}}{n_{\text{f}} \cos \varphi}. \tag{27}$$

In the case of Transverse Magnetic (TM) polarization, where the electrical field vector is in the incidence plane, the half phase-shift angle Φ_{TM} is given by

$$\tan \Phi_{\text{TM}} = \frac{n_{\text{f}}^2}{n_{\text{cl}}^2} \frac{\sqrt{n_{\text{f}}^2 \sin^2 \varphi - n_{\text{cl}}^2}}{n_{\text{f}} \cos \varphi}. \tag{28}$$

A three-layer dielectric slab waveguide consists of a film with refractive index n_{f} embedded between a cladding layer with refractive index n_{cl} and a substrate with refractive index n_{s}. The refractive indices of the substrate and cladding layers are lower than the index of the film. As shown in Fig. 19, a wave undergoes total reflection at both interfaces for large incidence angles φ when traveling through the film. In the absence of optical absorption in the material, the wave will be guided without energy losses in the two-dimensional film. Making the assumption that the wave vector $\boldsymbol{k} = (k_x, 0, k_z)$ has only components in the propagation direction z and the transverse direction x, the wave is reflected back and forth between the interfaces in the x direction. Such a wave can be described by the wave equation

$$\mathcal{E} = \mathcal{E}_0 \exp\left[\text{i} \left(\omega t \mp k_x x - k_z z \right) \right]. \tag{29}$$

The electric field vector $\mathcal{E}$ points in the y direction for a TE-polarized wave and is located in the x–z plane for TM polarization. The $\boldsymbol{k}$ vector of the wave

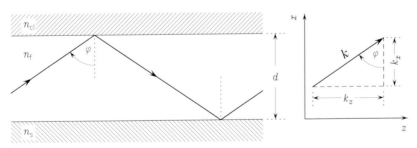

Fig. 19. Guided optical wave in a film with vertical thickness d being back and forth reflected at the interfaces to cladding layer and substrate. The wave vector $\boldsymbol{k}$ is split into a component k_z in the propagation direction and a component k_x in the vertical direction. To allow total reflection, the angle of incidence φ must be larger than the critical angle φ_{crit} defined by (26) at both interfaces

can be split into a component k_z which describes the propagation along the waveguide and a component k_x:

$$k_x = n_f k_0 \cos\varphi, \qquad k_z = n_f k_0 \sin\varphi, \qquad k_0 = \frac{2\pi}{\lambda_0}. \qquad (30)$$

The component k_z is used to define the effective refractive index n_{eff} for the traveling wave in the propagation direction

$$n_{\text{eff}} = n_f \sin\varphi, \qquad k_z = n_{\text{eff}} k_0. \qquad (31)$$

In the x direction, however, a standing wave has to build up, otherwise destructive interference will cause an extinction of the wave. In an asymmetric waveguide, the refractive index of the layers above and below the film may differ, e.g. $n_{\text{cl}} \leq n_s$. In order to allow total reflection at both interfaces, $\sin\varphi$ must be larger than n_{cl}/n_f and n_s/n_f, according to (26). On the other hand, $\sin\varphi$ is always equal to or smaller than 1.

$$
\begin{aligned}
n_{\text{cl}}/n_f \leq n_s/n_f \leq & \ \sin\varphi \ \leq 1, \\
n_{\text{cl}} \ \leq \ n_s \ & \leq n_f \sin\varphi \leq n_f, \\
n_{\text{cl}} \ \leq \ n_s \ & \leq \ n_{\text{eff}} \ \leq n_f.
\end{aligned}
\qquad (32)
$$

This equation limits the range for the effective refractive index n_{eff}.

To allow the formation of a standing wave in the x direction, the phase path must be a multiple of 2π after a roundtrip in the vertical cavity with thickness d. The phase-shift angles $2\Phi_{\text{cl}}$ and $2\Phi_s$ from (27) or (28) after the total reflection at the cladding layer and the substrate, respectively, must be taken into account.

$$2 d n_f k_0 \cos\varphi_m - 2\Phi_{\text{cl}} - 2\Phi_s = m \, 2\pi. \qquad (33)$$

m is the number of nodes of the standing wave and is defined as the order number of the transverse optical mode. The mode $m = 0$ is called the fundamental mode. For given geometry (d), refractive indices (n_f, n_s, n_{cl}), and wavelength λ_0, (33) determines the allowed values φ_m for the incidence angle φ depending of the mode number m.

3.2   Normalized Propagation Diagrams

In order to present calculated values for the effective refractive index n_{eff} for all kinds of geometries, refractive indices, and wavelengths, normalized parameters are usually defined.

$$V = k_0 d \sqrt{n_f^2 - n_s^2}, \qquad b = \frac{n_{\text{eff}}^2 - n_s^2}{n_f^2 - n_s^2}. \qquad (34)$$

V is called the normalized frequency since $k_0 = 2\pi/\lambda_0 = 2\pi\nu/c$ is proportional to the frequency ν of the wave. b is the normalized propagation parameter which is a measure for the effective refractive index n_{eff}. Asymmetric waveguides are taken into account by defining normalized asymmetry parameters a_{TE} and a_{TM} for transverse electric and magnetic waves:

$$a_{\text{TE}} = \frac{n_s^2 - n_{cl}^2}{n_f^2 - n_s^2}, \qquad a_{\text{TM}} = \frac{n_f^4}{n_{cl}^4} \frac{n_s^2 - n_{cl}^2}{n_f^2 - n_s^2}. \tag{35}$$

Using these normalized parameters, (33) can be transformed into

$$V\sqrt{1-b} = m\pi + \arctan\left(\sqrt{\frac{b}{1-b}}\right) + \arctan\left(\sqrt{\frac{b + a_{\text{TE}}}{1-b}}\right) \tag{36}$$

for TE-polarized waves. This dependency is plotted in Fig. 20 for the first three modes. There is always a solution to (36) for the fundamental mode ($m = 0$) in a symmetric waveguide ($a = 0$). In asymmetric waveguides, propagation is not possible below a cut-off frequency $V_0 = \arctan\sqrt{a_{\text{TE}}}$ for fundamental modes. For higher-order modes, the cut-off frequency is $V_m = V_0 + m\pi$.

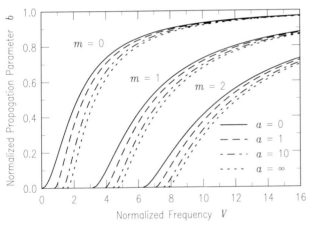

Fig. 20. Normalized propagation parameter b versus normalized frequency V at different normalized asymmetry parameters a for TE modes in a three-layer dielectric waveguide [26,27]

In semiconductor-diode lasers, fundamental-mode operation is desired in transverse and lateral directions. So, in both directions, the V parameter must be below the cut-off frequency $V_1 = \pi$ for higher-order modes. According to (34), for a given wavelength $\lambda_0 = 2\pi/k_0$ the V parameter is proportional to the product of the film thickness d and the refractive-index step $\sqrt{n_f^2 - n_s^2}$ between film and cladding layers. Since the waveguide in the transverse direction consists of epitaxially grown semiconductor material, small layer thicknesses (below $d = 1.5\,\mu\text{m}$) can be accurately controlled. Therefore, a relatively high refractive-index step is possible, which has the additional advantage of good carrier confinement since a high index step $\Delta n = n_f - n_{cl}$ is attributed to a high step in band-gap energy $\Delta E_g = E_{g,cl} - E_g$ (Fig. 4). For lateral index guiding, the width of the waveguide is defined by the ohmic contact or the ridge geometry (Fig. 9) which is patterned by a lithographic process allowing only inferior dimensional control. On the other hand, a small lateral waveguide will result in a high series resistance of the device, causing a lot of undesired ohmic heat generation. For these reasons, a relatively wide ($W > 3\,\mu m$) lateral waveguide is used together with a low effective-refractive-index step. To form a lateral-index waveguide, a stripe must be defined, where the effective refractive index of the transverse waveguide under the stripe is higher than outside it. This lateral control of the effective refractive index can be achieved in several ways:

- by a lateral change in the asymmetry $a = a(y)$, e.g. changing the thickness (ridge-waveguide laser) or the refractive index of one of the cladding layers,
- by a lateral change in the film thickness $d = d(y)$, or
- by a lateral change in the refractive index of the film $n_f = n_f(y)$ as in a buried-heterostructure laser (see Sect. 1.5).

3.3   Optical Near- and Far-Field Patterns

The amplitude of the electric-field vector $\mathcal{E}(x, y, z)$ in dielectric material with given refractive-index distribution $n(x, y, z)$ for sinusoidal time dependencies at a fixed angular frequency ω can be derived using the Helmholtz equation [28]

$$\left(\frac{\partial^2}{\partial x^2} + \frac{\partial^2}{\partial y^2} + \frac{\partial^2}{\partial z^2} \right) \mathcal{E}(x, y, z) + k_0^2 n^2(x, y, z)\, \mathcal{E}(x, y, z) = 0 \,. \tag{37}$$

In a film waveguide, the refractive-index profile $n(x, y, z) = n(x)$ only has a dependency in the x direction as shown in Fig. 19. For a TE-polarized wave propagating in the z direction, the electric field vector $\mathcal{E}$ is oriented in the y direction. Such a wave can be described by

$$\mathcal{E}_y(x, z) = \mathcal{E}_y(x) \exp(-\mathrm{i}\, n_{eff} k_0 z) \,. \tag{38}$$

Inserting (38), the Helmholtz equation (37) can be written as

$$\frac{\partial^2 \mathcal{E}_y(x)}{\partial x^2} + \left[k_0^2 n^2(x) - k_0^2 n_{eff}^2 \right] \mathcal{E}_y(x) = 0 \,, \tag{39}$$

which is a one-dimensional differential equation yielding the eigenvalue n_{eff} and the y-polarized vertical electric-field distribution $\mathcal{E}_y(x)$.

The field distribution and the effective refractive index of a one-dimensional dielectric waveguide can be approximated using the effective-index method, where the lateral effective-index variation $n_{eff}(y)$ is determined by solving (39) for the different lateral positions y. This effective-index distribution is then again inserted into a one-dimensional Helmholtz equation similar to (39) for a lateral waveguide, this time for a TM-polarized wave (magnetic-field vector orientated in the $-x$ direction) and yields the effective index of the one-dimensional waveguide and the field distribution in the lateral direction. This method is a rather good approximation for waveguides of edge-emitting diode lasers, which are thin in the vertical and broad in the lateral direction. As illustrated in Fig. 21 for a ridge-waveguide laser, the field distribution $\mathcal{E}(x,y)$ at the laser facet is called the near-field profile.

When the emitted light leaves the laser resonator and propagates from the waveguide into free space, the beam is broadened by diffraction. The diffracted pattern some distance away from the facet is called the far field. The transition occurs at a distance roughly W^2/λ_0, where W is the width of the near-field pattern. The far-field intensity $J(\Theta_x, \Theta_y)$ can be deduced from the Fresnel–Kirchhoff diffraction integral [10]. For the direction x, the far-field intensity distribution is given by

$$J(\Theta_x) \propto \cos^2 \Theta_x \left| \int_{-\infty}^{+\infty} \mathcal{E}(x) \exp(\mathrm{i}\, k_0 \sin \Theta_x x)\, dx \right|^2 . \tag{40}$$

Replacing x by y in (40) yields the same equation for the y direction. The factor $\cos \Theta_x$ in front of the integral is called the Huygens obliquity factor, which accounts for the fact that the intensity from a surface tilted by an angle Θ is reduced by $\cos^2 \Theta$. For very large off-axis angles, slight corrections must be added to the Huygens factor, which also depends on the polarization

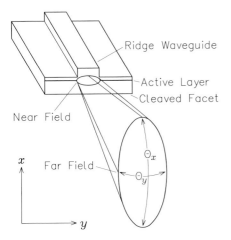

Fig. 21. The transition from the optical near-field profile $\mathcal{E}(x,y)$ at the laser facet to the far-field intensity pattern in free space $J(\Theta_x, \Theta_y)$ is described by diffraction theory

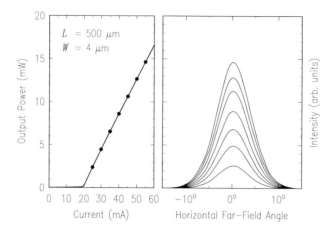

Fig. 22. Output characteristic and horizontal far-field patterns for a ridge-waveguide laser with a ridge width $W = 4\,\mu\text{m}$ and a cavity length $L = 500\,\mu\text{m}$ [29]. Single-mode behavior with perfectly Gaussian far-field patterns is obtained

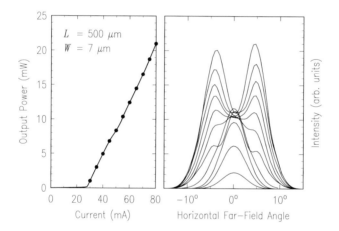

Fig. 23. Output characteristic and horizontal far-field patterns for a ridge-waveguide laser with a ridge width $W = 7\,\mu\text{m}$ and a cavity length $L = 500\,\mu\text{m}$. The ridge width of this laser is close to the limit, where higher-order modes can develop in the waveguide. When the laser is operated at higher power levels, the lateral effective refractive-index step is increased due to the higher temperature under the ridge allowing the propagation of a second lateral mode associated with a kink in the light-output characteristic

of the beam. For small angles, the far-field pattern is the Fourier transform of the near-field distribution. Thus, a narrow emitting aperture leads to a wide angular profile in the far field and vice versa. If the near-field function $\mathcal{E}(x, y)$ has a constant phase front at the facet, the far-field profile, being the Fourier transform of $\mathcal{E}(x, y)$, is always a symmetric function even if the emitted near field shows an asymmetric amplitude distribution. Waveguides with real index guiding have planar wavefronts in the propagation direction. In gain-guided lasers, the near field has a curved wave front originating from the optical absorption in the lateral cladding layers where no current is injected (see Sect. 1.5).

Figures 22 and 23 show lateral far-field patterns with increasing output powers of two ridge-waveguide lasers having different ridge widths W. Whereas for the small ridge width (Fig. 22, $W = 4\,\mu m$), nice Gaussian profiles are observed, a second-order mode develops for higher output powers in the laser with larger ridge width (Fig. 23, $W = 7\,\mu m$).

3.4   Fabry–Perot Resonator

An electromagnetic wave traveling in the z direction in a waveguide with an absorption coefficient α and an effective refractive index n_{eff} can be described using the wave equation

$$\mathcal{E} = \mathcal{E}_0 \exp\left(-\frac{\alpha}{2}\right) \exp\left[i\left(\omega t - k_0 n_{\mathrm{eff}} z\right)\right] . \tag{41}$$

The effective refractive index n_{eff} describing the phase evolution and the absorption coefficient α describing the amplitude evolution can be combined in a complex propagation number γ.

$$\mathcal{E} = \mathcal{E}_0 \exp\left(-\gamma z\right) \exp\left(i\omega t\right) , \tag{42}$$

$$\gamma = \frac{\alpha}{2} + i k_0 n_{\mathrm{eff}} , \qquad k_z = k_0 n_{\mathrm{eff}} = \frac{2\pi}{\lambda_0} n_{\mathrm{eff}} . \tag{43}$$

In a Fabry–Perot resonator, multiple reflections occur at the mirrors which have transmission coefficients t_1, t_2 and reflection coefficients r_1, r_2 for the amplitude of the wave. As illustrated in Fig. 24, the transmitted wave $\mathcal{E}_t$ can be expressed as a superposition of the contributions from these multiple reflections.

$$\mathcal{E}_t = t_1 t_2\, \mathcal{E}_i \exp\left(-\gamma L\right) \sum_{m=0}^{\infty} \left[(r_1 r_2)^m \exp\left(-2m\gamma L\right)\right] . \tag{44}$$

This geometric series can be transformed into

$$\frac{\mathcal{E}_t}{\mathcal{E}_i} = \frac{t_1 t_2 \exp\left(-\gamma L\right)}{1 - r_1 r_2 \exp\left(-2\gamma L\right)} , \tag{45}$$

which is called the Airy function for the Fabry–Perot resonator. At laser threshold, the denominator of (45) becomes zero, resulting in a transmitted wave having infinitely large amplitude.

$$1 = r_1 r_2 \exp\left(-2\gamma L\right) = r_1 r_2 \exp\left(-\alpha L\right) \exp\left(-i\frac{4\pi}{\lambda_0} n_{\mathrm{eff}} L\right) . \tag{46}$$

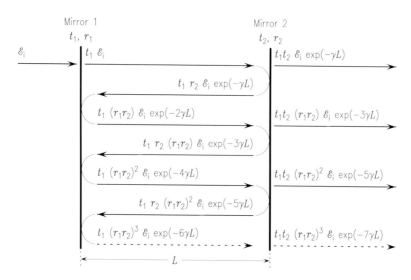

Fig. 24. Multiple reflections of a planar electromagnetic wave inside a Fabry–Perot resonator with a cavity length L. The mirrors have amplitude transmissions t_1, t_2 and amplitude reflectivities r_1, r_2

This equation can be split into two parts, yielding conditions for the phase and the amplitude.

$$1 = r_1 r_2 \exp\left(-\alpha L\right) , \tag{47}$$

$$1 = \exp\left(-\mathrm{i}\frac{4\pi}{\lambda_0} n_{\mathrm{eff}} L\right) . \tag{48}$$

The amplitude condition in (47) can be written as

$$1 = r_1 r_2 \exp(-\alpha L) ,$$
$$0 = \ln(r_1 r_2) - \alpha L ,$$
$$\alpha = \frac{1}{L} \ln(r_1 r_2) . \tag{49}$$

Replacing the absorption coefficient α by $\alpha_{\mathrm{i}} - \Gamma g_{\mathrm{th}}$ according to (22) for the threshold gain g_{th} and substituting the amplitude reflection factors r_1, r_2 by the intensity reflectivities $R_1 = r_1^2$ and $R_2 = r_2^2$, the equation can be further transformed into the laser-threshold condition known from (24).

$$\Gamma g_{\mathrm{th}} = \alpha_{\mathrm{i}} - \frac{1}{L} \ln(r_1 r_2)$$

$$= \alpha_{\mathrm{i}} + \frac{1}{L} \ln\left(\frac{1}{r_1 r_2}\right)$$

$$= \alpha_{\mathrm{i}} + \frac{1}{2L} \ln\left(\frac{1}{R_1 R_2}\right) . \tag{50}$$

Equation (48) yields the phase condition known from (19).

$$\frac{4\pi}{\lambda_0} n_{\text{eff}} L = m\, 2\pi\,, \qquad m = \frac{2L n_{\text{eff}}}{\lambda_0}\,,$$

$$L = m\, \frac{\lambda_0}{2n_{\text{eff}}}\,, \qquad m = 1,\, 2,\, 3,\, \dots\,. \tag{51}$$

As shown in Sect. 3.1, the effective refractive index n_{eff} of an optical waveguide is a function of the wavelength λ_0 even if there is no dispersion in the semiconductor material itself. The spectral separation between two neighboring modes can be derived by building the total differential of

$$m\, \lambda_0 = 2L\, n_{\text{eff}}\,. \tag{52}$$

$$m\, \partial\lambda_0 + \lambda_0\, \partial m = 2L\, \partial n_{\text{eff}}\,,$$

$$\frac{2L n_{\text{eff}}}{\lambda_0}\, \partial\lambda_0 + \lambda_0\, \partial m = 2L\, \frac{\partial n_{\text{eff}}}{\partial\lambda_0}\, \partial\lambda_0\,. \tag{53}$$

According to (52), a decrease of m by 1 ($\partial m \approx \Delta m = -1$) switches to the next-higher resonant value for λ_0. The wavelength separation $\Delta\lambda_{\text{FP}} \approx \partial\lambda_0$ between two optical modes is called the free spectral range of the Fabry–Perot resonator and results in

$$\frac{2L n_{\text{eff}}}{\lambda_0}\, \Delta\lambda_{\text{FP}} - \lambda_0 \approx 2L\, \frac{\partial n_{\text{eff}}}{\partial\lambda_0}\, \Delta\lambda_{\text{FP}}\,,$$

$$2L\, n_{\text{eff}}\, \Delta\lambda_{\text{FP}} - \lambda_0^2 \approx 2L\, \frac{\partial n_{\text{eff}}}{\partial\lambda_0}\, \lambda_0\, \Delta\lambda_{\text{FP}}\,,$$

$$\Delta\lambda_{\text{FP}} \approx \frac{\lambda_0^2}{2L\left(n_{\text{eff}} - \frac{\partial n_{\text{eff}}}{\partial\lambda_0}\lambda_0\right)} \approx \frac{\lambda_0^2}{2n_{\text{gr,eff}}L}\,, \tag{54}$$

which depends on the mode dispersion $\partial n_{\text{eff}}/\partial\lambda_0$.

$$n_{\text{gr,eff}} \approx n_{\text{eff}} - \frac{\partial n_{\text{eff}}}{\partial\lambda_0}\, \lambda_0 \approx n_{\text{eff}} + \frac{\partial n_{\text{eff}}}{\partial\omega}\, \omega \tag{55}$$

is the group effective index. $v_{\text{gr}} = c/n_{\text{gr,eff}}$ is the group velocity of the optical mode. The group effective index is typically 20–30% larger than the effective refractive index depending on the specific photon energy relative to band-gap energy.

3.5 Diode-Laser Spectrum

Figure 25 shows the spectrum of the longitudinal Fabry–Perot modes together with the modal gain at the laser threshold. When the peak of the modal gain at the wavelength λ_{p} is equal to the threshold gain necessary to overcome the intrinsic absorption and the mirror losses (according to (50)), the diode laser starts to operate at the mode which is in the closest spectral

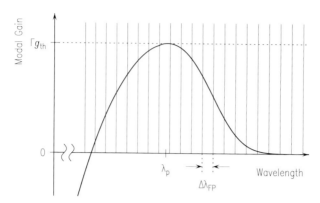

Fig. 25. Modal gain spectrum and Fabry–Perot modes of a diode laser at threshold. The modal gain $\Gamma g(\lambda_0)$ has its maximum value Γg_{th} at a peak wavelength λ_{p}. The spectral distance $\Delta\lambda_{\mathrm{FP}}$ of the Fabry–Perot modes is called the free spectral range

vicinity to the peak wavelength λ_{p}. As will be shown in Sect. 4.1, the modal gain is clamped above threshold at the threshold gain value Γg_{th}. Therefore only the mode closest to the peak gain is amplified, whereas for the other longitudinal modes, the losses are higher than the modal gain. This is shown in the spectrum of a diode laser above threshold in Fig. 26 where a suppression of the side modes of 31 dB is achieved. The emission wavelength of a diode laser changes when the temperature of the device is varied. Since the refractive index increases with temperature, the wavelength of a longitudinal optical mode at a given order number m increases according to (52). Additionally, the length of the laser cavity increases with temperature due to thermal expansion of the material, leading to a higher emission wavelength. The wavelength shift with temperature caused by these two effects is in the range $\Delta\lambda_0/\Delta T = 0.06\text{--}0.2\,\mathrm{nm/K}$. A stronger effect is the shift of the spectral-gain curve which is mainly determined by the decrease of the band-gap energy with temperature. The shift of the peak-gain wavelength λ_{p} with temperature T is approximately $\Delta\lambda_{\mathrm{p}}/\Delta T = 0.33\,\mathrm{nm/K}$. Both effects can be seen in Fig. 27 showing the variation of the emission wavelength of a laser with temperature.

3.6   Mirror Coatings

Dielectric mirror coatings are deposited on laser facets for two main reasons. Mirror coatings passivate and protect the extremely sensitive surfaces of the facets. Corrosion which results in device degradation and sudden failures can be reduced or even completely eliminated by a suitable coating process. On the other hand, the reflectivity of the mirrors can be changed, allowing the entire output power of the laser to be emitted through the front facet. Furthermore, by using a very low reflection at the output facet, the optical

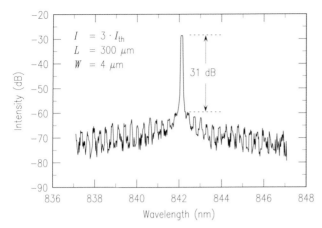

Fig. 26. Emission spectrum of an AlGaAs/GaAs single-mode diode laser with oxide aperture providing lateral current confinement and index guiding of the optical mode [30]. Since the cavity length is rather small ($L = 300\,\mu m$), the lateral modes can be resolved by the spectrometer. At an operating current $I = 3\,I_{th}$, a single longitudinal mode at a wavelength 842.1 nm dominates and a side-mode suppression of 31 dB is obtained

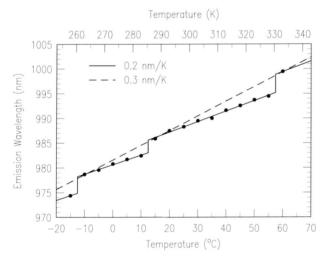

Fig. 27. The peak-emission wavelength of a diode laser increases with temperature. The device in this example shows a shift of 0.2 nm/K due to the change in refractive index and due to thermal expansion of the material. Since the peak gain shifts by approximately 0.33 nm/K, longitudinal mode hops to the next-lower-order number m occur with increasing temperature

density inside the cavity can be reduced, resulting in less filamentation of the lateral optical mode and an increase in the output power level where destruction of the facet occurs.

Figure 28 shows a typical facet coating for a high-power diode laser. Directly on both facet surfaces, Al_2O_3 layers are deposited. This material has a refractive index of approximately $n = 1.65$ and is well known for its good adhesion and passivation properties. The reflectivity R of the front facet depends on the layer thickness d and its refractive index n.

$$R(n,d) = \frac{(1 - n_{\text{eff}})^2 \cos^2(nk_0d) + \left(\frac{n_{\text{eff}}}{n} - n\right)^2 \sin^2(nk_0d)}{(1 + n_{\text{eff}})^2 \cos^2(nk_0d) + \left(\frac{n_{\text{eff}}}{n} + n\right)^2 \sin^2(nk_0d)}, \tag{56}$$

with $k_0 = 2\pi/\lambda_0$ and n_{eff} being the effective refractive index for the optical wave traveling in the waveguide between the laser mirrors. This is a periodic function with a periodicity in thickness of $\lambda_0/(2n)$ as illustrated on the left-hand side of Fig. 29. The minimum and maximum value R_{min} and R_{max} of the reflectivities are

$$R_{\text{min}} = \frac{\left(n_{\text{eff}} - n^2\right)^2}{\left(n_{\text{eff}} + n^2\right)^2}, \qquad R_{\text{max}} = \frac{(1 - n_{\text{eff}})^2}{(1 + n_{\text{eff}})^2}. \tag{57}$$

R_{max} is the natural reflectivity for an uncoated facet. The minimum reflectivity R_{min} is achieved at a layer thickness $d = \lambda_0/(4n)$. If the coating has a refractive index $n = \sqrt{n_{\text{eff}}}$, antireflection can be obtained.

(56) and (57) are only valid for a dielectric film on bulk material with a refractive index n_{eff} at normal incidence of the planar electromagnetic wave. If the wave is confined in a dielectric waveguide, significant modifications to these equations have to be considered [31,32].

Coatings for the back mirror having a high reflectivity consist of Bragg stacks. These are pairs of layers with high and low refractive index. The thickness of each layer is $\lambda_0/(4n)$. The constructive interference in the stacks provides a higher reflectivity which increases with the number of layer pairs

Fig. 28. Typical mirror coating for high-power edge-emitting lasers. The front facet is coated with a single layer of Al_2O_3 to reduce the reflectivity whereas on the back facet, a Bragg-mirror stack consisting of two pairs of Al_2O_3/Si layers is used to obtain a high reflectivity

and the difference in the refractive indices of the two materials. Standard coating materials for the Bragg reflector are Al_2O_3 with a refractive index $n = 1.65$ and Si with a refractive index in the range $n = 3.5$–4. The calculated spectral reflectivity of such a back mirror is plotted on the right-hand side of Fig. 29. With two layer pairs, a reflectivity of above 90% is obtained for a rather broad spectral range. The refractive index of Si strongly varies with wavelength and, for short wavelengths ($\lambda_0 < 700\,\mathrm{nm}$), the absorption becomes significant, leading to facet heating. For the short-wavelength range, the low-absorbing TiO_2 ($n = 2.45$) in combination with SiO_2 ($n = 1.45$) is a good alternative.

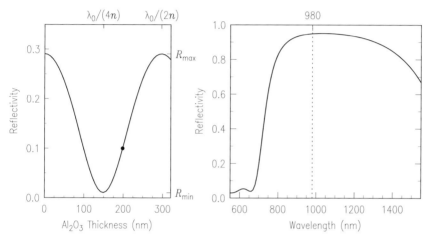

Fig. 29. Reflectivity R of the front and the back mirrors of a 980 nm laser having facet coatings as illustrated in Fig. 28. For the front mirror, an Al_2O_3 layer with a refractive index $n = 1.65$ is used. In the diagram on the *left-hand side*, the reflectivity versus Al_2O_3 layer thickness is plotted. At a thickness of 200 nm, 10% front-mirror reflectivity is achieved. A Bragg reflector with two pairs of Al_2O_3/Si layers is used to increase the back-mirror reflectivity above 90%. The reflectivity versus wavelength is plotted in the *right-hand-side* diagram. The data have been calculated using the transfer-matrix model [33]

4   Rate Equations and High-Power Operation

4.1   Rate Equations for Electronic Carriers and Photons

A phenomenological approach to describe the behavior of diode lasers during operation is the set of the rate equations. These are two coupled equations expressing balances for electronic carriers and photons. The rate equations provide relations between the external laser parameters extracted from the measured device characteristics and the internal physical effects.

The generation and recombination balance for the electronic carriers in the active region of a laser can be written as

$$\frac{dN}{dt} = \mathcal{R}_{\text{gen}} - \mathcal{R}_{\text{nr}} - \mathcal{R}_{\text{sp}} - \mathcal{R}_{\text{stim}} . \tag{58}$$

A change dN/dt of the carrier density N in the active region is attributed to the difference between the carrier-generation rate $\mathcal{R}_{\text{gen}}$ and the recombination rate $(\mathcal{R}_{\text{nr}} + \mathcal{R}_{\text{sp}} + \mathcal{R}_{\text{stim}})$. The generation rate for carriers is the number of electrons per time and volume which are injected into the active region. The number of the electrons per time is the electric current I divided by the elementary charge $q = 1.602\,189 \times 10^{-19}\,\text{C}$ of an electron. The volume of the active region is given by the product of the active-film thickness d and the current-injection area (LW). With the current density j being the current I divided by the injection area (LW), the carrier-generation rate can be written as

$$\mathcal{R}_{\text{gen}} = \frac{\eta_i\,j}{q\,d} , \tag{59}$$

where the internal efficiency η_i is the fraction of the current which generates carriers in the active region. The introduction of $\eta_i < 1$ takes into account that part $(1 - \eta_i)$ of the current does not enter the active region.

For the recombination rate of carriers, several processes must be considered, the nonradiative recombination $\mathcal{R}_{\text{nr}}$, spontaneous emission $\mathcal{R}_{\text{sp}}$, and stimulated emission $\mathcal{R}_{\text{stim}}$. For a spontaneous-recombination process, the presence of an electron and a hole is required. Therefore, the spontaneous-recombination rate $\mathcal{R}_{\text{sp}}$ is proportional to the product of the electron and hole densities. In undoped active regions, charge neutrality requires that both densities are equal; thus, the spontaneous-recombination rate is proportional to N^2.

$$\mathcal{R}_{\text{sp}} = B\,N^2 . \tag{60}$$

B is called the bimolecular recombination coefficient, and it has a value around $B \approx 10^{-10}\,\text{cm}^3/\text{s}$ for most III–V semiconductor materials.

In semiconductors, there are two major nonradiative recombination mechanisms. For the first mechanism, the recombination at point defects, the rate is proportional to the carrier density. The second mechanism is the Auger recombination, where the photon energy from a spontaneous recombination process is transferred to an electron in the form of kinetic energy (photo effect). Since this mechanism depends on the presence of three carriers, the rate is proportional to N^3.

$$\mathcal{R}_{\text{nr}} = A\,N + C\,N^3 . \tag{61}$$

In the absence of photons in the device, the carrier-recombination rate $\mathcal{R}_{\text{stim}}$ of the stimulated emission is negligible. In this case, the device behaves like a conventional LED. After switching off the external current, the generation of carriers is terminated and the carrier density in the device decays.

$$\frac{dN}{dt} = -\mathcal{R}_{\mathrm{nr}} - \mathcal{R}_{\mathrm{sp}} = -A\,N - B\,N^2 - C\,N^3. \tag{62}$$

To simplify the description of this decay, an exponential law with a carrier lifetime τ is used.

$$N = N_0 \exp\left(-\frac{t}{\tau}\right), \tag{63}$$

with N_0 being the initial carrier density. This exponential function is the solution of the differential equation

$$\frac{dN}{dt} = -\frac{N}{\tau}. \tag{64}$$

Compared to (62), this is an improper description of the nonradiative and spontaneous recombination in the device.

As shown in Sect. 1.3, the density of photons N_{ph} exponentially increases when light is traveling in a medium with optical gain originated from stimulated-emission processes.

$$N_{\mathrm{ph}}(z) = N_{\mathrm{ph}}(0) \exp[\,g(N,\lambda_0)z\,], \tag{65}$$

with the material gain $g(N,\lambda_0)$ depending on the carrier density N and the vacuum wavelength λ_0. This exponential function is the solution to the differential equation

$$\frac{dN_{\mathrm{ph}}}{dz} = g(N,\lambda_0)\,N_{\mathrm{ph}} \tag{66}$$

describing the photon-generation rate per unit length, which can be converted into the photon-generation rate per unit time by

$$\frac{dN_{\mathrm{ph}}}{dt} = \frac{dN_{\mathrm{ph}}}{dz}\frac{dz}{dt} = g(N,\lambda_0)\,N_{\mathrm{ph}}\,v_{\mathrm{gr}}. \tag{67}$$

v_{gr} is the group velocity of the photons in the active material. In the case of a dielectric waveguide, $v_{\mathrm{gr}} = c/n_{\mathrm{gr,eff}}$ is given by the group effective index $n_{\mathrm{gr,eff}} \approx n_{\mathrm{eff}} - (\partial n_{\mathrm{eff}}/\partial\lambda_0)\,\lambda_0$ as defined in (55). The photons which are generated in the active material are carrier losses by stimulated emission

$$\mathcal{R}_{\mathrm{stim}} = v_{\mathrm{gr}}\,g(N,\lambda_0)\,N_{\mathrm{ph}}. \tag{68}$$

Now the rate equation for the carriers is complete and can be written as

$$\frac{dN}{dt} = \frac{\eta_i\,j}{q\,d} - \frac{N}{\tau} - v_{\mathrm{gr}}\,g(N,\lambda_0)\,N_{\mathrm{ph}}. \tag{69}$$

It is common to approximate the dependency of the material gain $g(N, \lambda_0)$ as a linear function of the carrier density N.

$$g(N) = a_\mathrm{d} \left(N - N_\mathrm{tr} \right), \qquad a_\mathrm{d} = \frac{\partial g}{\partial N}, \qquad (70)$$

with a_d being the differential gain coefficient and N_tr the transparency carrier density. N_tr is the carrier density, where the material losses of the active medium are compensated by the optical gain, resulting in material which is optically transparent for light with vacuum wavelength λ_0.

The electronic carriers are confined in a layer with vertical thickness d. The photons are confined by a separate optical waveguide. The volume occupied by photons is therefore larger. The confinement factor Γ as defined in (21) can be regarded as a carrier–photon overlap factor, being the volume occupied by carriers divided by the volume occupied by photons. Since the volume for the photons is larger by the factor $1/\Gamma$ than the volume for the carriers, the rates per unit volume $\mathcal{R}$ are smaller by the factor Γ for the photons generated by spontaneous and stimulated emission.

$$\frac{dN_\mathrm{ph}}{dt} = \Gamma \, \mathcal{R}_\mathrm{stim} + \Gamma \, \beta_\mathrm{sp} \, \mathcal{R}_\mathrm{sp} - \frac{N_\mathrm{ph}}{\tau_\mathrm{ph}}, \qquad (71)$$

where $\beta_\mathrm{sp} < 1$ is the spontaneous-emission factor, which accounts for the fact that in contrast to the stimulated emission only the fraction β_sp of the spontaneously generated photons are emitted into the lasing mode. β_sp is approximately the reciprocal of the number of optical modes in the bandwidth of the spontaneous emission and therefore strongly depends on the cavity length and the geometry of the optical waveguide. While spontaneous and stimulated emission are photon-generation processes, the photon losses are taken into account by introducing a lifetime τ_ph for the photons in the laser cavity in analogy to (64). The differential equation (67) can be used to describe the exponential decay of the photon density with time in the absence of any photon-generation terms ($\mathcal{R}_\mathrm{stim} = \mathcal{R}_\mathrm{sp} = 0$).

$$\frac{dN_\mathrm{ph}}{dt} = -\frac{N_\mathrm{ph}}{\tau_\mathrm{ph}} = \frac{dN_\mathrm{ph}}{dz} \frac{dz}{dt} = -(\alpha_\mathrm{i} + \alpha_\mathrm{mirror}) \, N_\mathrm{ph} \, v_\mathrm{gr}. \qquad (72)$$

The two loss mechanisms for the photons are the intrinsic absorption α_i and the mirror loss α_mirror. The mirror loss of a Fabry–Perot resonator has been determined in Sect. 1.3 where (24) provides a condition for the laser threshold.

$$\alpha_\mathrm{i} + \alpha_\mathrm{mirror} = \alpha_\mathrm{i} + \frac{1}{2L} \ln \left(\frac{1}{R_1 R_2} \right) = \Gamma g_\mathrm{th}. \qquad (73)$$

A comparison of (73) and (72) yields a measure for the photon lifetime τ_ph.

$$\frac{1}{\tau_\mathrm{ph}} = v_\mathrm{gr} \left[\alpha_\mathrm{i} + \frac{1}{2L} \ln \left(\frac{1}{R_1 R_2} \right) \right] = v_\mathrm{gr} \, \Gamma g_\mathrm{th}. \qquad (74)$$

Inserting (68) into (71) yields the complete rate equation for the photons

$$\frac{dN_{\mathrm{ph}}}{dt} = v_{\mathrm{gr}} \, \Gamma \, g(N, \lambda_0) \, N_{\mathrm{ph}} + \Gamma \, \beta_{\mathrm{sp}} \, \mathcal{R}_{\mathrm{sp}} - \frac{N_{\mathrm{ph}}}{\tau_{\mathrm{ph}}} \,. \tag{75}$$

Both rate equations, (69) and (75), are strongly coupled by the stimulated emission process, which is a loss mechanism for the carriers and a generation process for the photons. The rate equations can be utilized to describe the static and dynamic behavior of diode lasers and LEDs.

4.2   Electrical and Optical Characteristics of Power Diode Lasers

For all further considerations, the contribution of the spontaneous emission to the photon density is neglected ($\beta_{\mathrm{sp}} = 0$), which is a good approximation for power diode lasers since these devices usually have large mode volumes resulting in rather low spontaneous-emission factors β_{sp}.

When a diode laser is operating in steady state, carrier and photon densities do not change,

$$\frac{dN}{dt} = 0, \qquad \frac{dN_{\mathrm{ph}}}{dt} = 0 \,. \tag{76}$$

So, the rate equations for the steady-state operation of diode lasers can be written as

$$0 = \frac{\eta_i \, j}{q \, d} - \frac{N}{\tau} - v_{\mathrm{gr}} \, g(N, \lambda_0) \, N_{\mathrm{ph}} \,, \tag{77}$$

$$0 = v_{\mathrm{gr}} \, \Gamma \, g(N, \lambda_0) \, N_{\mathrm{ph}} - \frac{N_{\mathrm{ph}}}{\tau_{\mathrm{ph}}} \,. \tag{78}$$

Rearranging (78) yields

$$N_{\mathrm{ph}} \left[v_{\mathrm{gr}} \, \Gamma \, g(N, \lambda_0) - \frac{1}{\tau_{\mathrm{ph}}} \right] = 0 \,, \tag{79}$$

and two solutions can be found to fulfill this equation.

$$N_{\mathrm{ph}} = 0 \,, \tag{80}$$

which is valid below laser threshold, where no photons are emitted with the exception of spontaneously generated photons, and

$$v_{\mathrm{gr}} \, \Gamma \, g(N, \lambda_0) = \frac{1}{\tau_{\mathrm{ph}}} \,, \tag{81}$$

which describes the situation above threshold. In this case, the gain is constant and according to (74) has a value $g(N, \lambda_0) = g_{\mathrm{th}}$. This behavior is called gain clamping. The carrier density $N = N_{\mathrm{th}}$ is also constant above threshold, since the gain $g(N, \lambda_0)$ is a function of N.

To derive the carrier density N below threshold, (80) is inserted into (77) giving

$$N = \frac{\eta_{\mathrm{i}}\, \tau}{q\, d}\, j \,. \tag{82}$$

The carrier density N below threshold is proportional to the density j of the injected current. In practice, this relation does not hold because (62) should be used to describe the carrier losses below threshold instead of (64). In this case, τ depends on the carrier density N.

$$\tau(N) = \frac{1}{A + B\,N + C\,N^2} \,. \tag{83}$$

To get the photon density above threshold, (81) is inserted into (77) giving

$$N_{\mathrm{ph}} = \Gamma \frac{\eta_{\mathrm{i}}\, \tau_{\mathrm{ph}}}{q\, d}\, j - \Gamma \frac{\tau_{\mathrm{ph}}}{\tau(N_{\mathrm{th}})}\, N_{\mathrm{th}} \,. \tag{84}$$

The photon density N_{ph} linearly increases with the injected current density j. The threshold current density j_{th} can be derived by setting $N_{\mathrm{ph}} = 0$.

$$j_{\mathrm{th}} = \frac{q\, d}{\eta_{\mathrm{i}}\, \tau(N_{\mathrm{th}})}\, N_{\mathrm{th}}. \tag{85}$$

Inserting this result into (84) yields

$$N_{\mathrm{ph}} = \eta_{\mathrm{i}} \frac{\Gamma\, \tau_{\mathrm{ph}}}{q\, d} (j - j_{\mathrm{th}}) = \eta_{\mathrm{i}} \frac{1}{q\, d\, v_{\mathrm{gr}}\, g_{\mathrm{th}}} (j - j_{\mathrm{th}}) \,. \tag{86}$$

The dependencies of carrier density N and photon density N_{ph} versus injected current density j are illustrated in Fig. 30 for the realistic case, where an additional spontaneous-emission process is considered and a nonlinear gain characteristic $g(N, \lambda)$ is assumed together with a nonlinear carrier-recombination mechanism according to (62).

Inserting the linear-gain approximation $g(N) = a_{\mathrm{d}}(N - N_{\mathrm{tr}})$ from (70) for $N = N_{\mathrm{th}}$ into (81) yields

$$N_{\mathrm{th}} = N_{\mathrm{tr}} + \frac{1}{v_{\mathrm{gr}}\, a_{\mathrm{d}}\, \Gamma\, \tau_{\mathrm{ph}}} = N_{\mathrm{tr}} + \frac{g_{\mathrm{th}}}{a_{\mathrm{d}}} \,. \tag{87}$$

N_{th} is larger than the transparency carrier density N_{tr}, since photon losses at the mirrors and inside the cavity have also to be compensated. These two loss mechanisms are combined in the photon lifetime τ_{ph}.

The volume occupied by electronic carriers in an edge-emitting laser is the product of the cavity length L, the lateral width W, and the active film thickness d. The volume occupied by photons is larger by a factor $1/\Gamma$ and therefore has a value $(LWd)/\Gamma$. Since each photon has a quantum energy $E_{\mathrm{ph}} = \hbar\omega$, the total optical energy E_{FP} stored in the Fabry–Perot resonator is

$$E_{\mathrm{FP}} = N_{\mathrm{ph}}\, \hbar\omega\, \frac{LWd}{\Gamma} \,. \tag{88}$$

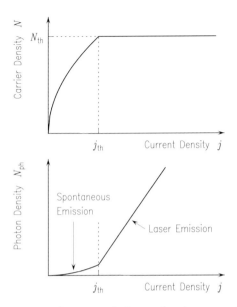

Fig. 30. Carrier and photon density versus injected current density of a diode laser as described by the rate equations. Below laser threshold, the device works like an LED with undoped recombination zone. Above threshold, the optical output power shows a linear increase with the injected current. The carrier density and thus the optical gain have a constant value. This behavior is called gain clamping

According to (72), the loss rate for photons through the laser mirrors is

$$\frac{dN_{\mathrm{ph}}}{dt} = -\alpha_{\mathrm{mirror}} \, v_{\mathrm{gr}} \, N_{\mathrm{ph}} \, . \tag{89}$$

The photons leaving the laser through the mirrors form the laser output. The output power P is the energy per time leaving the laser facets.

$$P = \alpha_{\mathrm{mirror}} \, v_{\mathrm{gr}} \, N_{\mathrm{ph}} \, \hbar\omega \, \frac{LWd}{\Gamma} \, . \tag{90}$$

The photon density N_{ph} above threshold is given by (86). Replacing the current densities j and j_{th} by the currents I and I_{th} divided by the injection area (LW) and inserting this equation into (90) yields

$$P = \eta_{\mathrm{i}} \, \frac{\alpha_{\mathrm{mirror}}}{\Gamma g_{\mathrm{th}}} \, \frac{\hbar\omega}{q} \, (I - I_{\mathrm{th}}) \, . \tag{91}$$

Now the threshold condition for the modal gain (Γg_{th}) (see (24) and (73))

$$\Gamma g_{\mathrm{th}} = \alpha_{\mathrm{i}} + \alpha_{\mathrm{mirror}} = \alpha_{\mathrm{i}} + \frac{1}{2L} \ln\left(\frac{1}{R_1 R_2}\right) \tag{92}$$

can be utilized to complete (91).

$$P = \eta_{\mathrm{i}} \frac{\alpha_{\mathrm{mirror}}}{\alpha_{\mathrm{i}} + \alpha_{\mathrm{mirror}}} \frac{\hbar\omega}{q} (I - I_{\mathrm{th}}) = \eta_{\mathrm{d}} \frac{\hbar\omega}{q} (I - I_{\mathrm{th}}) \ . \tag{93}$$

Above threshold current, the output power P linearly increases with current I. The differential efficiency η_{d} is defined as

$$\eta_{\mathrm{d}} = \frac{q}{\hbar\omega} \frac{dP}{dI} = \frac{(dP)/(\hbar\omega)}{(dI)/q} \ . \tag{94}$$

η_{d} is the differential increase in photons per time $(dP)/(\hbar\omega)$ emitted from the diode laser divided by the differential increase in injected electrons per time $(dI)/q$ above laser threshold. $(dP)/(dI)$ is the slope efficiency in W/A from the linear part of the laser characteristic above threshold.

In Fig. 31, the output-power characteristics of a real broad-area high-power diode laser are shown together with the current–voltage characteristic. The device shows an electrical-to-optical conversion efficiency of 63%. During operation of a high-power diode laser, 35–50% of the input power is dissipated as ohmic heat in the device. To avoid high device temperatures, a proper mounting is necessary, usually by using sophisticated junction-side-down soldering processes and water-cooled heat sinks.

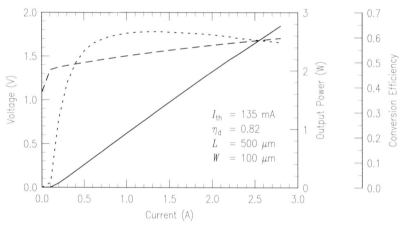

Fig. 31. Characteristics of a 950 nm InGaAs/AlGaAs broad-area diode laser with an active stripe width $W = 100\,\mu\mathrm{m}$ and a cavity length $L = 500\,\mu\mathrm{m}$ [34]. The optical output characteristic (*continuous line*) exhibits a threshold current $I_{\mathrm{th}} = 135\,\mathrm{mA}$ corresponding to a threshold current density $j_{\mathrm{th}} = 270\,\mathrm{A/cm^2}$ and a differential efficiency $\eta_{\mathrm{d}} = 0.82$. The current–voltage characteristic (*dashed line*) has a kink voltage 1.34 V which corresponds well to the photon energy $E_{\mathrm{ph}} = 1.305\,\mathrm{eV}$ for photons with a wavelength $\lambda = 950\,\mathrm{nm}$. From the slope of the characteristic, a differential resistance 163 mΩ can be derived. The electrical-to-optical conversion efficiency (*dotted line*) exceeds its maximum of 63% at an output power of 1 W. Even at an output power of 2.7 W, the conversion efficiency is above 55%

Internal laser parameters like the internal efficiency η_i and the intrinsic absorption α_i can be determined by plotting the inverse differential efficiencies n_d versus the cavity lengths L. From (93), the equation

$$\frac{1}{\eta_d} = \frac{1}{\eta_i} + \frac{\alpha_i}{\eta_i} \frac{2}{\ln\left(\frac{1}{R_1 R_2}\right)} L \tag{95}$$

can be derived. From the intersection of the linear fit with the $1/\eta_d$ axis extrapolated for $L = 0$, $1/\eta_i$ can be determined. Knowing the values for the facet reflectivities R_1 and R_2, α_i can be calculated using the slope of the fitted line.

Figure 32 shows an example of this type of plot for broad-area lasers with uncoated facets. Usually a rather simple device-fabrication process is employed and no device mounting is necessary to perform this characterization under pulsed conditions. Beside measurements of the threshold-current density, the characteristic temperature, the vertical far-field distribution, and the emission wavelength, such examinations have to be routinely performed on epitaxial laser material to control and verify the quality of the diode lasers.

A very important parameter for high-power lasers is the shift of the threshold-current density j_{th} with temperature T [35]. This behavior can phenomenologically be described by the equation

$$j_{th} \propto \exp\left(\frac{T}{T_0}\right), \tag{96}$$

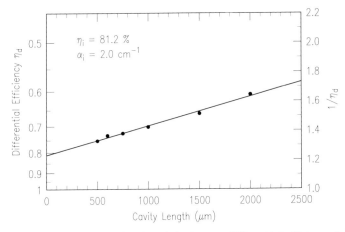

Fig. 32. A linear fit of a plot of the inverse differential efficiency $1/\eta_d$ versus the cavity length L gives access to the internal efficiency n_i and the intrinsic absorption α_i. Since the facets of the devices are uncoated, facet reflectivities $R_1 = R_2 = 0.29$ have been used for the calculations according to (95)

where the characteristic temperature T_0 is a measure of the temperature stability of the device. Figure 33 shows examples of two diode lasers with different characteristic temperatures.

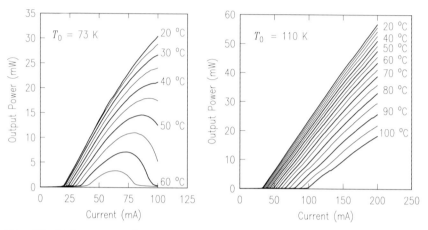

Fig. 33. Light-output characteristics of two single-lateral-mode semiconductor-diode lasers having different characteristic temperatures T_0

4.3   Design Considerations for High-Power Operation

When operating a diode laser at high output power, there are two main problems manufacturers and customers have to struggle with. The first problem is the distortion of the beam profile at high output levels. Devices with single-mode behavior in vertical and horizontal directions are available up to output power levels of approximately 200 mW. Above this power, the device shows beam distortions in the lateral direction. This phenomenon is called filamentation, which implies that there are hot regions inside the cavity where the refractive index is increased leading to parasitic optical waveguides, which destroy the lateral-mode profile. Especially in broad-area devices, filamentation is very pronounced. The main effect which causes filamentation, however, has a different origin. When the local optical intensity in a device is very high, the carrier density is reduced in this area. This behavior is called spatial hole burning. Due to the decrease in the local carrier density, the gain is also reduced and the refractive index is increased [36]. Taking into account the lateral-carrier diffusion and additional contributions from thermal effects, filamentation is a highly dynamic process which occurs on a picosecond time scale.

Another very important effect is Catastrophic Optical Mirror Damage (COMD). An example of this phenomenon is shown in Fig. 34. When a device is properly cooled, the output power limitation is caused by a destruction of the facet. A scanning electron micrograph of a facet after COMD is shown in Fig. 35. The dynamics of this mirror-degradation mechanism is described in detail in [37] and [38]. With the help of suitable facet coatings, the COMD effect can be drastically reduced or even completely eliminated [37].

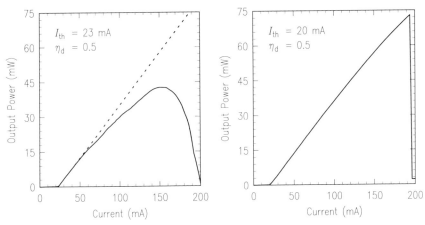

Fig. 34. Light-output power versus operating current for two single-lateral-mode diode lasers. The output power of the device on the *left-hand side* is limited by thermal rollover, whereas the device on the *right-hand side* is destroyed by Catastrophic Optical Mirror Damage (COMD)

The properties of high-power diode lasers with regard to device reliability and beam filamentation can be improved by lowering the optical density in the device. This measure will result in

- reduced spatial hole burning and thus better beam quality,
- more-reliable device operation since regions which increased optical density and increased temperature (hot spots) are avoided, and
- reduced localized power density at the laser facet, resulting in less device failures caused by COMD.

The optical density in the device can be reduced by a suitable vertical epitaxial structure [39]. An example of such a device structure is shown in Fig. 36. The core region of the vertical waveguide is rather narrow, resulting in a broad optical near-field intensity distribution. The near-field distribution shown in Fig. 36 has been calculated using the one-dimensional Helmholtz equation (39). According to (40), the far field is the Fourier transform of the near field. Therefore, a small vertical far-field angle is desired to have a low optical density in the optical waveguide. The small far-field angle has additional advantages, since due to the lower numerical aperture in the fast optical axis, lens and fiber coupling is easier. An alternative approach for the epitaxial structure of high-power laser uses a broad optical waveguide to

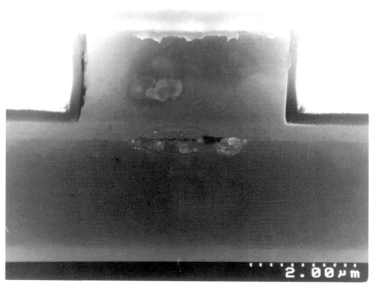

Fig. 35. Scanning electron micrograph of the facet region of a ridge-waveguide laser where the facet has been destroyed by catastrophic optical mirror damage. The pattern of the single-mode optical near-field distribution can be clearly identified

achieve a broad near-field intensity distribution. These Large-Optical-Cavity (LOC) structures are mainly used in the InGaAs/GaInAsP/GaInP material system (see Sect. 2) where a grading of the refractive index in the optical waveguide is hard to achieve [20,21].

A common way to reduce the optical intensity inside the cavity is the reduction of the facet reflectivity. To maintain a high efficiency, such devices have to be rather long and their epitaxial structure should be optimized for a low intrinsic absorption α_i.

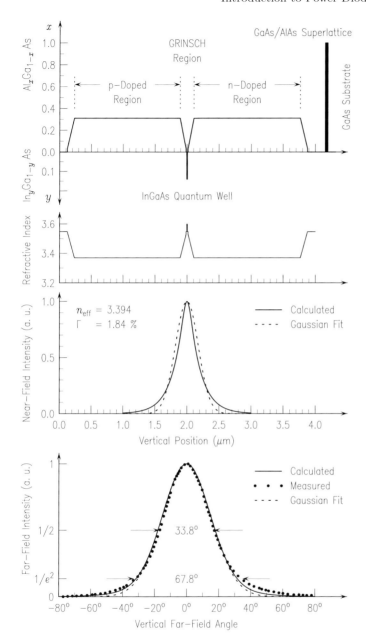

Fig. 36. Vertical design of a GRINSCH laser structure for high-power operation with weak optical confinement. The material composition is shown in the *top* diagram. The *center* diagrams show the vertical refractive-index profile n and the calculated electrical near-field intensity $\mathcal{E}$ together with a Gaussian fit. In the diagram at the *bottom*, calculated and measured far-field patterns and a Gaussian fit are plotted

List of Symbols

a, b	Normalized asymmetry and propagation parameters
a_d	Differential gain coefficient
A, B, C	Einstein coefficients, proportionality constants
d	Thickness in vertical direction, film thickness
D	Density of electronic states
E	Energy, energy levels
E_c, E_v	Conduction and valence band energy
E_F	Fermi-level energy
E_{Fc}, E_{Fv}	Quasi-Fermi-level energies for electrons and holes
E_g	Band-gap energy
$E_{ph} = \hbar\omega$	Photon energy
$\mathcal{E}$	Electric-field vector
f	Fermi function
f_c, f_v	Quasi-Fermi functions for electrons and holes
g, g_{th}	Optical gain, threshold gain
I, I_{th}	Electric current, threshold current
j_{th}, j_{tr}	Threshold and transparency current density
J	Optical intensity
$k = 2\pi/\lambda, \mathbf{k}$	Wavenumber, wave vector
$k_0 = 2\pi/\lambda_0$	Vacuum wavenumber
L	Cavity length
m	Order number for optical modes
m_e, m_h	Effective masses of electrons and holes
n, n_{eff}	Refractive index, effective refractive index
$n_{gr,eff}$	Group effective index
N, N_{ph}	Carrier and photon density
p	Momentum
P	Optical power
r_1, r_2	Amplitude reflectivity of mirrors 1 and 2
R_1, R_2	Intensity reflectivity of mirrors 1 and 2
$\mathcal{R}$	Transition rates
t	Time
t_1, t_2	Amplitude transmission of mirrors 1 and 2
T, T_0	Temperature, characteristic temperature
$u(\nu)\,d\nu$	Spectral energy density of blackbody radiation
U	Electric voltage
v, v_{gr}	Velocity, group velocity
V	Normalized frequency
W	Stripe width, ridge width, lateral width
x	Vertical or transverse direction
y	Horizontal or lateral direction
z	Propagation or axial direction

α	Optical intensity absorption coefficient
α_i	Intrinsic absorption
β_{sp}	Spontaneous-emission factor
$\gamma = \alpha/2 + i\,k_0 n_{eff}$	Complex propagation number
Γ	Confinement factor
$\eta,\ \eta_c$	Efficiency, conversion efficiency
$\eta_d,\ \eta_i$	Differential and internal efficiency
Θ	Far-field angle
$\lambda,\ \lambda_0$	Wavelength, vacuum wavelength
ν	Frequency
$\rho(\hbar\omega)\,d(\hbar\omega)$	Spectral photon density
$\tau,\ \tau_{ph}$	Carrier and photon lifetimes
φ	Angle of incidence
Φ	Phase-shift angle
$\omega = 2\pi\nu$	Angular frequency

List of Constants

$$
\begin{aligned}
c = 1/\sqrt{\epsilon_0\mu_0} &= 2.997\,925 \times 10^{10}\,\text{cm/s} & &\text{Speed of light in vacuum} \\
E_{ph}\,\lambda_0 &= 1.239\,852\,\mu\text{m\,eV} & &\text{Photon energy} \times \text{wavelength} \\
\hbar = h/(2\pi) &= 6.582\,173 \times 10^{-16}\,\text{eVs} & &\text{Reduced Planck constant} \\
k_B &= 8.617\,347 \times 10^{-5}\,\text{eV/K} & &\text{Boltzmann constant} \\
k_B T &= 25.852\,04\,\text{meV}\ (T = 300\,\text{K}) & &\text{Thermal energy at 300\,K} \\
m_0 &= 9.109\,534 \times 10^{-31}\,\text{kg} & &\text{Free-electron rest mass} \\
q &= 1.602\,189 \times 10^{-19}\,\text{C} & &\text{Elementary charge} \\
\epsilon_0 &= 8.854\,188 \times 10^{-14}\,\text{C/(Vcm)} & &\text{Dielectric constant} \\
\mu_0 &= 4\pi \times 10^{-9}\,\text{Vs}^2/(\text{Ccm}) & &\text{Permeability in vacuum}
\end{aligned}
$$

Abbreviations for Indices

B	Boltzmann	c	conduction
cl	cladding	crit	critical
d	differential	e	electron
eff	effective	f	film
F	Fermi	FP	Fabry–Perot
g	gap, band gap	gen	generation
gr	group	h	hole
i	internal, intrinsic, incident	nr	nonradiative
p	particle, peak	ph	photon
rt	roundtrip	s	substrate
sp	spontaneous	stim	stimulated
t	transmitted	TE	transversal electric
th	threshold	TM	transversal magnetic
tr	transparency	v	valence

References

1. H. Ibach, H. Lüth: *Solid-State Physics – An Introduction to Principles of Materials Science,* 2nd edn. (Springer, Berlin, Heidelberg 1995)
2. B. Sapoval, C. Hermann: *Physics of Semiconductors* (Springer, Berlin, Heidelberg 1995)
3. S. M. Sze: *Physics of Semiconductor Devices,* 2nd edn. (Wiley, New York 1981)
4. P. Bhattacharya: *Semiconductor Optoelectronic Devices,* 2nd edn. (Prentice-Hall, London 1997)
5. S. L. Chuang: *Physics of Optoelectronic Devices* (Wiley, New York 1995)
6. K. J. Ebeling: *Integrated Optoelectronics* (Springer, Berlin, Heidelberg 1993)
7. A. Yariv: *Optical Electronics in Modern Communications,* 5th edn. (Oxford Univ. Press, New York 1997)
8. G. P. Agrawal, N. K. Dutta: *Semiconductor Lasers,* 2nd edn. (Kluwer Academic Publishers, Dordrecht 1993)
9. N. W. Carlson: *Monolithic Diode-Laser Arrays* (Springer, Berlin, Heidelberg 1994)
10. H. C. Casey, M. B. Panish: *Heterostructure Lasers – Part A: Fundamental Principles – Part B: Materials and Operating Characteristics* (Academic Press, Orlando, Florida 1978)
11. W. W. Chow, S. W. Koch, M. Sargent III: *Semiconductor-Laser Physics* (Springer, Berlin, Heidelberg 1994)
12. L. A. Coldren, S. W. Corzine: *Diode Lasers and Photonic Integrated Circuits* (Wiley, New York 1995)
13. M. J. Adams, A. G. Steventon, W. J. Devlin, I. D. Henning: *Semiconductor Lasers for Long-Wavelength Optical-Fibre Communications Systems,* IEE Materials and Devices Series (Peter Peregrinus, London 1987)
14. M.-C. Amann, J. Buus: *Tunable Laser Diodes* (Artech House, Boston 1998)
15. R. Michalzik, M. Grabherr, K. J. Ebeling: High-power VCSELs: modeling and experimental characterization, in K. D. Choquette, R. A. Morgan (eds.): *Vertical-Cavity Surface-Emitting Lasers II,* Proc. SPIE **3286**, 206–218 (1998)
16. K. Iga, F. Koyama, S. Kinoshita: Surface emitting semiconductor lasers, IEEE J. QE **24**, 1845–1855 (1988)
17. K. J. Ebeling: Analysis of vertical cavity surface emitting laser diodes (VCSEL), in A. Miller, M. Ebrahimzadeh, D. M. Finlayson (eds.): *Semiconductor Quantum Optoelectronics – From Quantum Physics to Smart Devices* (SUSSP Publications and Institute of Physics Publishing, Bristol 1999) pp. 295–338
18. O. Madelung, R. Poerschke (eds.): *Data in Science and Technology, Semiconductors: Group IV Elements and III–V Compounds* (Springer, Berlin, Heidelberg 1991)
19. S. Nakamura, G. Fasol: *The Blue Laser Diode – GaN Based Light Emitters and Lasers* (Springer, Berlin, Heidelberg 1997)
20. D. Botez, L. J. Mawst, A. Bhattacharya, J. Lopez, J. Li, T. F. Kuech, V. P. Iakovlev, G. I. Suruceanu, A. Caliman, A. V. Syrbu: 66% wallplug efficiency from Al-free 0.98 μm-emitting diode lasers, Electron. Lett. **32**, 2012–2013 (1996)
21. L. J. Mawst, A. Bhattacharya, J. Lopez, D. Botez, D. Z. Garbuzov, L. DeMarco, J. C. Connolly, M. Jansen, F. Fang, R. F. Nabiev: 8 W continuous

wave front-facet power from broad-waveguide Al-free 980 nm diode lasers, Appl. Phys. Lett. **69**, 1532–1534 (1996)

22. P. Vettiger, M. K. Benedict, G.-L. Bona, P. Buchmann, E. C. Cahoon, K. Dätwyler, H.-P. Dietrich, A. Moser, H. K. Seitz, O. Voegeli, D. J. Webb, P. Wolf: Full-wafer technology – a new Approach to large-scale laser fabrication and integration, IEEE J. QE **27**, 1319–1331 (1991)

23. P. Unger, V. Boegli, P. Buchmann, R. Germann: Fabrication of curved mirrors for visible semiconductor lasers using electron-beam lithography and chemically assisted ion-beam etching, J. Vac. Sci. Technol. B **11**, 2514–2518 (1993)

24. P. Unger, V. Boegli, P. Buchmann, R. Germann: High-resolution electron-beam lithography for fabricating visible semiconductor lasers with curved mirrors and integrated holograms, Microelectron. Eng. **23**, 461–464 (1994)

25. P. Unger, G.-L. Bona, R. Germann, P. Roentgen, D. J. Webb: Low-threshold strained GaInP quantum-well ridge lasers with AlGaAs cladding layers, IEEE J. QE **29**, 1880–1884 (1993)

26. H. Kogelnik, V. Ramaswamy: Scaling rules for thin-film optical waveguides, Appl. Opt. **13**, 1857–1862 (1974)

27. H. Kogelnik: Theory of optical waveguides, in T. Tamir (ed.): *Guided-Wave Optoelectronics*, 2nd edn., Springer Series Electron. and Photon. **26**, 7–88 (Springer, Berlin, Heidelberg 1990)

28. D. Marcuse: *Theory of Dielectric Optical Waveguides*, 2nd edn. (Academic, Boston 1991)

29. P. Unger, P. Roentgen, G.-L. Bona: Junction-side up operation of (Al)GaInP lasers with very low threshold currents, Electron. Lett. **28**, 1531–1532 (1992)

30. J. Heerlein, M. Grabherr, R. Jäger, P. Unger: Single-mode AlGaAs-GaAs lasers using current confinement by native-oxide layers, IEEE Photon. Technol. Lett. **10**, 498–500 (1998)

31. T. Ikegami: Reflectivity of mode at facet and oscillation mode in double-heterostructure injection lasers, IEEE J. QE **8**, 470–476 (1972)

32. J. Buus: Analytical approximation for the reflectivity of DH lasers, IEEE J. QE **17**, 2256–2257 (1981)

33. H. A. Macleod: *Thin-Film Optical Filters,* 2nd edn. (Adam Hilger, Bristol 1986)

34. R. Jäger, J. Heerlein, E. Deichsel, P. Unger: 63% wallplug efficiency MBE grown InGaAs/AlGaAs broad-area laser diodes and arrays with carbon p-type doping using CBr$_4$, J. Crystal Growth **201/202**, 882–885 (1999)

35. P. M. Smowton, P. Blood: The differential efficiency of quantum-well lasers, IEEE J. Select. Topics QE **3**, 491–498 (1997)

36. Z. Dai, R. Michalzik, P. Unger, K. J. Ebeling: Numerical simulation of broad-area high-power semiconductor laser amplifiers, IEEE J. QE **33**, 2240–2254 (1997)

37. A. Moser: Thermodynamics of facet damage in cleaved AlGaAs lasers, Appl. Phys. Lett. **59**, 522–524 (1991)

38. A. Moser, E. E. Latta: Arrhenius parameters for the rate process leading to catastrophic damage of AlGaAs-GaAs laser facets, J. Appl. Phys. **71**, 4848–4853 (1992)

39. S. O'Brien, H. Zhao, A. Schoenfelder, R. J. Lang: 9.3 W CW (In)AlGaAs 100 μm wide lasers at 970 nm, Electron. Lett. **33**, 1869–1871 (1997)

Dynamics of High-Power Diode Lasers

Edeltraud Gehrig and Ortwin Hess

Theoretical Quantum Electronics, Institute of Technical Physics, DLR,
Pfaffenwaldring 38-40, D-70569 Stuttgart, Germany
edeltraud.gehrig@dlr.de
ortwin.hess@dlr.de

Abstract. This review gives an overview of the theory and discusses aspects of space–time modeling of high-power diode lasers. The dynamic interaction between the optical fields, the charge carriers, and the interband polarization are described on the basis of microscopic spatially resolved Maxwell–Bloch equations for spatially inhomogeneous semiconductor lasers. Thereby the influence of dynamic internal laser effects such as diffraction, self-focusing, scattering, carrier transport, and heating on the performance of broad-area or tapered amplifiers as well as the individual device properties (i.e. its epitaxial structure and geometry) are self-consistently considered.

Ever since its conception over thirty-five years ago, the semiconductor laser has attracted strong interest, not only for its experimental and technological significance but also for fundamental reasons. Over the years the semiconductor laser has matured from an exotic device which had to be operated in pulsed mode and at cryogenic temperatures with comparatively small output power to reliable high-power laser devices. It is, in particular, the recent improvements in material processing and laser technology which allow a realization of these large-area high-power diode lasers [1] with high spatial and spectral purity. However, this enlargement of the active area of the lasers not only increases the output power but also leads to instabilities and filamentation effects whose origin could be identified to lie in complex microscopic spatio-temporal interactions occurring on time-scales of pico- and nanoseconds [2,3]. The fact that the interplay of microscopic and macroscopic temporal and spatial dynamics strongly determines the overall behavior of high-power semiconductor lasers is vividly demonstrated by a direct comparison of the theoretical predictions [4,5,6] with experimental streak-camera measurements [7] revealing the characteristic spatio-temporal dynamics of broad-area lasers. This highlights that in the high-power semiconductor laser, indeed, it is necessary to fundamentally understand the complex interplay of microscopic material with macroscopic waveguide and device properties. A theoretical investigation and numerical modeling consequently has to include both the macroscopic external constraints imposed by a specific type of laser structure and a spatially resolved microscopic description of the interaction between the optical field and the active semiconductor medium. However, most theoretical studies are performed under the assumption of a steady-

R. Diehl (Ed.): High-Power Diode Lasers, Topics Appl. Phys. **78**, 55–81 (2000)
© Springer-Verlag Berlin Heidelberg 2000

state condition or, when dynamics are considered, characteristic and important semiconductor-laser properties such as the dependence of the gain and refractive index on the charge-carrier density, wavelength or temperature are usually omitted [8,9,10].

Thus, here we will give an overview of such a theory developed specifically for spatially inhomogeneous semiconductor lasers such as laser arrays and broad-area lasers [2,3] which, in contrast to the phenomenological modeling, includes the full space and momentum dependence of the charge-carrier distributions and the polarization on the basis of Maxwell–Bloch equations for spatially inhomogeneous semiconductor lasers [5,6,11]. We will particularly focus on the amplifier configuration, where a coherent optical signal is injected into the high-power semiconductor-amplifier. We will illustrate that this technologically extremely important configuration represents an additional degree of complexity with respect to the effects which have to be included in a simulation. It is the thermal effects which are known to play an important role in the performance of high-power semiconductor-amplifier systems. Indeed, in the high-power amplifier systems it is the nonlinear coupling of the optical field to the spatio-temporally varying temperature distribution which has significant influence on the overall laser properties [12].

In the theory discussed here, the microscopic dynamics of the carrier-distribution functions and the nonlinear polarization are self-consistently coupled with relevant macroscopic properties of the semiconductor device. These are e.g. the geometry, electronic contacts, epitaxial structure, and facet reflectivities. The direct and self-consistent consideration of the optical and thermal properties on a microscopic level, i.e. the spatial and temporal dynamics of the optical fields, the carriers, the interband polarization, and the carrier–carrier as well as the carrier–phonon scattering processes allows, an identification of the relevant physical processes such as dynamic local-carrier generation, carrier recombination by stimulated emission, diffraction, self-focusing, carrier diffusion and scattering, dynamic spatial and spectral gain as well as induced refractive index. In concert, these effects determine on a fundamental level the performance of high-power semiconductor lasers. The information on the spatio-temporally varying distributions of the light intensity and carrier density, dynamic spatial gain, and refractive index as well as characteristic temperature distributions which we obtain from our simulations thus allow an optimization of the structure and geometry of high-power semiconductor-laser devices with respect to spatial and spectral emission characteristics and output power.

1   Microscopic Spatio-Temporal Properties of Diode Lasers

In the following we will, after some introductory remarks on the importance of the main physical quantities, use a semiclassical approach to describe the physical processes in a semiconductor-diode laser. Thereby, the macroscopic optical fields and the polarization are described in a classical manner whereas the dynamics of the carrier distribution and the interband polarization are considered quantum-mechanically.

1.1 Role of Microscopic Spatio-Temporal Properties in Macroscopic Laser Characteristics

In order to optimize the performance of a high-power semiconductor laser one may vary its geometry or epitaxial structure. Thereby, it is necessary to consider the mutual influence of the light–matter dynamics within the active area and the macroscopic spatio-temporal behavior of the optical fields, carrier dynamics, spatial and spectral refractive index and gain as well as beam characteristics. Before deriving our microscopic theory we briefly list some of the main important physical quantities that either enter the theory as material parameters or are self-consistently calculated during the modeling. Along the way, we will briefly indicate their influence on the performance of high-power semiconductor lasers – tendencies which can be directly seen in the examples presented in Sect. 2.

- Optical fields E : the counterpropagating optical fields carry information on the spatio-temporal dynamics of the optical near field (both amplitude and phase information) within the laser and can therefore be used to calculate e.g. the temporally varying intensity distribution in the active area or snapshots of the near and far fields. Plotting the temporal variation of the near field, e.g. at the output facet of the laser, allows a direct comparison with experimental streak-camera images. Temporally averaged profiles of the intensity allow a comparison with more common experimental results.
- Carrier density N : the carrier density in the active area allows calculation of the spatio-temporal distribution of the inversion and may thus be of importance for the analysis of the efficiency of the gain reduction by optically injected light.
- Carrier distribution $f^{e,h}$: the distribution of electrons and holes provides the visualization of the dynamics of spatio-spectral carrier depletion, carrier accumulation and heating in dependence on macroscopic parameters such as carrier injection, temperature, material properties, geometry, and properties of an injected optical beam.
- Interband polarization p : the nonlinear interband polarization reveals via its real and imaginary parts the dynamic behavior of the spatially and spectrally dependent gain as well as the dynamic changes in the spatially and spectrally dependent refractive-index distribution. These parameters are important for the propagation of the light fields and thus for the spatial and spectral beam quality of the emitted radiation. They can be analyzed with respect to e.g. their dependence on carrier injection, temperature, material properties, geometry, and properties of an injected optical beam.

- Generation rate g : the spatially and spectrally dependent generation rate allows the localization of high- and low-gain areas in the active area in dependence on the corresponding light field and carrier distributions and also the analysis of spectral dependence of the gain dynamics.
- Pump rate Λ : the spatial injection of carriers followed by a carrier rearrangement in the conduction and valence bands can be analyzed with respect to the specific realization of the electrical contact (shape, location, etc.) and boundary conditions of the active layer as well as the influence of a nonuniformity in the carrier injection on the spatio-temporal carrier–light field-dynamics.
- Relaxation times τ_i : they can be calculated for various material systems in order to reveal their relevance for different performances (spatio-temporal carrier–light field-dynamics, spectra) of semiconductor-diode lasers of different emission wavelengths.
- Band gap $\mathcal{E}_g$ and transition frequency ω_T : these parameters depend microscopically on the carrier dynamics in the conduction and valence bands. Thereby Coulomb enhancement, thermalization and carrier–phonon interactions are self-consistently included. Thus the microscopic carrier dynamics in the dynamic bands can be used to extract spectral properties such as emission frequency or lineshape – also in their dependence on e.g. injection current or input-beam characteristics.
- Temperature T : the dependence of parameters such as e.g. emission wavelength or laser threshold on temperature can be compared to experimental observations. The self-consistently included thermal interactions can also be used to analyze critical aspects such as the formation of bulk damage.
- Background refractive index n_l and index variations included in η : the lateral and vertical index structure of the active area and the surrounding layers plays an important role for the propagation and guiding of the light fields and determines the spatial beam characteristics as well as the emission wavelength.
- Absorption coefficient α : the dynamic absorption coefficient allows the determination of loss in dependence on the dynamic band-gap structure. Thereby its variation with temperature is included.
- Confinement factor Γ : a variation of the confinement factor allows an analysis of its influence on the spatio-temporal carrier light-field dynamics and on the beam properties.
- Facet reflectivities R_1 and R_2 : the facet reflectivities play an important role, in particular for the filament formation and thus the spatial and spectral beam quality of optically injected semiconductor lasers.

1.2   Optical-Field Dynamics

Starting from Maxwell's equations, a wave equation for the optical fields $\boldsymbol{E}$ in a dielectric medium can be derived [2]

$$\frac{1}{\epsilon_0}\nabla\nabla\boldsymbol{P} + \nabla^2\boldsymbol{E} - \frac{1}{c^2}\frac{\partial^2}{\partial t^2}\boldsymbol{E} = \mu_0\frac{\partial^2}{\partial t^2}\boldsymbol{P}\,. \tag{1}$$

Splitting the polarization $\boldsymbol{P}$ into a linear ($\boldsymbol{P}_1$) and a nonlinear part ($\boldsymbol{P}_{\mathrm{nl}}$),

$$\boldsymbol{P} = \boldsymbol{P}_1 + \boldsymbol{P}_{\mathrm{nl}} = \varepsilon_0\chi_1\boldsymbol{E} + \boldsymbol{P}_{\mathrm{nl}} \tag{2}$$

leads after substitution in (1) to

$$\chi_1\nabla\nabla\boldsymbol{E} + \nabla^2\boldsymbol{E} - \frac{\varepsilon_1}{c^2}\frac{\partial^2}{\partial t^2}\boldsymbol{E} = -\frac{1}{\varepsilon_0}\left[\nabla\nabla\boldsymbol{P}_{\mathrm{nl}}\right]. \tag{3}$$

Since typical geometries of the active area in high-power semiconductor-diode lasers favor the propagation of light fields perpendicular to the facets (in the following: z coordinate) it is a common method [2] to divide the counterpropagating optical fields $\boldsymbol{E}$, the polarization $\boldsymbol{P}$ and the differential operator ∇ into longitudinal and transverse parts with the ansatz

$$\boldsymbol{E} = e^{\mathrm{i}\beta z - \mathrm{i}\omega t}\left(\boldsymbol{E}_{\mathrm{T}} + \boldsymbol{e}_z E_z\right) \tag{4}$$

$$\boldsymbol{P} = e^{\mathrm{i}\beta z - \mathrm{i}\omega t}\left(\boldsymbol{P}_{\mathrm{T}} + \boldsymbol{e}_z P_z\right) \tag{5}$$

$$\nabla = \nabla_{\mathrm{T}} + \boldsymbol{e}_z\frac{\partial}{\partial z}, \tag{6}$$

where $\beta = k_z + \mathrm{i}\alpha/2 + \delta\beta$ (k_z denotes the unperturbed propagation wavenumber of the optical fields with $k_z = n_1 k_0$), α is the linear absorption, $\delta\beta = \delta n\, k_0$ denotes the changes in the propagation wavenumber due to carrier- and temperature-induced refractive-index changes, and ω is the frequency. This results in a transverse wave equation

$$\chi_1\,\nabla_{\mathrm{T}}\left(\nabla_{\mathrm{T}}\boldsymbol{E}_{\mathrm{T}} + \left(\mathrm{i}\beta + \frac{\partial}{\partial z}\right)E_z\right) + \left(-\beta^2 + 2\mathrm{i}\beta\frac{\partial}{\partial z} + \frac{\partial^2}{\partial z^2} + \nabla_{\mathrm{T}}^2\right)\boldsymbol{E}_{\mathrm{T}}$$

$$+ \frac{\varepsilon_1}{c^2}\left(\omega^2 + 2\mathrm{i}\omega\frac{\partial}{\partial t} - \frac{\partial^2}{\partial t^2}\right)\boldsymbol{E}_{\mathrm{T}}$$

$$= -\frac{1}{\varepsilon_0}\left[\nabla_{\mathrm{T}}\left(\nabla_{\mathrm{T}}\boldsymbol{P}_{\mathrm{T,nl}} + \left(\mathrm{i}\beta + \frac{\partial}{\partial z}\right)P_{z,\mathrm{nl}}\right)\right.$$

$$\left. + \frac{1}{c^2}\left(\omega^2 + 2\mathrm{i}\omega\frac{\partial}{\partial t} - \frac{\partial^2}{\partial t^2}\right)\boldsymbol{P}_{\mathrm{T,nl}}\right] \tag{7}$$

and a longitudinal wave equation

$$\chi_1\left(\mathrm{i}\beta + \frac{\partial}{\partial z}\right)\nabla_{\mathrm{T}}\boldsymbol{E}_{\mathrm{T}} + \varepsilon_1\left(-\beta^2 + 2\mathrm{i}\beta\frac{\partial}{\partial z} + \frac{\partial^2}{\partial z^2}\right)E_z + \nabla_{\mathrm{T}}^2 E_z$$

$$+\frac{\varepsilon_1}{c^2}\left(+\omega^2 + 2\mathrm{i}\omega\frac{\partial}{\partial t} - \frac{\partial^2}{\partial t^2}\right) E_z$$

$$= \frac{1}{\varepsilon_0}\left[-\left(\mathrm{i}\beta + \frac{\partial}{\partial z}\right)\nabla_\mathrm{T}\boldsymbol{P}_{\mathrm{T,nl}} + \left(\beta^2 - 2\mathrm{i}\beta\frac{\partial}{\partial z} - \frac{\partial^2}{\partial z^2}\right)P_{z,\mathrm{nl}}\right.$$

$$\left.-\frac{1}{c^2}\left(\omega^2 + 2\mathrm{i}\omega\frac{\partial}{\partial t} - \frac{\partial^2}{\partial t^2}\right)P_{z,\mathrm{nl}}\right]. \tag{8}$$

The wave equations are now nondimensionalized to the characteristic width w of the device, the diffraction length $l = k_w^2 = (2\pi/\lambda)w^2$ and the characteristic propagation time $\tau_R = ln_e/c$, during which an optical beam covers that distance:

$$x = w\tilde{x}$$
$$y = w\tilde{y}$$
$$z = l\tilde{z} \tag{9}$$

and

$$t = \tau_R\tilde{t}$$
$$w = \frac{\tilde{w}}{\tau_R}. \tag{10}$$

In the next step we expand the transverse and longitudinal wave equations in powers of the dimensionless number $f = w/l = 1/k_z w \ll 1$ [2] and, correspondingly, the transverse and longitudinal parts of the optical fields and the nonlinear polarization

$$\boldsymbol{E}_{\tilde{\mathrm{T}}} = \boldsymbol{E}_{\tilde{\mathrm{T}}}^{(0)} + f^2\boldsymbol{E}_{\tilde{\mathrm{T}}}^{(2)} + s$$
$$E_{\tilde{z}} = fE_{\tilde{z}}^{(1)} + f^3 E_{\tilde{z}}^{(3)} + s$$
$$\boldsymbol{P}_{\tilde{\mathrm{T}}} = f^2\boldsymbol{P}_{\tilde{\mathrm{T}}}^{(2)} + f^4\boldsymbol{P}_{\tilde{\mathrm{T}}}^{(4)} + s$$
$$P_{\tilde{z}} = f^3 P_{\tilde{z}}^{(3)} + f^5 P_{\tilde{z}}^{(3)} + s . \tag{11}$$

Combining these expressions with the wave equations results in longitudinal and transverse equations in the orders $O(f), O(f^3), \ldots$. Finally one obtains in first order from the longitudinal equation the expression

$$\mathrm{i}\nabla_{\tilde{\mathrm{T}}}\boldsymbol{E}_{\tilde{\mathrm{T}}}^{(0)} = E_{\tilde{z}}^{(1)}. \tag{12}$$

This can be inserted into the transverse equation, which then reads in first order

$$\left(\frac{\partial}{\partial\tilde{z}} - \frac{\mathrm{i}}{2}\nabla_\mathrm{T}^2 + \left(\frac{\alpha}{2}l + \mathrm{i}\eta\right) + \frac{\partial}{\partial\tilde{t}}\right)\boldsymbol{E}_\mathrm{T}^{(0)} = \frac{\mathrm{i}}{2}\frac{1}{n_l^2\varepsilon_0}\boldsymbol{P}_{\mathrm{T,nl}}^{(2)}. \tag{13}$$

In first order, Maxwell's wave equation for the optical fields $\boldsymbol{E}(\boldsymbol{r},t)$ and the polarization $\boldsymbol{P}_{\mathrm{nl}}(\boldsymbol{r},t)$ is purely transverse, but it may nevertheless be transversely dependent.

In (13) α is the linear absorption and η includes static and dynamic changes in the permittivity affecting both refractive index and propagation wavenumber. The static waveguiding structure is a consequence of lateral and vertical confinement of the active area. Its spatial dependence can be obtained from static perturbation theory and then be included in effective parameters such as the background refractive index n_l of (13). Dynamic changes arise from the time-dependent carrier and polarization dynamics as well as thermal interactions. Their influence on the permittivity ε and refractive index are deduced from the microscopic carrier dynamics and considered in η whereas the change of the emission frequency is included in the microscopic equations, which will be described in Sect. 1.3.

Since the active area of a bulk semiconductor-diode laser is formed by a Fabry–Perot resonator the description of the light propagation in the active layer requires the simultanous consideration of counterpropagating optical fields. The optical fields and the polarization are thus composed of two waves, E^+ and E^-, which counterpropagate in the positive ('+') and the negative ('−') z-direction in the optical resonator formed by the active area ('c.c.' denotes the complex conjugate-expression).

$$\boldsymbol{E} = \frac{1}{2}(E^+ e^{ik_z - i\omega t} + E^- e^{-ik_z - i\omega t} + \text{c.c.})$$

$$\boldsymbol{P}_{\mathrm{nl}} = \frac{1}{2}(P_{\mathrm{nl}}^+ e^{ik_z - i\omega t} + P_{\mathrm{nl}}^- e^{-ik_z - i\omega t} + \text{c.c.}),$$

where k_z denotes the unperturbed wavenumber. Substituting (14) in (13) we thus finally obtain in dimensional quantities the following paraxial wave equation for the counterpropagating optical fields $E^\pm(\boldsymbol{r}, t)$ [5]

$$\pm \frac{\partial}{\partial z} E^\pm(\boldsymbol{r}, t) + \frac{n_l}{c} \frac{\partial}{\partial t} E^\pm(\boldsymbol{r}, t) =$$

$$\frac{\mathrm{i}}{2k_z} \frac{\partial^2}{\partial x^2} E^\pm(\boldsymbol{r}, t) - \left(\frac{\alpha}{2} + \mathrm{i}\eta\right) E^\pm(\boldsymbol{r}, t) + \frac{\mathrm{i}}{2} \frac{\Gamma}{n_l^2 \epsilon_0 L} P_{\mathrm{nl}}^\pm(\boldsymbol{r}, t), \qquad (14)$$

where

$$P_{\mathrm{nl}}^\pm(\boldsymbol{r}, t) = 2V^{-1} \sum_{k} d_{\mathrm{cv}}(k)\, p_{\mathrm{nl}}^\pm(k, \boldsymbol{r}, t) \qquad (15)$$

is the polarization which depends on the microscopic interband polarization $p_{\mathrm{nl}}^\pm(k, \boldsymbol{r}, t)$. It is the source of the optical fields and generally describes nonlinear spatio-temporal variations of the gain and refractive index. k is the carrier wavenumber, the position vector $\boldsymbol{r} = (x, z)$ denotes the lateral (x) and longitudinal (z) directions, V is the normalization volume of the crystal and $d_{\mathrm{cv}}(k)$ is the optical dipole matrix element. In (14) k_z denotes the (unperturbed) wavenumber of the propagating fields, n_l is the background refractive index of the active layer, and L the length of the structure. The parameter α considers the linear absorption and η includes lateral (x) and vertical (y) variations of the refractive index due to the waveguide structure as well as dynamic induced changes due to the carrier dynamics and thermal interactions. Γ denotes the optical confinement factor.

In typical high-power diode-laser devices (Fig. 1) the active layer imposes the lateral and longitudinal boundary conditions for the optical fields. The light propagating in the longitudinal direction is partially reflected and partially transmitted, in dependence on the device reflectivities, R_1 and R_2, leading to:

$$E^+(x, z = 0, t) = -\sqrt{R_1} E^-(x, z = 0, t)$$
$$E^-(x, z = L, t) = -\sqrt{R_2} E^+(x, z = L, t), \tag{16}$$

(a)

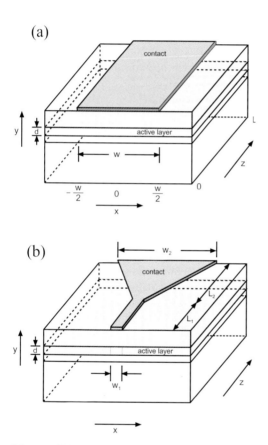

Fig. 1. Schematic of the geometry of a broad-area (**a**) and a tapered (**b**) semiconductor-diode laser. Charge carriers injected through the contact stripe at the top of the device recombine in the active zone. The active GaAs layer (*shaded light*) is located between two cladding layers of $Al_xGa_{1-x}As$ (*white*). Light, generated by stimulated emission and amplification, propagates in the longitudinal z-direction

where L denotes the length of the device. At the lateral edges, the active area is surrounded by (unpumped) layers of semiconductor media which strongly absorb the fields (absorption coefficient α_w). This is represented by the lateral boundary conditions

$$\frac{\partial E^\pm}{\partial x} = -\alpha_w E^\pm \tag{17}$$

at $x = w/2$ (w is the full lateral width of the laser structure) as well as

$$\frac{\partial E^\pm}{\partial x} = +\alpha_w E^\pm \tag{18}$$

at $x = -w/2$.

1.3 Physics of the Active Semiconductor Medium

The dynamics of the nonlinear polarization, is determined by the microscopic interband polarizations $p_{\mathrm{nl}}^\pm(k, \boldsymbol{r}, t)$. The polarizations in turn, depend on the dynamics of the distributions of electrons (e) and holes (h) $f^{\mathrm{e,h}}(k, \boldsymbol{r}, t)$. Their interplay and the coupling with the optical field are governed by the semiconductor Bloch equations

$$\frac{\partial}{\partial t} f^{\mathrm{e,h}}(k, \boldsymbol{r}, t) = g(k, \boldsymbol{r}, t) \tag{19}$$

$$-\tau_{\mathrm{e,h}}^{-1}(k, N)\left[f^{\mathrm{e,h}}(k, \boldsymbol{r}, t) - f_{\mathrm{eq}}^{\mathrm{e,h}}(k, \boldsymbol{r}, t)\right] + \Lambda^{\mathrm{e,h}}(k, \boldsymbol{r}, t)$$

$$-\Gamma_{\mathrm{sp}}(k, T_1) f^{\mathrm{e}}(k, \boldsymbol{r}, t) f^{\mathrm{h}}(k, \boldsymbol{r}, t) - \gamma_{\mathrm{nr}} f^{\mathrm{e,h}}(k, \boldsymbol{r}, t)$$

$$\frac{\partial}{\partial t} p_{\mathrm{nl}}^\pm(k, \boldsymbol{r}, t) = -\left[\mathrm{i}\bar{\omega}(k, T_1) + \tau_{\mathrm{p}}^{-1}(k, N)\right] p_{\mathrm{nl}}^\pm(k, \boldsymbol{r}, t) \tag{20}$$

$$+\frac{1}{\mathrm{i}\hbar} d_{\mathrm{cv}}(k) E^\pm(\boldsymbol{r}, t) \left[f^{\mathrm{e}}(k, \boldsymbol{r}, t) + f^{\mathrm{h}}(k, \boldsymbol{r}, t)\right] .$$

The microscopic generation rate is given by

$$g(k, \boldsymbol{r}, t) = -\frac{1}{4\mathrm{i}\hbar} d_{\mathrm{cv}}(k) \left[E^+(\boldsymbol{r}, t) p_{\mathrm{nl}}^{+*}(k, \boldsymbol{r}, t) + E^-(\boldsymbol{r}, t) p_{\mathrm{nl}}^{-*}(k, \boldsymbol{r}, t)\right]$$

$$+\frac{1}{4\mathrm{i}\hbar} d_{\mathrm{cv}}^*(k) \left[E^{+*}(\boldsymbol{r}, t) p_{\mathrm{nl}}^+(k, \boldsymbol{r}, t) + E^{-*}(\boldsymbol{r}, t) p_{\mathrm{nl}}^-(k, \boldsymbol{r}, t)\right] . \tag{21}$$

$f_{\mathrm{eq}}^{\mathrm{e,h}}(k, \boldsymbol{r}, t)$ denote the carrier distributions in thermal equilibrium with the lattice, each given by the corresponding Fermi distribution. In (20) $\Gamma_{\mathrm{sp}}(k, T_1)$ is the spontaneous-recombination coefficient and γ_{nr} denotes the rate due to nonradiative recombination. The microscopic pump term includes the pump-blocking effect

$$\Lambda^{\mathrm{e,h}}(k, \boldsymbol{r}, t) = \Lambda(\boldsymbol{r}, t) \frac{f_{\mathrm{eq}}^{\mathrm{e,h}}(k, \boldsymbol{r}, t)\left(1 - f^{\mathrm{e,h}}(k, \boldsymbol{r}, t)\right)}{V^{-1}\sum_k f_{\mathrm{eq}}^{\mathrm{e,h}}(k, \boldsymbol{r}, t)\left(1 - f^{\mathrm{e,h}}(k, \boldsymbol{r}, t)\right)}, \tag{22}$$

where the macroscopic pump term

$$\Lambda(\boldsymbol{r},t) = \frac{\eta_{\text{eff}}}{ed}\, \mathcal{J}(\boldsymbol{r},t) \tag{23}$$

depends on the space and time dependent injection current density $\mathcal{J}(\boldsymbol{r},t)$, with η_{eff} being the quantum efficiency and d the thickness of the active layer. The scattering rates $\tau_{\text{e,h}}^{-1}(k,N)$ and $\tau_{\text{p}}^{-1}(k,N)$ in (19) are microscopically determined [5] and include carrier–carrier scattering mechanisms and the interaction of carriers with optical phonons (LO phonons). Both depend on wavenumber and carrier density and thus vary spatio-temporally within the active semiconductor layer. The detuning between the cavity frequency ω and the transition frequency ω_{T}

$$\hbar\overline{\omega}(k,T_{\text{l}}) = \mathcal{E}_{\text{g}}(T_{\text{l}}) + \frac{\hbar^2 k^2}{2m_{\text{r}}} + \delta\mathcal{E}(N,T_{\text{l}}) - \hbar\omega$$
$$= \hbar(\omega_{\text{T}} - \omega) \tag{24}$$

contains via $\mathcal{E}_{\text{g}}$ and $\delta\mathcal{E}$ the dependence of the semiconductor band gap on the density of charge carriers N and on the lattice temperature T_{l}. The variation of the band gap with carrrier density is given by [14]

$$\delta\mathcal{E}(N,T_{\text{l}}) = \mathcal{E}_0 \frac{-a\left(Na_0^3\mathcal{E}_0^2\right)^{1/4}}{\left(Na_0^3\mathcal{E}_0^2 + b^2\left(k_{\text{B}}T_{\text{l}}\right)^2\right)^{1/4}}, \tag{25}$$

with the exciton binding energy $\mathcal{E}_0 = m_{\text{r}}e^4/(2\epsilon_0^2\hbar^2)$, the exciton Bohr radius $a_0 = \hbar^2\epsilon/(e^2 m_{\text{r}})$, the lattice temperature T_{l}, Boltzmann's constant k_{B}, the effective mass m_{r}, and the numerical factors $a = 4.64$ and $b = 0.107$. The dependence on the lattice temperature can be expressed as [13]

$$\mathcal{E}_{\text{g}}(T_{\text{l}}) = \mathcal{E}_{\text{g}}(0) - v_1 T_{\text{l}}^2/(T_{\text{l}} + v_2) \tag{26}$$

where $\mathcal{E}_{\text{g}}(0)$ ($= 1.519$ eV for GaAs) is the band gap at $T_{\text{l}} = 0$ K, and v_1, v_2 are two material parameters given by $v_1 = 5.405 \times 10^{-4}$ eV/K and $v_2 = 204$ K for GaAs, respectively.

The Wigner distributions, describing the microscopic semiconductor dynamics, are functions of space and momentum. As has been discussed in [5], the spatial transport of charge carriers usually occurs on a much slower time-scale in the picosecond up to the nanosecond regime (10 ps–10 ns) than the $\boldsymbol{k}$-space relaxation of the microscopic variables towards their local quasi-equilibrium values occurring on a femtosecond time-scale. This typical separation of time-scales between the $\boldsymbol{k}$-space and the $\boldsymbol{r}$-space dynamics allows us to treat both regimes separately, with the influence of spatial gradients on the $\boldsymbol{k}$-space dynamics often being negligible. Transport of carriers thus effectively occurs as ambipolar transport of electrons and holes allowing one to derive a macroscopic transport equation for the carrier density $N(\boldsymbol{r}, t)$ [5] and a corresponding relaxation equation for the energy densities $u^{e,h}(\boldsymbol{r}, t)$ [12]

$$\frac{\partial}{\partial t} N(\boldsymbol{r}, t) = \nabla \left(D_f \nabla N(\boldsymbol{r}, t) \right) + \Lambda(\boldsymbol{r}, t) + G(\boldsymbol{r}, t)$$
$$- \gamma_{nr} N(\boldsymbol{r}, t) - W(\boldsymbol{r}, t) \tag{27}$$

$$\frac{\partial}{\partial t} u^{e,h}(\boldsymbol{r}, t) = \Lambda_u^{e,h}(\boldsymbol{r}, t) + G_u^{e,h}(\boldsymbol{r}, t)$$
$$- \gamma_{nr} u^{e,h}(\boldsymbol{r}, t) - W_u^{e,h}(\boldsymbol{r}, t) - R^{e,h}(\boldsymbol{r}, t) \tag{28}$$

with spatially structured carrier injection (Λ and $\Lambda_u^{e,h}$), carrier recombination by stimulated emission (G and $G_u^{e,h}$), nonradiative recombination ($\gamma_{nr} N$ and $\gamma_{nr} u^{e,h}$), spontaneous emission (W and $W_u^{e,h}$) and carrier–phonon relaxation ($R^{e,h}$). The recombination of charge carriers, which occurs at the lateral edges of the structure, is modeled by using characteristic values of the surface recombination velocity coefficient v_{sr} via the boundary conditions

$$\frac{\partial N}{\partial x} = -v_{sr} N \tag{29}$$

at $x = w/2$ and

$$\frac{\partial N}{\partial x} = +v_{sr} N \tag{30}$$

at $x = -w/2$. In (27), D_f is the ambipolar diffusion coefficient given by

$$D_f = \frac{\sigma^h D^e + \sigma^e D^h}{\sigma^e + \sigma^h} \,, \tag{31}$$

with the conductivities of electrons and holes, $\sigma^{e,h}$ [5]. The macroscopic gain is given by

$$G = \chi'' \epsilon_0 / 2\hbar \left(\left| E^+ \right|^2 + \left| E^- \right|^2 \right) - 1/4\hbar \, \text{Im} \left[E^+ P_{nl}^{+*} + E^- P_{nl}^{-*} \right], \tag{32}$$

where Im[...] denotes the imaginary part and the spontaneous emission reads

$$W(\boldsymbol{r}, t) = V^{-1} \sum_k \Gamma_{sp}(k, T_l) f^e(k, \boldsymbol{r}, t) f^h(k, \boldsymbol{r}, t), \tag{33}$$

with a phenomenological rate of spontaneous emission [15]. The corresponding expressions in the heat equations are, with $\epsilon^{e,h}(k) = \hbar^2 / 2m_{e,h}$,

$$\Lambda_u^{e,h}(\boldsymbol{r}, t) = V^{-1} \sum_k \epsilon^{e,h}(k) \Lambda^{e,h}(k, \boldsymbol{r}, t) \tag{34}$$

$$G_u^{e,h}(\boldsymbol{r}, t) = -\frac{1}{2\hbar^2 V} \left[\left| E^+(\boldsymbol{r}, t) \right|^2 + \left| E^-(\boldsymbol{r}, t) \right|^2 \right]$$
$$\times V^{-1} \sum_k \epsilon^{e,h}(k) \frac{\tau_p^{-1}(k, N)}{\overline{\omega}^2(k, T_l) + \tau_p^{-2}(k, N)} \left| d_{cv}(k) \right|^2$$
$$\times \left[f^e(k, \boldsymbol{r}, t) + f^h(k, \boldsymbol{r}, t) - 1 \right] \tag{35}$$

$$W_u^{e,h}(\boldsymbol{r}, t) = V^{-1} \sum_k \epsilon^{e,h}(k) \Gamma_{sp}(k, T_l) f^e(k, \boldsymbol{r}, t) f^h(k, \boldsymbol{r}, t). \tag{36}$$

The relaxation term

$$R^{e,h}(\boldsymbol{r},t) = \sum_{ph} \frac{1}{\tau_{ph}^{e,h}} \left(u^{e,h}(\boldsymbol{r},t) - u_{eq}^{e,h}(\boldsymbol{r},t) \right) \tag{37}$$

$$+ V^{-1} \sum_{k} \epsilon^{e,h}(k) \frac{1}{\tau_{po}^{e,h}(k,N)} \left(f^{e,h}(k,\boldsymbol{r},t) - f_{eq}^{e,h}(k,\boldsymbol{r},t) \right)$$

includes carrier–phonon interactions, where, in particular, the polar optical phonons ($\tau_{po}^{e,h}$) are considered in dependence on density and wavenumber. On the basis of a grand canonical ensemble interpretation having both energy and particle contact and exchange with the environment, the carrier density N and the carrier energies $u^{e,h}$ can be expressed as functions of the independent variables of the chemical potentials $\mu^{e,h}$ and the plasma temperatures $T_{pl}^{e,h}$. Solving the resulting set of equations for the plasma temperatures $T_{pl}^{e,h}$ leads to [12]:

$$\dot{T}_{pl}^{e,h}(\boldsymbol{r},t) = J_u^{e,h}(\boldsymbol{r},t)\dot{u}^{e,h}(\boldsymbol{r},t) - J_N^{e,h}(\boldsymbol{r},t)\dot{N}(\boldsymbol{r},t), \tag{38}$$

where $[\dot{\ldots}]$ denotes the temporal derivativ. and the Jacobian derivates $J_u^{e,h}$ and $J_N^{e,h}$ are given by

$$J_u^{e,h}(\boldsymbol{r},t) = \frac{\partial N^{e,h}}{\partial \mu^{e,h}} \left(\frac{\partial u^{e,h}(\boldsymbol{r},t)}{\partial T_{pl}^{e,h}(\boldsymbol{r},t)} \frac{\partial N^{e,h}(\boldsymbol{r},t)}{\partial \mu^{e,h}(\boldsymbol{r},t)} - \frac{\partial u^{e,h}(\boldsymbol{r},t)}{\partial \mu^{e,h}(\boldsymbol{r},t)} \frac{\partial N^{e,h}(\boldsymbol{r},t)}{\partial T_{pl}^{e,h}(\boldsymbol{r},t)} \right)^{-1} \tag{39}$$

$$J_N^{e,h}(\boldsymbol{r},t) = \frac{\partial u^{e,h}}{\partial \mu^{e,h}} \left(\frac{\partial u^{e,h}(\boldsymbol{r},t)}{\partial T_{pl}^{e,h}(\boldsymbol{r},t)} \frac{\partial N^{e,h}(\boldsymbol{r},t)}{\partial \mu^{e,h}(\boldsymbol{r},t)} - \frac{\partial u^{e,h}(\boldsymbol{r},t)}{\partial \mu^{e,h}(\boldsymbol{r},t)} \frac{\partial N^{e,h}(\boldsymbol{r},t)}{\partial T_{pl}^{e,h}(\boldsymbol{r},t)} \right)^{-1}. \tag{40}$$

The dynamics of the lattice temperature is given by

$$\dot{T}_l(\boldsymbol{r},t) = -\gamma_a \left[T_l(\boldsymbol{r},t) - T_a \right] + \sum_{ph} \frac{1}{\tau_{ph}^e} \left[T_{pl}^e(\boldsymbol{r},t) - T_l(\boldsymbol{r},t) \right]$$

$$+ J_u^e(\boldsymbol{r},t)\pi^{-2}V^{-1} \sum_{k} \epsilon^e(k) \frac{1}{\tau_{po}^e} \left(f^e(k,\boldsymbol{r},t) - f_{eq}^e(k,\boldsymbol{r},t) \right)$$

$$+ \sum_{ph} \frac{1}{\tau_{ph}^h} \left[T_{pl}^h(\boldsymbol{r},t) - T_l(\boldsymbol{r},t) \right]$$

$$+ J_u^h(\boldsymbol{r},t)\pi^{-2}V^{-1} \sum_{k} \epsilon^h(k) \frac{1}{\tau_{po}^h} \left(f^h(k,\boldsymbol{r},t) - f_{eq}^h(k,\boldsymbol{r},t) \right)$$

$$+ \hbar\omega\gamma_{nr}N(\boldsymbol{r},t) + \frac{\mathcal{J}^2 A^2 R}{c_q V_{BAL}} \tag{41}$$

where $-\gamma_a$ describes the relaxation to the ambient temperature T_a, c_q is the specific heat, R denotes the total resistance, A is the cross section, and V_{BAL} the volume of the active zone.

2 Spatio-Temporal Dynamics of High-Power Diode Lasers

The system of equations (14) and (19) describing the microscopic and macro-scopic carrier–light field-dynamics in the active area of semiconductor lasers is the basis for a numerical modeling of the performance of high-power semi-conductor lasers based on microscopic principles. This allows, in particular, an analysis of the temporal behavior of the optical fields and the carriers as well as the spatio-temporal variation of the gain, the emitted wavelength, the induced refractive index, and the spatial beam characteristics (far field, near field).

The approach taken here incorporates the full microscopic spatio-temporal dynamics together with the relevant macroscopic properties of typical high-power semiconductor lasers. It thus represents a firm and well-founded basis for the analysis of relevant physical processes which influence the macro-scopic operating characteristics of the devices. It also generally applies to semiconductor-laser systems with a large variety of geometries and active-material systems. To be specific, we will in the following focus on typical III–V semiconductor-material systems and use the relevant parameters for the GaAs-AlGaAs system [16].

In the following sections we will discuss typical examples which illus-trate the way the microscopic dynamics and the material properties di-rectly influence and determine the macroscopic performance of high-power semiconductor-laser systems with optical injection.

2.1 Optical Injection

The microscopic modeling allows one to consider an optical injection in a very general and detailed way: not only the input power serves as an input parameter, but also the spatial and spectral shape and angle of incidence are self-consistently transferred to the internal dynamics.

The laterally dependent injected light field at the possible injection po-sitions $z = 0$ or $z = L$ may conveniently be expressed via the boundary conditions

$$E^+(x, z = 0, t) = -R_1 E^-(x, z = 0, t) + E_{\text{inj}}(x, z = 0, t)$$
$$E^-(x, z = L, t) = -R_2 E^-(x, z = L, t) + E_{\text{inj}}(x, z = L, t), \tag{42}$$

where

$$E_{\text{inj}}(x, z = 0, L, t) = E_{\text{inj}}^0(x, z = 0, L, t)$$
$$\times \exp\left[-\mathrm{i} \tan^{-1}\left(\frac{z}{z_0}\right)\right] \exp\left[\mathrm{i}k \frac{(x - x_0)^2}{2R(z)}\right] \tag{43}$$

with

$$E_{\text{inj}}^0(x, z = 0, L, t) = \mathcal{E}_{\text{inj}}^0(x, z = 0, L, t)$$

$$\times T_{1,2} \exp\left[-i\omega_{\text{inj}}(z = 0, L)t\right] \exp\left[-\frac{(x - x_0)^2}{w_0^2}\right]$$

$$\times \exp\left[\frac{i2\pi}{\lambda}(x - x_0)\sin(\alpha_{\text{inj}})(z = 0, L)\right]$$

$$R(z) = z + \frac{z_0^2}{z}. \tag{44}$$

In (43), we have assumed a Gaussian-shaped monofrequent optical injection. $T_{1,2} = 1 - R_{1,2}$ are the transmissions of the front and back facets and $x - x_0$ is the lateral distance from the lateral injection position x_0. $\omega_{\text{inj}}(z = 0, L) = 2\pi c/(\lambda_{\text{inj}}(z = 0, L))$ and $\alpha_{\text{inj}}(z = 0, L)$ are the frequency and the injection angle of the injected light field, respectively, $z_0 = \pi w_0^2/\lambda_{\text{inj}}$ is the Rayleigh range, $R(z)$ is the radius of curvature of areas of constant phase and $2w_0$ the beam waist of the injected beam. Since propagation after entering the active layer occurs automatically via Maxwell's wave equation, no term $e^{ik_z z}$ which is usually a part of the formula for Gaussian beams has to be considered here. Equation 44 describes the laterally varying amplitude of the injected optical field after transmission in the active layer, including the phase change between different lateral positions due to the angle of incidence. Thus, (43) very generally represents the dynamic spatio-temporally resolved injection of an optical beam or pulse and allows one to optimize the injection (e.g. width of the injected light beam, angle of incidence, frequency with respect to the gain bandwidth of the amplifier) or to analyze the physical effects during the interaction of one or several light fields with the carrier dynamics, such as in the case of dynamic wave mixing when using injection of various different light beams from one or opposite sides.

2.2 Influence of Laser Geometry and Facet Reflectivities

Figure 1 shows typical structures of broad-area (a) and tapered (b) devices with an active layer (here: GaAs) sandwiched between cladding layers (here $Al_x Ga_{1-x} As$).

The broad-area structure is usually gain-guided, whereas the tapered structure may have a special index structure in the narrow part serving as an optical waveguide. The geometry and the boundary conditions (e.g. the reflectivities of the facets) of the device are important for the propagation and reflection of the light in the active layer [17].

2.2.1 Broad-Area Amplifier

Broad-area amplifiers, used in a Master-Oscillator–Power-Amplifier (MOPA) configuration [18], are very appropriate for the generation of high output

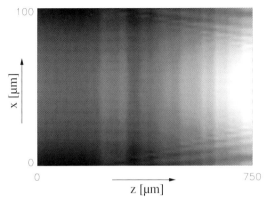

Fig. 2. Amplification and longitudinal self-structuring of an injected optical CW signal within the active layer of a broad-area semiconductor laser. The plot displays a snapshot of the intra-cavity intensity $I(\boldsymbol{r}, t_0) \sim \left(\mid E^+(\boldsymbol{r}, t_0) \mid^2 + \mid E^-(\boldsymbol{r}, t_0) \mid^2 \right)$. The power of the injected optical CW signal is $P_{\mathrm{inj}} = 10$ mW. Dark shading corresponds to low intensity and bright colours to areas of high intensity. The optical signal is injected from the *left* and the outcoupling facet is located at the *right-hand side* of the laser. Its transverse stripe width is $w = 100$ μm. The longitudinal extension corresponds to 750 μm

powers in an almost-diffraction-limited beam. As an example, Fig. 2 shows a snapshot in the active area of an anti-reflection-coated broad-area laser at an amplifier current of 1 A with an injected optical power (from the left in Fig. 2) of 10 mW, 500 ps after injection. The longitudinal and transverse intensity modulations arise from internal microscopic carrier–carrier and carrier–phonon scattering processes and charge-carrier transport leading to combined self-focusing and diffraction. This self-structuring is characteristic for high-power diode lasers, and occurs even if the pump profile and the shape of the input beam are assumed to be uniform. The diffusion of charge carriers with a typical diffusion length of 18 μm leads, together with the density-dependent momentum relaxation of the carriers on time-scales of the order of ≈ 50–100 fs, to the formation of a multitude of intensity maxima in the active layer. The partial reflection at the facets – the reflectivities were chosen as 10^{-4} in the example shown – and the interaction with the diffraction grating built up by the density distribution increase the counterpropagation effects, so that the formation of structures depends very critically on the injection current as well as the reflectivity of the facets. Our simulations generally allow a systematic variation of these parameters. Figure 3 shows for example an intensity snapshot for the analogue-amplifier configuration with a rather high reflectivity of 10^{-2}. As can be seen from Fig. 3, the increased counterpropagation effects lead to stronger dynamic changes in the carrier–light field-dynamic and thus with a higher degree to intensity structures and filament formation. For growing reflectivity an increasing percentage of the

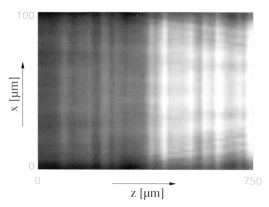

Fig. 3. Same configuration as Fig. 2 but with facet reflectivities $R = 1 \times 10^{-2}$

optical fields is back-reflected into the resonator, enhancing the formation of dynamically varying longitudinal and transverse structures in the intensity distribution. The interaction of the counterpropagating optical fields reduces the influence of the input beam and thus deteriorates the spatial and spectral quality of the amplified radiation. The spatio-temporal variation of the intensity is accompanied with corresponding dynamical changes of the carrier density, the macroscopic gain, the refractive index and temperature which reveal further information on the spatio-temporal dynamics of high-power semiconductor lasers [17].

2.2.2   Tapered Amplifier

The tapered-amplifier geometry has recently attracted growing theoretical [6,19,20], experimental [21] and technological [22] attention. A detailed discussion of a corresponding high-power diode-laser structure is presented in Chap. 7. Here, the tapered geometry is briefly discussed in the context of this chapter's objective. The tapered geometry can partly suppress the dynamical transverse intensity structures in the amplifier and leads – due to the small lateral and vertical dimension of the waveguide – to a high small-signal gain. However, the tapered amplifier may also exhibit dynamic intensity structures and the ideal situation strongly depends on the particular operating conditions and the laser geometry.

Figure 4 displays a characteristic snapshot of the intensity distribution of a tapered amplifier with $w_1 = 5$ μm, $w_2 = 100$ μm (Fig. 1b), and an overall cavity length $L = 1.5$ mm, about 500 ps after injection of the optical beam. From comparing the snapshot in Fig. 4 with the corresponding intensity distributions in the broad-area amplifier (Fig. 2) it follows that the particular geometry of the tapered amplifier is more appropriate for the amplification and guiding of the diverging propagating light, and the transverse

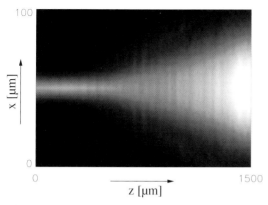

Fig. 4. Snapshot of the intensity distribution in the active layer of a tapered amplifier for an injected CW signal of 1 mW

structures in the intensity distribution are suppressed more efficiently. The residual reflectivities of the facets (which were again chosen to be 10^{-4}) and the interaction with the density distribution in the active area still lead to spatio-temporally varying intensity distributions. In comparison with the linear broad-area amplifier (Fig. 2), the 'beam shaping' by the tapered shape of the amplifying section helps to avoid transverse structures and filamentation. Thus, for given experimental situations, such as amplifier current and input power, a systematic analysis of the influence of the amplifier geometry (overall length, length of the preamplifier, angle of the tapered section) can be achieved by a variation of parameters and calculation of the respective light fields and carrier distributions, which may be decisive for the spatial and spectral stability of the emitted beam. Usually, one can extract that the preamplifier should have a length of several 100 μm in order to provide preamplification and mode selection, and the angle of the tapered section should coincide with the diffraction angle of the propagating Gaussian beam. The quality of the AR coating turns out to be particularly important in the case of the tapered amplifier. As an example Fig. 5 shows a snapshot of the intensity distribution in a tapered amplifier with a facet reflectivity of 10^{-2} illustrating the increased counterpropagation of the optical fields. In combination with the geometry of the active layer, this leads to the formation of strong and irregular intensity fluctuations. The internal spatio-temporal dynamics dominates over the amplification of the injected input beam, and the amplifier shows multi-mode lasing behavior.

2.3   Dynamics of Optical-Emission Characteristics

Typical dimensions of the active layer of high-power semiconductor lasers range from 100 μm to 3000 μm, while the diffusion and diffraction length,

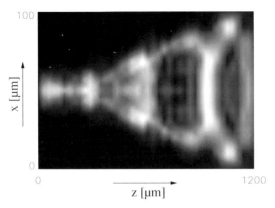

Fig. 5. Snapshot of the intensity distribution in the active layer of a highly pumped tapered amplifier ($I = 1.2$ A) with a facet reflectivity $R = 1 \times 10^{-2}$

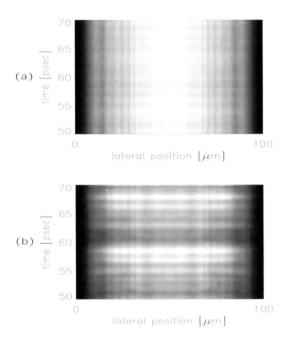

Fig. 6. Temporal behavior of the near field at the position of the output facet for a reflectivity of (**a**) 1×10^{-5} and (**b**) 1×10^{-2}

or the internal coherence length, all lie in the μm regime. As a consequence, spontaneous formation of optical filaments has for quite some time been observed. Recently, it has been shown that in broad-area lasers the optical structures vary dynamically both in transverse [7] and in longitudinal directions. Thus, in a MOPA system also, the spatio-temporal modulations within the broad-area amplifier interact with the propagating beam and contribute significantly to the spatial and spectral quality of the emitted light. As the spatio-temporal dynamics in high-power semiconductor lasers is also generally strongly dependent on material properties (epitaxial structure, waveguiding properties, residual reflectivity of the facets, carrier injection) and operating characteristics (injection current, power and spatio-spectral quality of the injected beam) a profound understanding of the dynamical behavior is of great interest. As an example, Fig. 6 shows the temporal behavior of the intensity distribution, $I(x, z = L) \sim | E^+(x, z = L) |^2 + | E^-(x, z = L) |^2$, at the output facet for an amplifier with low (Fig. 6a, $R = 10^{-5}$) and with high (Fig. 6b, $R = 10^{-2}$) facet reflectivity. The temporal behavior of the amplifier with a low reflectivity of 10^{-5} (Fig. 6a) is rather stable in time. Only very slight intensity fluctuations arise due to the residual reflectivity and the diffraction in the density pattern. In addition, small transverse structures can be seen due to the microscopic scattering and relaxation processes. In the case of the amplifier with high a reflectivity of 10^{-2} (Fig. 6b) strong intensity modulations appear in space and time due to the increased counterpropagation.

The influence of this tendency on the spatial beam characteristics can be seen in Fig. 7a–d. The temporally averaged near-field and far-field profiles show fewer structures in the case of the good anti-reflection coating (Fig. 7a,b). There, most of the amplified light is contained in the diffraction-limited central lobe of the far-field distribution. Note that although the injected beam was assumed to have a uniform Gaussian shape, the small transverse modulations of the emitted radiation cannot be suppressed completely, because the light-field and carrier dynamic lead to temporal variations on a picosecond time-scale and spatial variations in the μm regime, both giving rise to the structures in the temporal averaged values.

2.4   Spatial and Spectral Carrier Dynamics

In the simulation, the dynamics of the microscopic properties – which determine the macroscopic variables discussed so far – are self-consistently calculated at every spatial location. The microscopic variables can very often reflect macroscopic properties. An important way to analyse the spatio-temporal behavior of high-power semiconductor lasers is for example to look at the deviation of the carrier distributions of electrons and holes from their quasi-equilibrium values and at the interband polarization. These Wigner distributions include on a microscopic level the dynamic spatial and spectral carrier–light field-dynamics.

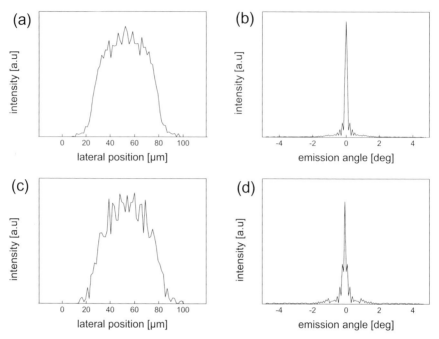

Fig. 7. Temporal average spatial beam characteristics for a broad-area amplifier with injection of a Gaussian CW beam: (**a**) near field for $R = 1 \times 10^{-5}$, (**b**) far field for $R = 1 \times 10^{-5}$, (**c**) near field for $R = 1 \times 10^{-2}$, (**d**) far field for $R = 1 \times 10^{-2}$

In the following we analyse examples of the Wigner distributions for the situation of a broad-area amplifier ($I = 1.2$ A, $L = 750$ μm, $w = 100$ μm, $R_1 = R_2 = 10^{-4}$) with optical injection of a Gaussian CW beam with a total power of 20 mW.

Figure 8a–f shows for three characteristic values of λ_{inj} snapshots of the dependence of the electron (left column) and hole (right column) Wigner distributions $\delta f_z^{\mathrm{e}}(k, z, t) = f^{\mathrm{e}}(k, 0, z, t) - f_{\mathrm{eq}}^{\mathrm{e}}(k, 0, z, t)$ and $\delta f_z^{\mathrm{h}}(k, z, t) = f^{\mathrm{h}}(k, 0, z, t) - f_{\mathrm{eq}}^{\mathrm{h}}(k, 0, z, t)$ on the wavenumber k and the longitudinal position z in the center $x = 0$ of the broad-area laser. The propagating beam has, after a number of passes, significantly led to a deviation of the carrier distributions from their equilibrium values each given by Fermi–Dirac statistics. In principle, the distributions of electrons and holes show a similar behavior. However, due to the band structure and to the difference in effective masses, the absolute values are not identical. In Fig. 8a,d the wavelength of the injected beam has been set to 815 nm, the maximum of the gain bandwidth of the laser. Thus, the spectral depletion of the carrier distribution is very high, and the longitudinally increasing but spectrally well-confined kinetic trench lies within the gain maximum. The dependence on λ_{inj} becomes apparent when comparing Fig. 8a,d with Fig. 8b,e and Fig. 8c,f, where λ_{inj} is

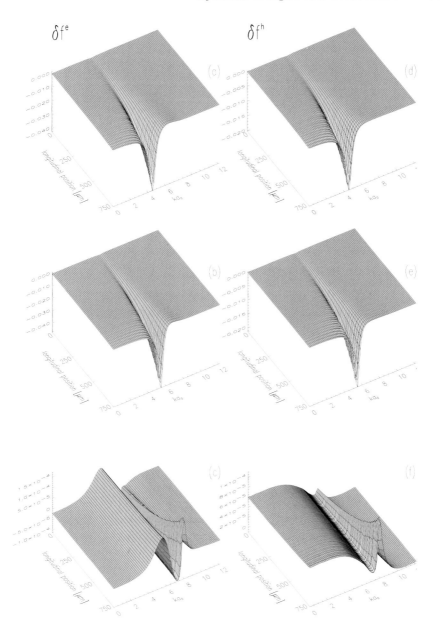

Fig. 8. Snapshots of the nonequilibrium Wigner distributions of electrons $\delta f_x^c(k, z, t_0)$ (*left column*) and holes $\delta f^h(k, z, t_0)$ (*right column*). The injection of the optical beam into the active layer of the broad-area laser occurs (**a,d**) at resonance with $\lambda_{\mathrm{inj}} = \lambda = 815$ nm, (**b,e**) at a wavelength detuning $\delta\lambda = 20$ nm, (**c,f**) at a strongly nonresonant excitation ($\lambda_{\mathrm{inj}} = 765$ nm)

815 nm, 795 nm, and 765 nm, respectively. As the input beam determines the progression and the position of the kinetic trench in the generation rate, a shift of λ_{inj} towards shorter wavelengths results in a corresponding spectral drift. In addition, high detuning towards absorbing states leads, due to interaction with the longitudinal optical phonons, to the situation depicted in Fig. 8c,f where a significant proportion of carriers has accumulated in the heated nonequilibrium states at $k \sim 6a_0^{-1}$. Stimulated recombination at resonance, however, is prevented by the particular distribution of the generation rate and refractive index (see Fig. 9c,f). As a consequence, the laser adjusts its emission wavelength accordingly and the depletion of carriers takes place at wavelengths corresponding to $k \sim 7a_0^{-1}$.

2.5   Spatial and Spectral Refractive-Index and Gain Dynamics

As already mentioned in Sect. 1.1, the spatio-temporal and spectrally resolved modeling can be used to extract parameters such as the carrier-induced refractive index and the spatio-spectral gain. Both are important for the propagation of the light fields and thus the spatial and spectral beam quality. In Fig. 8 the shape of the kinetic trench burnt into the carrier distributions by the optical fields is considerably smoother than both the real (Fig. 9a–c) and imaginary part (Fig. 9d–f) of the interband polarization. In the polarization (Fig. 9), longitudinal structures appear on length scales which span a regime from 10 μm to 100 μm. These nonlinear structures reflect the density-dependent microscopic relaxation of the polarization [5] which, due to the propagation of the optical beam, is directly transformed to a characteristic length scale. Consequently, the microscopic spatio-spectral dynamics of the interband polarization being governed by the fast carrier–carrier and, in particular, by the carrier–optical-phonon scattering processes, determine both spectral and spatial scales of the light–matter interaction within the active area: it is these modulations in the polarization which act as an induced macroscopic distributed feedback grating and which are the seeds of the longitudinal variance of the intensity distribution (see Fig. 2). Together with the spatio-temporal variation of the carrier density they lead to the nonstationary spatio-temporal dynamics of semiconductor-diode lasers.

The combined spatial and spectral variation of the interband polarization displayed in Fig. 9 demonstrates the influence of the microscopic spatio-spectral dynamics on the nonlinear gain and refractive-index variations. The real part of the nonlinear polarization shown in Fig. 9a–c visualizes the spectral dependence and longitudinal variation of the generation rate and thus the spatio-spectral distribution of the optical gain. Note that in Fig. 9a,d negative values of $p_z'(k, z, t_0)$ represent positive local gain. The corresponding spatial variation of the dispersion of the nonlinear induced refractive index δn can on the other hand be directly deduced from the k-dependence of $p_z''(k, z, t_0)$ (Fig. 9d–f). Noting, that generally $-\delta n \sim p_z''$, the longitudinal

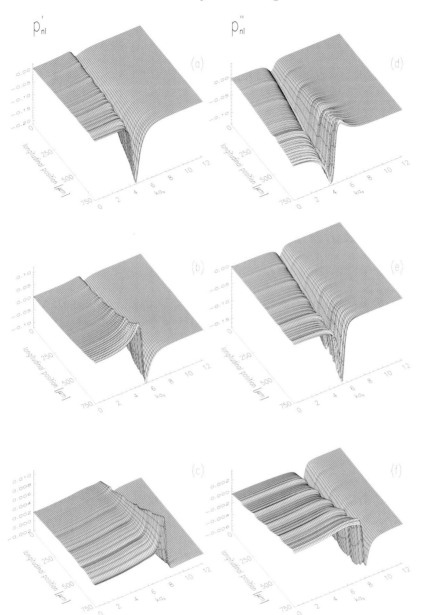

Fig. 9. Snapshots of the nonequilibrium Wigner distributions of the real $(p'_z(k, z, t_0)$, *left column*) and imaginary parts $(p''_z(k, z, t_0)$, *right column*) of the interband polarization. The injection of the optical beam into the active layer of the broad-area laser occurs (**a,d**) at resonance with $\lambda_{\mathrm{inj}} = \lambda = 815$ nm, (**b,e**) at a wavelength detuning $\delta\lambda = 20$ nm, (**c,f**) at strongly nonresonant excitation $(\lambda_{\mathrm{inj}} = 765$ nm)

shape of $p_z''(k, z, t_0)$ (Fig. 9d–f) reveals the density dependence of the induced nonlinear refractive index. If the optical beam is injected at resonance (Fig. 9a,d) the propagating light field gains in amplitude and thus leads to a rising depletion of the density of charge carriers and to an increase in gain. This, in turn, causes an increase of the refractive index within the cavity from $z = 0$ to $z = L$. With increasing wavelength detuning, Fig. 9b,e and Fig. 9c,f show that the carrier heating effects gain influence while at the same time the carrier depletion by stimulated emission is reduced: fewer and fewer electron–hole pairs are locally available at momentum states (wavelengths) suitable for recombination. Thus, with decreasing wavelength, the real and the imaginary parts of the interband polarization (Fig. 9) change their sign towards absorption.

3   Conclusion

We have given an overview of the theory of high-power semiconductor lasers and have selectively discussed the macroscopic and microscopic spatio-temporal dynamics of broad-area and tapered diode amplifiers. Our theory and the numerical simulations are based on a microscopic spatially, spectrally, and temporally resolved Maxwell–Bloch model for semiconductor lasers which self-consistently includes, in particular, opto-electronic and thermal properties of the active semiconductor-laser area and generally allows an investigation of diode-amplifier systems with variable geometries, as well as specific laser-material systems. It therefore can reveal typical relevant properties which influence the overall performance of the device: the spatio-temporal distribution of the intensity within the laser cavity, the charge-carrier density, the local nonlinear gain as well as the carrier-induced refractive index.

Since transverse and longitudinal dynamically varying structures and filaments caused by the carrier–light-field interactions within the active layer strongly influence the optical properties of high-power semiconductor-diode lasers, the analysis and prediction of the spatio-temporal dynamics is extremely important when designing semiconductor-diode-laser systems. The simulation results discussed in this chapter have recently lead to an improved interpretation of experimental data obtained at the University of Kaiserslautern (c.f. chapter "Properties and Frequency Conversion of High-Brightness Diode-Laser-Systems" by K.-J. Boller et al. of this book). The flexibility of the modeling allows us to vary and eventually optimize typical device geometries and operating conditions. By revealing the internal spatio-temporal processes within the semiconductor-laser, the self-consistent microscopic theoretical description provides a new way to describe and analyse the physical effects which determine the spatio-temporal behavior of semiconductor MOPA devices.

Acknowledgement

A substantial part of this work has been done at the University of Kaiserslautern. We would like to thank Prof. Wallenstein for his continuing support and interest.

List of Symbols

E	Optical fields
N	Carrier density
$f^{e,h}$	Carrier distribution (e: electron, h: hole)
p	Interband polarization
g	Generation rate
Λ	Pump rate
L	Cavity length
w	Stripe width of the broad-area laser
d	Thickness of active layer
n_c	Refractive index of the cladding layers
n_l	Refractive index of active layer
n	Effective index
ε	Permittivity
λ	Laser wavelength
R_1	Front facet-mirror reflectivity
R_2	Back facet-mirror reflectivity
γ_{nr}	Nonradiative recombination rate
Γ_{sp}	Spontaneous-recombination coefficient
a_0	Exciton Bohr radius
m_0	Mass of the electron
m_e	Effective electron mass
m_h	Effective hole mass
$\mathcal{E}_g$	Band gap energy
ω_T	Transition frequency
α	Absorption coefficient
η_i	Injection efficiency
Γ	Confinement factor
α_w	Absorption
v_{sr}	Surface recombination velocity
k_z	Wavenumber of the propagating field
d_{cv}	Opticle dipole-matrix element
k	Carrier wavenumber
$\tau_{e,h}^{-1}$	Carrier relaxation rate (carrier–carrier and carrier–phonon (LO) scattering)
τ_{ph}^{-1}	Carrier–phonon scattering rate
γ_a	Relaxation rate to the ambient temperature
$u^{e,h}$	Carrier energy
T_l	Lattice temperature
$T_{pl}^{e,h}$	Plasma temperature of electons (e) and holes (h)

References

1. D. Botez, D. R. Scrifres: *Diode-Laser Arrays* (Cambridge Univ. Press, Cambridge 1994)
2. O. Hess: *Spatio-Temporal Dynamics of Semiconductor Lasers* (Wissenschaft und Technik, Berlin 1993)
3. O. Hess, T. Kuhn: Spatio-temporal dynamics of semiconductor lasers: theory, modeling and analysis, Prog. Quant. Electron. **20**, 85–179 (1996)
4. O. Hess: Spatio-temporal complexity in multi-stripe and broad-area semiconductor lasers, Chaos, Solitons & Fractals **4**, 1597–1618 (1994)
5. O. Hess, T. Kuhn: Maxwell-Bloch equations for spatially inhomogeneous semiconductor lasers I: Theoretical description, Phys. Rev. A **54**, 3347–3359 (1996)
6. O. Hess, T. Kuhn: Maxwell-Bloch equations for spatially inhomogeneous semiconductor lasers II: Spatio-temporal dynamics, Phys. Rev. A **54**, 3360–3368 (1996)
7. I. Fischer, O. Hess, W. Elsäßer, E. Göbel: Complex spatio-temporal dynamics in the near field of a broad-area semiconductor laser, Europhys. Lett. **35**, 579–584 (1996)
8. C. M. Bowden, G. P. Agrawal: Maxwell-Bloch formulation for semiconductors: effects of Coulomb exhange, Phys. Rev. A **51**, 4132–4139 (1995)
9. J. Yao, G. P. Agrawal, P. Gallion, C. M. Bowden: Semiconductor-laser dynamics beyond the rate-equation approximation, Opt. Commun. **119**, 246–255 (1995)
10. S. Balle: Effective two-level model with asymmetric gain for laser diodes, Opt. Commun. **119**, 227–235 (1995)
11. O. Hess, S. W. Koch, J. V. Moloney: Filamentation and beam propagation in broad-area semiconductor lasers, IEEE J. QE **31**, 35–43 (1995)
12. E. Gehrig, O. Hess: Nonequilibrium spatio-temporal dynamics of the Wigner distributions in broad-area semiconductor lasers, Phys. Rev. A **57**, 2150–2163 (1998)
13. Y. P. Varshni: Temperature dependence of the energy gap in semiconductors, Physica **34**, 149–154 (1967)
14. R. Zimmermann: Nonlinear optics and the Mott transition in semiconductors, Phys. Stat. Sol. B **146**, 371–384 (1988)
15. G. H. B. Thompson: *Physics of Semiconductor Laser Devices* (Wiley, New York 1980)
16. S. Adachi: Material parameters for use in research and device applications, J. Appl. Phys. **58**, R1–R29 (1985)
17. E. Gehrig, O. Hess, R. Wallenstein: Modeling of the performance of high-power diode amplifier systems with an opto-thermal microscopic spatio-temporal theory, IEEE J. QE **35**, 320–331 (1999)
18. E. Gehrig, B. Beier, K.–J. Boller, R. Wallenstein: Experimental characterization and numerical modelling of an AlGaAs oscillator broad-area double-pass amplifier system, Appl. Phys. B **66**, 287–293 (1998)
19. R. J. Lang, A. Hardy, R. Parke, D. Mehuys, S. O'Brien, J. Major, D. Welch: Numerical analysis of flared semiconductor laser amplifiers, IEEE J. QE **29**, 2044–2051 (1993)
20. J. V. Moloney, R. A. Indik, C. Z. Ning: Full space-time simulation for high-brightness semiconductor lasers, IEEE Photon. Techn. Lett. **9**, 731–733 (1997)

21. J. N. Walpole: Semiconductor amplifiers and lasers with tapered gain regions, Opt. Quant. Electron. **28**, 623–645 (1996)
22. S. O'Brien, R. Lang, R. Parker, D. F. Welch, D. Mehuys: 2.2 W continuous-wave diffraction-limited monolithically integrated master oscillator power-amplifier at 854 nm, IEEE Photon. Techn. Lett. **9**, 440–442 (1997)

Epitaxy of High-Power Diode Laser Structures

Markus Weyers, Arnab Bhattacharya, Frank Bugge, and Arne Knauer

Ferdinand-Braun-Institut für Höchstfrequenztechnik (FBH),
Albert-Einstein-Straße 11, D-12489 Berlin, Germany
weyers@fbh-berlin.de

Abstract. Excellent semiconductor-material quality is an essential prerequisite for the fabrication of high-power diode lasers and laser bars. This review discusses issues in the epitaxial growth of semiconductor materials and layer sequences that form the basis for diode lasers. First, an overview of the material systems used for diode lasers with emission wavelengths extending from the far-infrared to the blue range of the spectrum is presented. The following sections then concentrate on materials that have, until now, been used for high-power diode lasers: the GaAs-based, and to a lesser extent also the InP-based members of the III–V family (Al, Ga, In)(As, P).

Different growth techniques are described, with stress on the two modern methods – Molecular-Beam Epitaxy (MBE) and Metalorganic Vapor-Phase Epitaxy (MOVPE) – that are currently used for the growth of diode-laser structures. An attempt is made to compare the relative strengths and weaknesses of these epitaxial growth techniques.

The issues of purity, ordering (observed for example in GaInP), phase separation (occurring for certain compositions of GaInAsP) and p- and n-type doping for the constituent III–V materials found in high-power diode lasers are discussed in detail. The sequential growth of layers of controlled thicknesses, composition and doping profiles to build up the desired heterostructures requires careful optimization of the growth processes, especially at the heterointerfaces. While most of the layers within the heterostructure are lattice-matched to the underlying substrate, the active, light-emitting layer usually consists of one or more strained quantum wells (QWs). The advantages of such strained layers along with the corresponding implications from the growth viewpoint are highlighted. Finally, an overview of the different material combinations used and the state-of-the-art for 600–1060 nm emitting GaAs-based diode lasers is presented.

The first reports on semiconductor-diode lasers [1] date back to 1962. Since then these devices have been realized over a wide range of emission wavelengths in a variety of material systems. Since the first demonstration of the semiconductor laser, threshold-current densities have fallen by more than three orders of magnitude, due to various breakthroughs made possible primarily by advancements in crystal-growth technologies. Improvements in Liquid-Phase Epitaxy (LPE) in the early 1970s enabled the realization of the Double Heterostructure (DH) concept and Continuous-Wave (CW) operation of GaAs/AlGaAs lasers, while progress in growth technologies like Molecular-Beam Epitaxy (MBE) and Metalorganic Vapor-Phase Epitaxy (MOVPE) were responsible for the development of quantum-well lasers, which are the

R. Diehl (Ed.): High-Power Diode Lasers, Topics Appl. Phys. **78**, 83–120 (2000)

basis of most high-power diode lasers today. This review describes and compares the different epitaxial growth methods used to fabricate diode lasers in various material systems and examines important growth-related issues such as doping, ordering and homogeneity that are essential for the successful fabrication of high-performance lasers.

The basic transverse structure of most edge-emitting diode lasers is quite similar (Fig. 1). In the Separate Confinement Heterostructure (SCH) design, light generated by band-to-band recombination in the Quantum-Well (QW) active region is confined in the transverse (vertical) direction by means of the waveguide formed by the cladding and confinement layers. In the longitudinal direction the laser cavity is defined by the two mirror facets formed by cleaving along crystallographic planes of the substrate crystal. Confinement in the lateral direction is achieved by gain- or index-guiding, the latter typically accomplished using etched ridges or through buried structures fabricated via multi-step epitaxy.

Lasers with emission in the far infrared are primarily of interest for monitoring and measurement applications. Using lead salt materials like PbS, PbSnTe or PbEuSe, wavelengths between 30 μm and 2.5 μm can be achieved [2]. All these devices work only at cryogenic temperatures with output powers in the mW range. In the range 4 μm to 2.5 μm GaSb-based devices have shown better performance than the lead salt-based ones [3], including room-temperature operation at 2.8 μm [4]. While diode lasers rely on recombination between conduction and valence bands, quantum cascade lasers [5] based on intersubband transitions have also been used to reach the mid-IR

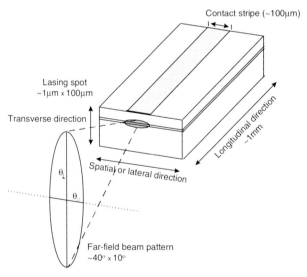

Fig. 1. Schematic representation and far-field beam pattern of a typical broad-area-type edge-emitting semiconductor laser

wavelength range. As the transition energy in these structures is determined only by the multi-quantum-well structure design and not by the fundamental band gap, they offer a high degree of flexibility in wavelength selection by tailoring the depth and width of GaAs/AlGaAs or GaInAs/AlInAs QWs. Further, these structures are based on well-established GaAs or InP technology. Intensive research has resulted in quantum cascade lasers outperforming lead salt and other diode lasers in the $3.5\,\mu m$ to $17\,\mu m$ wavelength range: lasers emitting at $8\,\mu m$ have demonstrated peak power $> 0.5\,W$ and average power of $14\,mW$ at room-temperature [6] with more than $1\,W$ CW being reported [7] at $30\,K$.

The wavelength range of 1900–1100 nm can be covered with the InP-based materials GaInAs(P) and (Al)GaInAs. The most-prominent applications here are lasers for optical-fiber transmission around 1300 nm and 1550 nm [8,9]. While most fiber-communication applications do not require high power, research efforts in this wavelength range towards enhanced output power are driven by emerging areas such as illumination, range-finding, free-space communications and marking.

GaAs-based devices cover the wavelength range from around 1060 nm to 630 nm. Strained InGaAs QWs are typically used in the active region for devices operating in the infrared above 850 nm. Adding P allows one to shift the wavelength down to approximately 730 nm with quaternary GaInAsP QWs. Even shorter wavelengths down to 715 nm can be achieved using GaAsP or AlGa(In)As active regions. Visible (red) wavelengths in the 630–690 nm region are realized with GaInP QWs. These active regions can be combined with AlGaAs/AlGaAs (above 730 nm), GaInP/AlGaInP, GaInAsP/GaInP (above 800 nm) or GaInP/AlGaAs waveguide and cladding layers, and are discussed in detail in this review.

Blue-green laser diodes have been fabricated using II–VI materials, where lattice matching to GaAs substrates can be achieved using the ZnMgSSe-based system [10]. While laser operation in the 470–530 nm range has been demonstrated, device lifetimes are still limited. Despite the absence of a lattice-matched substrate, GaN-based diode lasers have taken the lead in the short-wavelength (blue and green) region with an amazing speed. It took only three years from the first reports on blue LEDs made from Ga(In)N until CW operation at room-temperature from diode lasers was reported [11].

Among the materials discussed above, only the GaAs-based ones are at present commercially used for high-power diode lasers with powers greater than 1 W in multimode operation or greater than about 100 mW single mode. The market for devices with $> 1\,W$ power in the 780–980 nm range is the most rapidly expanding section of the diode-laser market with 1997 revenues of US \$131 million [12]. The driving force behind these developments mostly comes from the requirements in pumping of fiber amplifiers and various solid-state lasers as well as applications in printing, graphics and medicine. Another reason for the dominance of GaAs-based devices is the relative maturity of

GaAs technology which is pushed by both microelectronic and optoelectronic applications. GaAs substrates with the low dislocation density necessary for reliable high-power operation are readily available today. Further, for GaAs-based materials, the epitaxial growth methods discussed in this chapter have been developed to a level not yet achieved for other materials. Similarly, GaAs processing has reached the level of consistency and reproducibility needed for mass production. In the range 730 nm to 1000 nm, CW output powers higher than 7 W from a 100 μm aperture have been achieved using different material combinations [13,14,15,16,17].

Currently, the only other material system used for high-power diode lasers is the InP-based one. Here, output powers around 5 W CW from a 100 μm apertures have been achieved for emission around 1500 nm [18]. Very recently, high CW powers have been reported at longer wavelengths with 11 W CW at 1870 nm from a 1 cm-wide bar [19] and 218 W quasi-CW operation [20] from 2D arrays of devices at 1830 nm being demonstrated. If a certain application requires a specific wavelength in the range 1300 nm to 1900 nm, InP-based laser structures have to be used. Additionally, eye safety is brought forward as an argument for wavelengths above 1400 nm. However, at the extreme power densities encountered in the use of high-power diode lasers the perception of wavelengths above 1400 nm being less harmful to the eye than shorter wavelengths appears questionable.

GaN is also a potential candidate for high-power applications. The material quality of GaN grown on non-lattice-matched SiC or sapphire substrates is relatively poor with dislocation densities above $10^8\,\mathrm{cm}^{-3}$. Surprisingly enough, this crystal quality allows for the production of high-brightness LEDs [11]. However, laser diodes require lower dislocation densities which can be achieved by pseudo-substrates grown via Epitaxial Lateral OverGrowth (ELOG). In this technique, growth starts from seed areas and proceeds over masked areas where the growing film is not bonded to the substrate. With this approach laser diodes with 420 mW CW output power have been demonstrated [21]. It will be interesting to follow the future developments in this field. At the present stage, however, GaN-based laser technology is not mature enough for production of high-power diode lasers.

Since only GaAs-based high-power diode lasers are currently in large-scale production, the rest of this contribution is focused on this material system. Additionally, issues in the epitaxy of InP-based devices are discussed where appropriate. This review lays stress on epitaxial growth methods and their application to high-power diode lasers.

1   Growth Methods

Most high-power diode lasers are based on QW structures where layer thicknesses of just a few nm in the active region and the desired compositions have to be accurately controlled with good homogeneity over a 2 in (in future possibly 3 in) wafer. At the same time the waveguide and cladding layers, making the bulk of the 3–5 µm thick structure have to be grown at a reasonable growth rate to make the process economically feasible. All this has to be achieved maintaining high crystalline quality with minimal defects and with a tight control of the doping profile. While a variety of epitaxial growth techniques, described in the following section, have been used to fabricate III–V diode lasers, we focus primarily on the techniques Molecular-Beam Epitaxy (MBE) and Metalorganic Vapor-Phase Epitaxy (MOVPE) that are used in the production of high-power diode lasers.

Historically, Liquid-Phase Epitaxy (LPE) was the first technique used to grow diode lasers [22]. In LPE, layer growth proceeds by precipitation of the semiconductor material from the melt or from a solution onto a substrate. A well-established process with machines capable of growing on several tens of substrates at high rates make this method cheap and LPE is still widely used for the mass-production of LEDs. However, while LEDs are usually based on thick layers, diode lasers require thin QW structures with well-controlled, homogeneous compositions and thicknesses, a requirement LPE is not capable of fulfilling, primarily due to melt-back effects while switching between different compositions at the high growth rates used. Consequently, this method no longer plays a role in the growth of the active region of diode lasers.

Similarly, Chloride- (ClVPE) and Hydride-Vapor-Phase Epitaxy (HVPE) have been used for the growth of DH (Double Heterostructure) lasers [23]. In this method group-III metals are supplied via reaction of a melt with chlorine in the source area with the reverse reaction occurring on the substrate. In the case of ClVPE, trichloride group-V sources ($AsCl_3$, PCl_3) are passed over the metal melt generating the chlorine necessary for transport and simultaneously providing the group-V species. In HVPE, arsine and phosphine act as group-V sources and HCl is independently injected to transport the metal. Etching reactions are significant in these techniques, and the particularly corrosive nature of $AlCl_3$ (which etches quartz) makes AlGaAs growth virtually impossible. The metal-transport mechanism makes rapid switching from one composition to another and thus the growth of thin layers difficult. This is why VPE is also no longer used in diode-laser production.

The two predominant techniques used for the growth of high-power diode lasers are MBE and its variant Gas-Source MBE (GSMBE) and MOVPE, also termed Metalorganic Chemical Vapor Deposition (MOCVD). Both methods are used in the commercial production of III–V devices on a multiwafer scale.

1.1 Molecular-Beam Epitaxy and Its Variants

In Molecular-Beam Epitaxy (MBE) directed thermal beams of atoms or molecules that are evaporated from heated effusion-cells react on the clean surface of a substrate held at high temperature under Ultra-High Vacuum (UHV) conditions to form an epitaxial film. The effusion-cell temperature controls the beam flux, and mechanical shutters in front of the cells permit rapid switching of beam species and thus abrupt changes in layer composition and doping. A schematic of this process is shown in Fig. 2. A complete treatment of the MBE process can be found in texts on the subject [24,25]. To form molecular beams the mean free path between gas-phase collisions has to be greater than the effusion-cell to substrate distance, requiring the process to be carried out under UHV conditions. The stainless steel growth chamber is equipped with ion pumps, cryopumps and titanium sublimation pumps yielding base pressures below 10^{-10} torr. In addition, liquid-nitrogen-filled cryoshrouds surround the effusion-cells and the substrate. Thus the substrate sees only cold walls onto which most impurities and re-evaporated layer constituents condense.

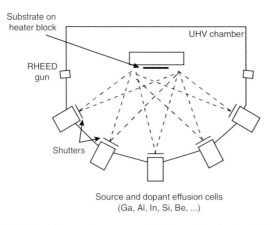

Fig. 2. Schematic representation of a typical Molecular-Beam Epitaxy (MBE) system

Loading and unloading the growth chamber proceeds via UHV buffer chambers and load-lock systems that minimize the introduction of impurities like oxygen and water. Such oxidizing agents are particularly incompatible with the hot metal filaments and the melt in the effusion-cells. If the vacuum in the growth chamber has to be broken for maintenance, extensive bake-out procedures are necessary to re-establish high purity of the growth system and the grown layers, especially those containing aluminum.

MBE started as a research tool to study crystal growth. The UHV environment makes it ideally suited for combination with analytical techniques like electron diffraction RHEED (Reflection High-Energy Electron Diffraction), LEED (Low-Energy Electron Diffraction) or mass spectrometry. Optical techniques like reflectometry have also been successfully adapted to MBE machines. However, these research tools are not usually found on production machines geared for high throughput and high up-time. Driven by microelectronic applications, e.g. growth of AlGaAs/InGaAs/GaAs PHEMTs (Pseudomorphic High Electron Mobility Transistors), machines capable of growth on multiple 4 in wafers with good homogeneity over a single wafer and also from wafer to wafer are available. The next generation of MBE machines capable of handling multiple 6 in wafers is expected to be in operation in 1999. This allows for a wafer output of several 10.000 4 in wafers per year just from the merchant epi vendors, not including the capacity installed and used for captive production. Thus, MBE is prepared for the very strong demand foreseen for high-power diode lasers.

The majority of basic growth studies as well as production applications in MBE are in the AlGaAs material system. AlGaAs diode lasers for CD players with emission at 780 nm are grown by MBE in large quantities. Reliable 980 nm high-power single-mode diode lasers for pumping of Er-doped fiber amplifiers are also commercially grown by MBE [26]. Based on the longstanding experience in this material system, growth recipes for different applications (e.g. diode lasers require different growth conditions than transistor structures) have been derived. High-purity material with controlled p- and n-doping can easily be achieved. Homogeneity has improved over the years due to improvements in system geometry and temperature uniformity on the substrate holders and has now reached a very high level. Transfers between the growth chamber and the UHV buffer chamber can be performed at elevated temperatures leading to reduced cycle times. For extended uptime at high throughput, large-capacity effusion-cells have been developed. Technical improvements in cell design have resulted in improved beam-flux stability. As the flux from an effusion-cell depends on the level of the melt within it, MBE users have developed strategies to correct for such depletion effects by varying the temperature set-point of the cell depending on its filling level.

In MBE the growth of layers containing phosphorus is more difficult than that of arsenide layers. Elemental arsenic primarily evaporates as As_4 with a small fraction of As_2 yielding a stable beam pressure at a given effusion-cell temperature. In contrast to this, depending on the temperature and pressure, phosphorus exists in many allotropes, each having a different vapor pressure. This results in an ill-defined beam composition and unstable beam fluxes. Since P_4 has a negligible sticking coefficient on the surface, only the dimer P_2 contributes to the growth and the effective P flux to the surface is very difficult to control using conventional effusion-cells. This makes the controlled growth of mixed group-V compounds like GaInAsP more or less impossible. Additionally, the high vapor pressure of P makes it difficult to outgas the source material by heating in the vacuum chamber. A group-V source thus containing trace amounts of residual water can seriously compromise system purity.

To overcome these difficulties the group-V sources are introduced as their gaseous hydrides, arsine and phosphine, in a modification of the MBE technology called Gas-Source MBE (GSMBE) [27]. Beam-flux control for the group-V species is achieved by controlling the gas flux into the growth chamber. The stability of AsH_3 and PH_3 makes pre-cracking upon introduction into the growth chamber necessary, which is done in a cracker cell at elevated temperatures. Since the cracker temperature has no impact on the beam flux it can be independently optimized to enhance the generation of the dimer species As_2 and P_2 over that of the tetramers. This process can be supported by catalysts like tungsten or tantalum. The hydrogen background produced by hydride cracking has to be removed from the growth chamber by means of turbomolecular pumps and/or cryopumps. GSMBE has successfully been used for the growth of GaInAsP/GaAs high-power diode-laser structures [28,29].

The use of high-pressure hydride sources is not common in an MBE laboratory where typically solid sources are handled; the additional safety precautions to handle the toxic hydrides and the exhaust gases are expensive and require experienced personnel. This has prompted the development of sources using solid phosphorus. Such sources have become available with the invention of valved cracker cells, where evaporation, flux control and control of the species leaving the cell (i.e. dimers) are done in separate areas. The phosphorus is kept in a heated furnace to provide the necessary vapor pressure and introduced into the chamber via a valve that is heated to avoid condensation. Behind the valve a cracker cell like that used in GSMBE is located. With this type of cell a reproducible supply of P appears possible. The source material can be outgassed by heating the crucible while keeping the valve closed, improving system purity. Additionally, source materials can be loaded without breaking the growth-chamber vacuum. Since the group-V elements are always required in larger amounts than the metals Al, Ga and In this remote loading option reduces system down-time. Consequently, this approach has also been applied to As sources and is now frequently used for AlGaAs growth. GaInAsP/GaAs diode lasers grown by solid-source MBE yield results comparable to those obtained using GSMBE [30].

For reasons of completeness a further variant of MBE, Metalorganic MBE (MOMBE) [31], also named Chemical Beam Epitaxy (CBE) is mentioned. Here the supply of the group-V materials is as in GSMBE. Additionally, the group-III compounds are supplied as metalorganic compounds from external sources and introduced into the chamber via a gas injector. MOMBE can thus be considered as MOVPE in the absence of gas-phase reactions. It offers advantages in selective growth and also in the fabrication of buried structures. The complicated surface chemistry involved in the breakdown of the metalorganic molecules requires a tight control especially of the substrate temperature. While a variety of device structures have been demonstrated, MOMBE has still not found the way from research laboratories into large-scale production.

1.2   Metalorganic Vapor-Phase Epitaxy

In MOVPE [32,33,34,35] metalorganic (MO) precursors, usually the trimethyl compounds of Al, Ga and In like $Ga(CH_3)_3$ (TMGa, also named TMG) react with sources of As and P, usually the hydrides AsH_3 and PH_3, on the heated substrate to form the epitaxial layer. A schematic depiction of this process is shown in Fig. 3a. The reaction is carried out in a gas flow with hydrogen being the usual carrier gas at atmospheric or reduced pressures. Parts of the overall reaction already take place in the gas phase. The final steps, i.e. release of the constituent elements and incorporation into the lattice, happen on the semiconductor surface. The local growth rate and layer composition are determined by the local temperature and the supply of the starting materials from the gas phase. To reduce unwanted pre-reactions and avoid turbulent flow, reduced pressures are used with 100 hPa being a typical value. The higher flow velocity at reduced pressure and constant total flow lead to enhanced homogeneity of material supply. This also reduces the time

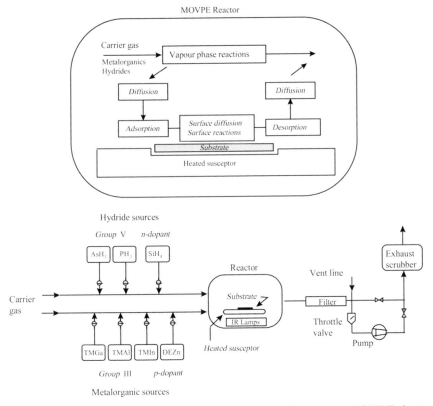

Fig. 3. *Top*: schematic depiction of reactions taking place during MOVPE; *bottom*: schematic layout of a typical MOVPE system

necessary for an exchange of the gas phase in the reactor, enabling abrupt switching between epilayers of different compositions.

A typical MOVPE growth system (Fig. 3b) consists of three main parts – a gas-mixing system, the reactor, and the pump and exhaust handling system. Despite differences in detail the general layout of the gas-mixing system is very similar for most MOVPE systems. In the gas-supply part the MO precursors are kept in stainless steel cylinders at a precisely controlled temperature and at a fixed pressure. Flowing hydrogen through the liquid TMG(a), TEG(a), TMA(l), DMZn or solid TMI(n) precursors in the bottle results in a controlled transport of the saturated vapor into the reactor. The fluxes of hydrogen and the other gaseous species like AsH_3, PH_3 or SiH_4 for n-doping are metered by mass-flow controllers. These devices measure the thermal conductivity of the flowing gas. The pressure is controlled by capacitance or piezoelectric pressure transducers. Additional purge flows introduced downstream of the dosing devices increase the flow velocity of the reactants and thus reduce transient effects that can occur when fluxes are changed [36]. The reactants are switched into the reactor or into a bypass directly into the exhaust by switching manifolds [37]. These valves are constructed for rapid switching operation and minimized dead volumes that could lead to memory effects.

For the reactor design a variety of different concepts are being used. A common design found in research and also in low-volume production systems is the horizontal linear reactor [38] where the gas stream enters on one side, is passed over the lying, or in some cases under the hanging [39] substrate and leaves the reactor on the other side. This type of reactor is used for single-wafer operation up to 4 in with machines for one 2 in wafer being the most-common ones. For the production of high-power diode-laser structures by MOVPE currently reactors of horizontal type are mostly employed.

The vertical-reactor type with a lying substrate that was used in the fabrication of the first MOVPE-grown AlGaAs diode lasers [40] is still used in some laboratories. Barrel reactors capable of multiwafer operation [41] are in use at several companies especially for the fabrication of AlGaAs solar cells. In this vertical-reactor type the substrates are not lying but stand against the substrate holder.

A common feature of all the above reactor types is that the temperature of only the susceptor (heated either by infrared lamps or inductively) is controlled. The temperature of the reactor walls is not controlled but results from radiative heating from the susceptor and heat removal by the flowing carrier gas, heat conduction or heat radiation into the environment.

In the close-spaced or showerhead reactor [42] gases are introduced through an actively cooled injector distributing the flow over the substrate placed close to this injector. In this reactor the substrate is heated by a resistive heating element within the reaction chamber. The Turbodisc^{TM} reactor [43] is of similar design. The gases are supplied through separate inlets for the group-III and group-V materials in the cooled reactor top with a mesh acting as flow distributor. The substrates are placed at a much larger distance than in the close-spaced reactor and a homogeneous distribution of the reactants is achieved by a very fast (1000 rpm) rotation of the substrate holder that also suppresses convection. This reactor has also been up-scaled for multiwafer operation up to 38×2 in wafers per run, a capacity that currently is primarily needed for solar-cell and LED applications.

Another reactor concept is the Planetary^{TM} type [44]. In this reactor the source materials are injected in the center of the radial symmetric reactor and flow towards its periphery. The substrates are placed at some distance from the inlet. To compensate for the radial expansion of the gas and the source consumption, the wafers are individually rotated. Also the main substrate holder rotates to average out any deviation from radial symmetry leading to a planetary motion of the substrates. The quartz reactor top located around 20 mm above the substrates is actively temperature-controlled by a cooling-gas mixture. This reactor is available in different diameters with capacities from 5×3 in over 5×6 in up to 95×2 in and is widely used for the production of LED and solar-cell structures.

For each of these geometries optimization of the flow pattern and the temperature field in the reactor is necessary. Since different processes often need different optimum growth temperatures, no standard set of growth parameters is available. Instead, for each application the crystal grower has to find specific optimum conditions with respect to material quality and homogeneity. The latter criterion is usually neglected in research but is crucial for a successful transfer into production. In the optimization of reactor geometry and increasingly also of the growth parameters, numerical modeling [45,46] has been applied to supplement the usual empirical approach. This has resulted in excellent homogeneity in composition and also emission wavelength over a wafer and also from wafer to wafer.

Using different reactor types MOVPE has successfully been applied for the fabrication of AlGaAs/GaAs, GaInAsP/GaAs and AlGaInP/GaAs-based high-power diode lasers and currently is the method of choice for InP-based lasers.

Diode-laser structures are grown by low-pressure MOVPE. Downstream, after the reactor, a pumping system follows that consists of a rotary pump, a throttle valve for pressure control and traps for particles or condensable materials. After leaving the pump the effluent gas is treated in wet or dry chemical scrubbers or passed through a burner to remove the toxic materials in a quantitative manner. Treatment of the exhaust gases is especially troublesome when large amounts of phosphorus-containing materials have to be grown. The self-igniting nature of P is a hazard that has to be controlled. The huge demand for AlGaInP-based LEDs has led to various approaches to tackle phosphorus deposition in the exhaust system. Cold traps can be used to facilitate P condensation but bear the risk of clogging. Intentional oxidation after every few runs has also been tried to prevent accumulation of large amounts of phosphorus.

1.3   Comparison of MBE and MOVPE

When comparing different growth methods one is often confronted with the question as to which is the better one. If there was a simple answer to that question a comparison would probably no longer be necessary because the better method would be chosen and the inferior one abandoned!

Taking device performance as the criterion both MBE and MOVPE offer, to the best of our knowledge, equal performance. The AlGaAs-material system has been the mainstay of MBE and the accumulated knowledge of many years of experience is available. High-purity material is standard for many years due to the availability of extremely pure elemental sources and minimized contamination of the growth chamber. MOVPE of AlGaAs has long been plagued by an insufficient purity of the starting materials [47,48]. With the improvement in precursor quality and leak tightness of the growth systems, material of high quality can be grown. Threshold-current densities, internal losses and efficiency yield similar values for similar device structures grown by both MBE and MOVPE [26,49]. On the other hand the P-containing materials have long been the realm of MOVPE while MBE results were less encouraging. With the availability of P sources for MBE either as PH_3 cracker cells or as valved phosphorus cracker cells this gap has been closed and laser performance similar to that in MOVPE has been demonstrated by MBE [29,30]. An important point in device performance, reliability, is very difficult to assess from the literature. Device reliability can be affected by the presence of native defects like vacancies or interstitials [50]. The concentration of such point defects may depend on the growth parameters, which are different for the two methods. Currently, no indication can however be found that one method is superior to the other in this respect.

When looking at productivity, yield from one wafer and wafer throughput both have to be considered. The yield from one wafer is determined by the area that results in devices within the specifications, especially for the emission wavelength. Here also both methods can perform equally well. In MBE homogeneity is a matter of system design. The user has little chance to alter the geometry or change the temperature profile of the substrate holder. Together with an optimized growth system the user thus buys homogeneity. Of course, in MOVPE too the system has to be designed to allow for homogeneous growth. However, depending on the application, the user has to optimize the growth parameters in order to achieve the desired homogeneity. After having solved this task both systems again yield very similar levels of homogeneity. Extended defects can severely reduce yield, especially when producing laser bars with their large area of around 1 mm by 10 mm. With respect to particles both systems again show similar performance with particle densities of some 10 per cm^{-2}.

Wafer throughput is affected by capacity per run, cycle time and system up-time. Although driven by different applications both MBE and MOVPE now offer capacities per run that are sufficient for the current demand for high-power diode lasers. The use of vacuum-transfer chambers and load-locks in MBE reduces the time required for loading of the substrates. This feature is usually not common in commercial MOVPE systems but equipment makers are working on and in some cases already implementing solutions for reduced loading times. On the other hand MOVPE allows for higher growth rates, above 2 μm/h, leading to a shorter growth time in comparison to MBE where rates are typically 1 μm/h. MOVPE systems require cleaning of quartz parts and of the exhaust system on a frequent basis. Often the parts can simply be exchanged causing no additional down-time. If the gas-supply system is opened to air, extensive purging and leak testing can cause down-time of one or two days. The exchange of a bubbler of metalorganic source material takes the same amount of time but is only necessary at large time intervals. If no component fails the up-time of an MBE system is governed by the replenishment of the sources. In modern systems this can also be done without breaking the vacuum for the group-V sources. In the case of an empty effusion cell for the metals or a problem within the UHV chamber bake-out procedures over a week or even longer are necessary. According the throughput criterion, no clear preference can thus be given.

The prices of multiwafer growth machines are similar for MBE and MOVPE. While MOVPE requires substantial investment in gas cabinets for the hydrides and the safety equipment associated with their use, MBE needs a supply system for liquid nitrogen. The necessary investment thus more depends on the situation in a given facility than on the choice of the growth method.

The decision for a growth method currently is and probably will continue to be driven by the experience at hand. If for example high-power diode lasers are to be grown in an environment where LEDs, especially AlGaInP ultra-high-brightness LEDs, are fabricated, MOVPE will be the system of choice since MBE is not competitive in LED growth due to its lower growth rates. If ample expertise in MBE is available in a laboratory or production site this method will be the first option. Only very specific applications result in demands that one method cannot fulfill. One example is selective-area growth which is used for the production of diode lasers for wavelength multiplexing where different emission wavelengths are required from one growth run [51]. Since selective growth in MBE is not possible at usual growth conditions MOVPE has to be used for this application.

In the field of electronic devices based on epitaxial layers, e.g. High Electron Mobility Transistors (HEMTs) or Heterojunction Bipolar Transistors (HBTs), a growing trend to second-sourcing or even complete outsourcing of the layer production to specialized merchant epitaxial growth companies can be seen. It is likely that a similar trend will appear in the epitaxy of diode lasers once the production volume starts to increase.

2   Materials for High-Power Diode Lasers

High-power diode lasers ($> 1\,\mathrm{W}$ optical output power) are almost exclusively based on material systems grown on a GaAs substrate with layer structures using a variety of binary, ternary and quaternary materials of the (AlGaIn)(AsP) III–V compound semiconductor family. Material properties of AlGaAs and GaInAsP are available in the literature [52,53]. The thick waveguiding and cladding layers have to be grown lattice-matched to the substrate to avoid formation of misfit dislocations that are often sources of very rapid degradation of diode-laser performance. In addition, the crystalline quality of the layers can be adversely affected by ordering as in the case of GaInP or by phase separation as observed for certain compositions of quaternary GaInAsP. Further, the concentration of point defects which act as centers for nonradiative recombination and thus losses in efficiency have to be kept low. These requirements call for optimal growth conditions for each material.

Unfortunately, the optimum conditions for the different materials within a heterostructure may be different. This may require compromises to avoid long growth interruption for the adjustment of the growth parameters when switching from one composition to the other. Depending on the materials to be combined, optimization of the switching sequence can be a challenge to the skills of the crystal grower.

The substrates used for devices have a (100)-oriented surface. Misorientation from the exact (100) orientation by some degrees leads to a larger step density with the step structure being dependent on the miscut direction. Since the attachment of the atoms to step edges is a very important step in the growth process, step orientation and density can have a strong impact on the crystalline properties of the grown layers. In case of GaIn(As)P ordering which may affect the band gap is dependent on the miscut direction [54,55]. Although less obvious, depending on the growth conditions the surface morphology of AlGaAs can be significantly different for different substrate orientations.

2.1   GaAs and AlGaAs

(Al)GaAs is the most intensively investigated and also most widely used compound semiconductor-material system. The primary reason is the small lattice mismatch $\Delta a/a$ of only 1.3×10^{-3} between AlAs and GaAs. This makes the fabrication of heterostructures easy since the requirements on composition control are not very tight. While electronic devices usually require only thin $\mathrm{Al}_x\mathrm{Ga}_{1-x}\mathrm{As}$ layers of low Al content ($x < 0.3$) where the lattice mismatch can be easily accommodated elastically, the notion of AlGaAs being lattice-matched to GaAs is clearly not true for diode lasers having thick cladding and waveguiding layers with a high Al content. According to the theory of *Matthews* and *Blakeslee* [56] the critical-layer thickness beyond which formation of misfit dislocations is expected for $Al_{0.5}Ga_{0.5}As$ is only 500 nm and for AlAs only 220 nm. This fact has to be kept in mind, although AlGaAs turns out to be more immune against relaxation of strain than theoretically predicted. Another reason for the dominance of AlGaAs is the fact that As is much easier to handle than P, which has a much higher vapor pressure and occurs in different modifications some of which can spontaneously self-ignite.

2.1.1 Purity

With the available precursor materials, GaAs of very high purity can be grown by MBE as well as by MOVPE. Background doping levels below $10^{14}\,cm^{-3}$ and very sharp and well-resolved excitonic transitions in photoluminescence are reported for both methods.

If Al is added this situation changes and impurity incorporation increases with increasing Al content due to the high reactivity of this element. The reaction rate for oxide formation from the metal and water vapor is several orders of magnitude higher for Al than for Ga and In. In MBE residual gases like water and CO are sources for oxygen incorporation. Maintaining a clean growth system and avoiding any introduction of such contaminants is thus a precondition for clean layers. In MOVPE too a leak-tight growth system is a must. In addition to leaks, starting materials can be sources for oxygen contamination. From the hydrogen carrier gas such contamination is removed by diffusion through a Pd membrane. The purity of arsine has significantly improved over the past few years. Additionally, point-of-use filtering or purification is usually employed. The most significant sources for oxygen are the metalorganic group-III precursors. Any contact to air or residual moisture can lead to the formation of alkoxides which have oxygen between the metal and the organic ligand. Over a couple of years a number of alternative precursors has been studied to reduce the risk of O incorporation [35]. However, in device production a conservative approach prevails using the standard trimethyl and for Ga also triethyl compounds at the highest available purity.

The tendency for oxide formation decreases with increasing temperature and also with increasing V/III ratio [57]. Consequently high growth temperatures are favored for reduced oxygen uptake. AlGaAs for laser structures is grown hot, i.e. at temperatures around $750°$ C both in MBE as well as in MOVPE.

Carbon is the other impurity usually found in AlGaAs where it acts as an acceptor. Like the metal–oxygen bond the metal–carbon bond strength increases in the order In, Ga and Al [32]. In MOVPE such bonds are already present in the starting materials and have to be broken to release the metal in the growth process. For the Ga precursors this is easily achieved and background doping levels below $10^{14}\,\mathrm{cm}^{-3}$ are obtained. In the presence of Al this level increases [58]. Pyrolysis of TMAl in the absence of hydrogen or arsine even yields aluminum carbide. In the presence of arsine as a reducing agent and at the temperatures employed for the growth of laser structures carbon incorporation in AlGaAs is reduced to levels in the range around $10^{16}\,\mathrm{cm}^{-3}$. This purity is sufficient to allow for controlled doping at the levels employed for diode lasers (i.e. above several $10^{17}\,\mathrm{cm}^{-3}$).

The AlGaAs, material quality obtained in both growth methods allows for the growth of diode-laser structures with excellent device performance even with high Al-containing waveguide and cladding layers. Addition of Al into the active region causes a degradation of the device performance due to the introduction of centers for nonradiative recombination [59,60]. In limiting the wavelength range accessible with AlGa(In)As-active regions, this constraint is more severe than the fact that AlGaAs is an indirect semiconductor for an Al content above $x = 0.43$.

2.2   GaInP and AlGaInP

(Al)GaInP lattice-matched to GaAs is less well established than AlGaAs. Lattice matching is obtained at $x = 0.52$ for $\mathrm{Ga}_x\mathrm{In}_{1-x}\mathrm{P}$ and at $x = 0.51$ for $\mathrm{Al}_x\mathrm{In}_{1-x}\mathrm{P}$. Ordering on the group-III sublattice [61] that can occur on (111)B planes makes these materials interesting from a material-science point of view and challenging with respect to device manufacturing. GaInP has been mainly used as active-region material in red LEDs and diode lasers with emission in the visible wavelength range 700 nm to 630 nm that is not accessible with AlGaAs structures. More recently it has also found interest as a direct competitor to AlGaAs in waveguiding or cladding layers for longer-wavelength devices [62]. GaInP has a lower surface recombination velocity and a much lower tendency for oxidation than AlGaAs, and thus promises a higher stability of the laser facets against Catastrophic Optical Mirror Damage (COMD) [62]. Additionally, regrowth processes are easier in the absence of Al on the surface on which the overgrowth is initiated, making GaInP advantageous for structures requiring two-step epitaxy [63]. GaInP is mostly grown by MOVPE though MBE and GSMBE have found increasing interest over the last few years [30].

2.2.1   Purity

GaInP grown by MOVPE usually has an n-type background due to intrinsic defects in the range of $10^{16}\,\mathrm{cm}^{-3}$ over a wide range of growth temperatures. This is sufficient for laser applications. As in the case of AlGaAs, the addition of Al leads to enhanced oxygen incorporation and consequently undoped AlInP tends to be semi-insulating. Increasing the growth temperature is again effective in suppressing oxygen incorporation. However, due to In re-evaporation very high temperatures are not ideal for In-containing alloys, thus making the optimal temperature window for AlGaInP quite narrow [64]. High temperatures are also chosen for other reasons discussed in the section on ordering.

Oxygen incorporation is also influenced by the phosphorus overpressure during growth, with high V/III ratios minimizing oxygen incorporation [64]. However, the use of high PH$_3$ flows leads to heavy loading of excess phosphorus in the reactor exhaust and associated disposal problems.

Carbon in (Al)GaInP is amphoteric. Under usual growth conditions there seems to be some preference for incorporation as a donor. However, not much has been published on the role of C in AlGaInP. It is likely that it is also partly responsible for the semi-insulating nature of undoped AlInP but proof for this assumption is still lacking.

2.2.2   Ordering

Due to the different covalent radii of Ga and In there is a driving force for an arrangement of these atoms that deviates from a random distribution on the group-III lattice. This arrangement known as CuPt ordering leads to alternating preferential occupation of (111)B planes by Ga and In (Fig. 4). In the extreme case, never observed up to now, a natural superlattice of alternating InP and GaP planes would be formed in this crystal direction. This superlattice leads to a reduction of the symmetry and hence to a reduction of the band gap energy of this alloy. Thus no strict correlation of band gap and composition exists. Since band gap, band-offsets and also refractive index depend on the degree of ordering, an understanding of this property is important for the design of diode lasers.

Ordering appears to be a process taking place on the surface of the growing layer by preferential incorporation of Ga and In at certain sites that mini-

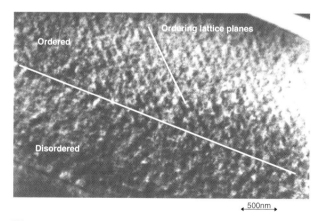

Fig. 4. TEM picture of GaInP layer: the *lower part* grown at 720° C is disordered, the *upper part* grown at 650° C shows ordering on (111) planes

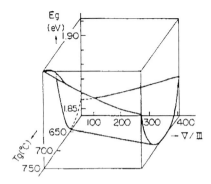

Fig. 5. Schematic depiction of GaInP band gap in dependence on MOVPE growth parameters: disordered material has a band gap of 1.91 eV, strongly ordered material has a band gap of 1.82 eV at 300 K

mize the local strain energy. This process strongly depends on the growth conditions. Figure 5 sketches the general trends observed for the band gap of MOVPE-grown material when growth temperature and V/III ratio are varied. The growth rate also plays a role since it determines the time available for an atom to find a certain site. High rates and high V/III ratios favor growth of ordered material. Layers grown at intermediate temperatures around 650° C are ordered, while at high and low temperatures the ordering is reduced.

Ordering also depends on substrate misorientation and is promoted if steps in the direction of the ordering superlattice are present [54]. On an exact (100) substrate these steps are present as upward and downward B steps. Due to the locally varying residual misorientation of such an exact surface both ordering variants can occur. If the substrate is misoriented to (111)B one variant is favored. A strong miscut towards (111)A ($> 5°$) helps to suppress ordering.

Ordering is not a homogenous process. In ordered material domains with different degrees of ordering coexist with a size distribution that depends on the growth conditions and substrate misorientation [65]. The boundaries of these domains break the translational crystal symmetry and give rise to special features in the photoluminescence of ordered material. Locally indirect transitions over domain boundaries result in excitation-dependent moving emission, while band gap fluctuations lead to broadened PL peaks. The ordered domains have different surface structures similar to different growth facets. This facetting leads to the observed trend that ordered material has a higher microscopic surface roughness than disordered material grown under proper conditions. This surface roughness can have an impact on the quality of subsequently grown layers and on the abruptness of interfaces. Structural imperfections brought about by ordering are potential sources for device degradation and have to be considered when trying to optimize device performance.

For the quaternary AlGaInP a study of the effect of ordering on the band gap is difficult since knowledge of the lattice constant alone is not sufficient to determine the composition. Thus there is less data available on the dependence of ordering on growth conditions. Nevertheless Transmission Electron Microscopy (TEM) studies [66,67] suggest that the general trends observed for GaInP also hold for AlGaInP. Depending on the composition, ordering extends to higher growth temperatures. To reduce the effects of both ordering and also the tendency for O incorporation, laser structures incorporating AlGaInP are also grown at high temperatures of around 750° C similar to the case of AlGaAs.

2.3   GaInAsP on GaAs

The quaternary GaInAsP lattice-matched to GaAs is of interest for active regions for emission around 800 nm and also for waveguiding layers at longer wavelengths. Thermodynamic calculations predict a miscibility gap in this material system [32,68]. Additionally, ordering effects like in GaInP are observed [69]. With respect to material purity it behaves like GaInP, with n-type background doping levels between $10^{15}\,\mathrm{cm}^{-3}$ and $10^{16}\,\mathrm{cm}^{-3}$ being obtained depending on growth conditions and composition. Again this level of purity is sufficient for laser applications.

2.3.1   Ordering

For compositions close to GaInP the quaternary $Ga_x In_{1-x} As_y P_{1-y}$ shows a very similar ordering behavior. For an As content $y = 0.1$, a composition found for example in the waveguide layer of an 800 nm-emitting Al-free diode laser, ordering has been observed in two variants on exact (100) substrates and in only one variant on substrates misoriented to (111)B [69]. For higher As content, ordering is also observed but not intensively studied. It appears that the degree of ordering is lower and the growth-parameter window in which ordering occurs smaller as the composition moves away from GaInP.

2.3.2   Phase Separation

The miscibility gap theoretically predicted [32,68] for GaInAsP shows up as an inferior material quality [70] for compositions with an As content between $y = 0.2$ and $y = 0.5$. Broadening of luminescence spectra and X-ray rocking curves in this compositional range indicate a deterioration in crystalline quality. This broadening is asymmetric. TEM studies reveal that platelets form in the growth direction due to columnar growth. These platelets are elongated along the steps on the growing surface. Neighboring domains are slightly mismatched against each other. If one assumes that all the strain is due to a different group-III composition the In content differs by $\Delta x = 0.02$ [71]. Although no detailed study is available as yet it is very likely that such inhomogeneous material could adversely affect device reliability. It has also been observed that the quality of GaInAs degrades when deposited onto a decomposed quaternary layer [72,73]. Although GaInAs itself does not show such behavior, phase separation from an underlying layer can apparently be inherited and lead to composition modulation in GaInAs QWs.

2.4   InP, GaInAs(P) and AlGaInAs

InP-based diode lasers currently play only a small role for high-power applications. However, since this material system is important for long-wavelength applications and high-power diode lasers are being investigated in some laboratories, InP-based materials are briefly discussed. P-containing layers are mostly grown by MOVPE. MBE is primarily used for GaInAs/AlInAs heterostructures that do not require the use of a P source.

2.4.1   Purity

In MOVPE, InP can be grown with high purity and background-doping levels in the range of $10^{14}\,cm^{-3}$ using either the hydride phosphine or the liquid source TertiaryButyl-Phosphine (TBP, $C_4H_9PH_2$). The same holds true for GaInAs(P) although the background doping due to intrinsic defects tends to be higher in the ternary and quaternary. Again the addition of Al leads to higher impurity uptake. AlGaInAs is used in cladding and waveguide layers to enhance the barrier height in the conduction band and consequently reduce leakage current. Compared to GaAs the typical growth temperatures for InP-based structures are lower due to the lower thermal stability of the substrate material. This leads to relatively high oxygen incorporation in AlGaInAs and especially AlInAs. The solid In source TMIn is suspected to be an important source for O contamination as it is more difficult to purify than the liquid TMAl or TMGa due to its large surface area. Using high-purity precursors epi-material of sufficiently high quality for diode-laser application can be grown.

2.4.2   Ordering and Phase Separation

In contrast to compositions lattice-matched to GaAs, quaternary GaInAsP lattice-matched to InP is outside the miscibility gap. However, some strain-compensated Multi-Quantum-Well (MQW) active regions used to achieve better temperature characteristics employ strained layers lying inside the miscibility gap. This leads to roughening of the growth front and thickness modulation with increasing number of quantum wells [74]. By using compositions outside the miscibility gap this problem can be avoided without sacrificing laser performance. Similar to phase separation, that though observed, plays a minor role for InP-based structures, some tendency for ordering is reported for example for GaInAs, but this does not impede the fabrication of diode lasers [75].

3 Doping

The operation of a diode laser requires the presence of a p-n junction. For usual applications the lower cladding and sometimes also the waveguide layers are grown n-type on an n-type substrate and the upper cladding and waveguide layers are doped p-type. The location of the p-n junction, the series resistance especially in the p-type layers and the losses due to free-carrier absorption depend on the doping profile and the doping level. Thus a good control of these features is necessary.

3.1 n-Type Doping

The most common n-type dopant for AlGaAs is Si which is incorporated on a Ga (or Al) site. In MBE it is supplied by evaporation from an effusion-cell. In MOVPE silane or the thermally less stable disilane are used [76]. n-type doping up to $5 \times 10^{18}\,\mathrm{cm}^{-3}$ in GaAs can be obtained. Above this level the amphoteric nature of Si leads to self-compensation by incorporation on As-sites or as an interstitial. However, doping levels of mid $10^{18}\,\mathrm{cm}^{-3}$ are usually sufficient for laser applications. In $\mathrm{Al}_x\mathrm{Ga}_{1-x}\mathrm{As}$ with $x > 0.23$ the presence of DX-centers acting as electron traps can make the determination of the doping level problematic. While these DX-centers have deleterious effects on the transport properties, for example of PHEMT devices, they are apparently not harmful to laser operation.

Si is also the most widespread n-type dopant for (Al)GaInP and GaInAsP. The doping levels of low $10^{18}\,\mathrm{cm}^{-3}$ required for cladding layers can easily be achieved. In some MOVPE laboratories Se (supplied as H_2Se) is also used, for which a smoothing effect on the surface has been reported [77]. The big Se atom floats on the growing surface and, as a surfactant, affects the diffusion of Ga and In and their attachment to steps. The accumulation of Se on the surface may bear the risk of doping transients. For special applications it may also be important that the group-VI atom Se can be incorporated up to $10^{19}\,\mathrm{cm}^{-3}$ without problems with self-compensation like those discussed for the group-IV atom Si.

Similarly, Si as well as the group-VI dopants Se or S are used in MOVPE of InP-based structures. These dopants can be incorporated electrically active into both InP and GaInAs up to $10^{19}\,\mathrm{cm}^{-3}$. The group-VI dopants are reported to have a smoothing effect as already discussed for Se doping of GaInP. Due to their higher vapor pressure they are however more susceptible to memory effects.

3.2 p-Type Doping

Whilst Si is the most common n-type dopant in MBE and MOVPE for all III–V materials, with Se and S being additionally used in MOVPE especially for the In-containing materials, a more heterogeneous picture arises for p-type doping.

In MBE Be is the p-dopant of choice for all III–V materials. The contenders Zn and Mg have considerably higher vapor pressures increasing the risk of memory effects. In comparison to these two elements Be also has a smaller diffusion coefficient, making the controlled growth of doping profiles easier. Due to the small amounts that are used and the containment in the steel growth chamber the high toxicity of Be is tolerated.

Organic Be precursors like diethyl-beryllium have also occasionally been used in MOVPE [78]. The higher vapor pressure in comparison to elemental Be in combination with its high toxicity are concerns for safety and Be doping has not found wide acceptance in the MOVPE community. Instead, Zn introduced as dimethyl- or diethyl-Zn is the usual p-dopant for the GaAs- and InP-based compounds. However, Zn easily diffuses especially at high doping levels [79]. This diffusion has to be taken into account when optimizing the growth process to obtain the desired doping profile. An additional problem is the re-evaporation of Zn that reduces the incorporation efficiency at high temperatures. The highly doped top p-type GaAs contact layer of GaAs-based devices for example is usually grown at reduced temperatures to achieve the desired doping level of $2 \times 10^{19}\,\mathrm{cm}^{-3}$. Further, this avoids excessive outdiffusion into the underlying cladding layer.

P-type doping of AlGaInP in MOVPE is difficult due to a number of reasons. As the Al fraction and hence the band gap increases, the ionization energy of Zn becomes larger. Consequently it is more difficult to ionize incorporated acceptor atoms. Further, at high-Al content, more oxygen is incorporated as well, leading to compensation. Additionally, the vapor pressure of Zn over Zn_3P_2 is much higher than over Zn_3As_2 and thus re-evaporation of Zn from AlGaInP is much easier than from AlGaAs. This is demonstrated in Fig. 6 which shows data on Zn doping into different materials in the same reactor. At the high growth temperatures needed for AlGaInP laser structures the DMZn partial pressure necessary to reach the desired doping level may even exceed those of the group-III precursors. Zn incorporation can be enhanced by providing a high step density where Zn can form double bonds which are more stable. In addition to the effect on ordering, enhanced Zn incorporation is another reason for choosing substrates highly misoriented towards (111)A for AlGaInP-based red diode lasers. In as-grown layers not all Zn atoms are necessarily electrically active as acceptors. Hydrogen remaining in the lattice can form complexes neutralizing the Zn acceptors [64]. By annealing under nitrogen the hydrogen can be removed. This process can also be carried out in the MOVPE reactor by switching off the arsine soon after the GaAs p-type contact layer growth and cooling under nitrogen from near growth temperature. Despite this procedure in AlInP the highest p-type doping level obtained with Zn is $< 5 \times 10^{17}\,\mathrm{cm}^{-3}$ at usual growth tempera-

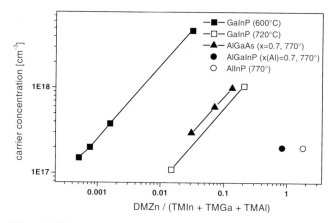

Fig. 6. Hole concentration versus DMZn partial pressure for different materials grown in the same MOVPE reactor

ture. This level can be increased by reducing the temperature to 700° C and compromising on the material quality.

The other p-dopant used in MOVPE for AlGaInP is Mg. This dopant is notorious for transients caused by adsorption of the precursor molecules in the stainless steel tubing and on the reactor walls and also for memory effects. It also needs post-growth activation to drive out passivating hydrogen. The advantage it offers is its higher incorporation efficiency. So for AlGaInP the choice is between two p-type dopants that both have their respective drawbacks.

In section 2.1.1 carbon has been identified as a prominent residual impurity in GaAs and AlGaAs. This acceptor has a much lower diffusivity than Zn and can be incorporated up to very high levels onto arsenic sites. This can either be achieved by growing at low V/III ratios or by adding additional sources for carbon. For GaAs intrinsic C doping to levels of 10^{19} cm^{-3} requires growth temperatures below 600° C and V/III ratios around 1. Lower doping levels are also possible at higher temperatures. Depending on the Al content AlGaAs can be intrinsically doped even at the usual high growth temperatures but again low V/III ratios have to be employed. In this approach a change in the doping level may also be accompanied by a change in the concentration of intrinsic defects like vacancies. To allow for a more independent control of doping and intrinsic defects carbon sources like CCl$_4$, CBr$_4$ or trimethyl-As can be added. Halogenated C precursors have been successfully used for the growth of surface, emitting diode lasers (VCSELs) where the p-type (Al)GaAs/Al(Ga)As top Bragg mirrors require a well-controlled doping profile with δ-doping spikes at high doping levels [80]. While for such new structures new doping schemes are being applied, for edge emitters the more conservative approach using Zn still dominates for AlGaAs devices. In some cases this approach is combined with a C-doped p$^+$-contact layer grown at reduced temperatures.

4   Heterostructures

Diode lasers require a number of layers of different compositions to be stacked on top of each other as shown in Fig. 1. Doing so with (Al)GaAs is an easy and straightforward task. With a properly designed growth system nearly atomically flat interfaces between GaAs and AlGaAs can be generated. For devices like (P)HEMTs that rely on electron transport along such a Ga(In)As/AlGaAs interface roughness on the monolayer scale may be important. For laser structures, however, where transport is across the interface such roughness simply leads to a grading of the potential. An abrupt interface would be associated with a potential spike that could give rise to additional series resistance. To avoid this, interfaces between (Al)GaAs layers with significantly different Al content usually are graded over a distance of several nm. In MOVPE this is achieved by simply ramping the flux of the Al or Ga precursor. Due to the thermal mass of the effusion-cells in MBE a quick ramp of the flux by a change in cell temperature is difficult to control, and a pulsed supply using mechanical shutters can be employed.

In the waveguiding layers grading has little effect on the optical properties and helps with respect to the series resistance as already mentioned. A graded profile between QW and barrier changes the position of the subbands in the QW. Although this does not necessarily adversely affect laser performance, grading at this position usually is not introduced voluntarily. As will be discussed later it can often hardly be avoided.

The situation is less well controllable if the group-V component changes at an interface like for InP/GaInAs and/or if the materials have a considerably different In-content like GaInP/(Al)GaAs. A graded interface in such cases is associated with strain that can have a negative impact on device performance. Further, parasitic QWs can be formed if, for example, a graded InGaAs layer forms between InGaP and GaAs. This can even result in emission at a wavelength completely different from the intended one!

Growth is usually performed under an excess supply of group-V species resulting in a surface covered with adsorbed group-V atoms or molecules, with the coverage depending on the growth temperature and the group-V supply. When switching from an arsenic-containing layer to a phosphorus-containing one, the excess arsenic has to be removed from the surface to avoid incorporation into the subsequent layer. Usually the excess arsenic is desorbed during a growth interruption. The length of this growth pause has to be adjusted depending on the temperature. In MOVPE the time necessary for an exchange of the gas phase has also to be taken into account. The switching scheme applied is a matter of individual optimization and often considered as proprietary knowledge.

Phosphorus has a much higher vapor pressure than arsenic. Consequently it is more easily removed from the surface. When switching from a P-containing to an As-containing layer P can desorb from the first mono-layers of the grown layer and can be replaced by the less volatile As. Again a process-dependent optimization of the switching scheme is necessary.

In MOVPE polycrystalline material is usually deposited on the substrate holder. Among the group-III elements In has the highest vapor pressure. If large amounts of In-containing material are deposited upstream of the substrate there is a risk of desorption or transport by chemical reactions resulting in In incorporation into the subsequently grown layer. Such effects have to be reduced by both optimizing the susceptor-temperature profile and also by optimized growth interrupts.

5   Strained Quantum Wells

In some cases important emission wavelengths (for example 980 nm needed for pumping Er-doped fiber amplifiers) are not accessible with active regions lattice-matched to GaAs (or InP). In such cases lattice-mismatched layers are the only choice. Such layers can be employed as long as the strain re-sulting from the mismatch is accommodated elastically without formation of misfit dislocations [56]. The beneficial consequences that strain-induced band-structure modifications might introduce in QW lasers were first sug-gested independently by *Yablonovitch* and *Kane* [81] and *Adams* [82]. The introduction of strain in a typical zincblende-type semiconductor like GaAs splits the degeneracy at the valence-band-edge maximum, and separates the heavy-hole and light-hole subbands as is schematically depicted in Fig. 7. The change in the band structure leads to a reduction of the density of states at the edge of the valence band and also the effective mass of electrons and holes. This reduces the threshold-current density of strained QWs in comparison to unstrained ones. Additionally, the lifted degeneracy leads to a preferential polarization of the emitted light. For tensile strain, transitions involving the light holes have the lowest energy leading to polarization in the QW plane (TM mode); for compressive strain, transitions between electrons and heavy holes lead to polarization perpendicular to the QW plane (TE mode). Due to these positive effects, strain is nowadays introduced even if the same emission wavelength is, in principle, possible using unstrained material.

5.1   Pseudomorphic Growth and Strain Relaxation

In pseudomorphic growth a mismatched epitaxial layer adopts the lattice con-stant of the substrate in the growth plane while it adjusts the lattice constant in the perpendicular direction according to its elastic properties. For a cubic crystal this results in tetragonal distortion. This deviation from the equilib-rium lattice constants leads to excess energy. If this energy exceeds a certain

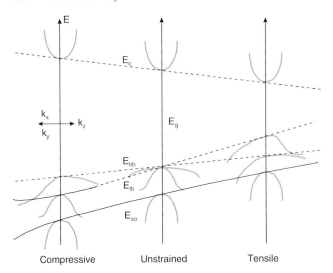

Fig. 7. Schematic representation of the changes in band structure caused by introduction of tensile and compressive strain

limit the layer can relax to an energetically more favorable state by formation of misfit dislocations. Depending on the type of structure (i.e. single or double heterostructure) and the type of dislocations involved the maximum strain that can be accommodated elastically, i.e. without relaxation by dislocations, has been theoretically predicted. The original calculations by *Matthews* and *Blakeslee* [56] have been modified by others [83,84] to determine more accurately the critical thickness up to which layers of a certain mismatch can be grown.

While these theories provide a good guideline for an estimation of a critical layer thickness they are not accurate enough to give quantitative results. Experimentally it has been found that the critical layer thickness also depends on the growth temperature [85] and is different for dislocations running in different crystallographic directions. It has also been found that depending on the growth conditions point defects can be formed before extended defects are generated [86]. While extended defects are known to lead to rapid device degradation, point defects are suspected to be the cause of gradual degradation [50]. This means that the device performance is the criterion that determines the maximum permissible strain in a QW. This maximum strain is not necessarily a quantity determined only by the lattice constants but may depend on the growth conditions.

In the case of GaInAs/GaAs QWs used as active regions for emission wavelengths in the range of below 850 nm to above 1100 nm other factors related to strain have to be taken into account. The compressive strain in such layers creates a barrier against the incorporation of the big In atom. This leads to a reduction in the In uptake into strained layers in comparison to relaxed ones [87]. This reduction of incorporation efficiency results in an increased coverage of the growing surface with In. At the beginning of the growth of the QW this coverage gradually builds up. This can lead to a graded lower interface. At the upper interface the In floating on the surface can be incorporated into the subsequent layer, again resulting in a graded interface. At high coverage even In droplets can form which can lead to three-dimensional growth of In-rich crystallites on top of the QW and consequently rough interfaces. Above a certain coverage incorporation into these In-rich areas can be preferred over incorporation into the QW. This leads to a reduction of the In content in the QW if the In supply is increased over a critical limit. The maximum In content observed at $750°$ C for example is $x = 0.30$. This limit can be pushed to higher values by using high growth rates and/or low temperatures. The limitation of the In uptake during growth has to be taken into account in addition to the limitation given by strain relaxation when trying to push the operation of lasers with strained InGaAs QWs to longer wavelengths. Although less-well studied, similar observations have been made for the growth of compressively strained, In-rich GaInP QWs emitting in the visible red region.

Tensile-strained GaAsP QWs are used as an active region for emission wavelengths in the range 800 nm to 715 nm [16,88]. Such QWs have not been as intensively studied as compressively strained InGaAs QWs. The phenomena discussed above are related to the accumulation of In on the growing surface. Such accumulation is not possible for highly volatile materials like P. Consequently, no limitation of P incorporation due to strain effects has been observed. In MOVPE, due to the different thermal stability of PH_3 and AsH_3, control of the GaAsP composition is difficult. At low temperatures the P content is very sensitive to the AsH_3/PH_3 ratio. This situation is improved at higher temperatures ($\approx 750°$ C) as are used in any case for the AlGaAs waveguide layers.

5.2 Strain Compensation

In MQW structures the net strain accumulates, reducing the maximum allowable strain in a single QW. InP-based devices in particular often use stacks of several QWs to reduce the leakage current and obtain better temperature characteristics [89]. This is possible by using strain-compensating barriers which have the opposite strain of the QW. Each QW thus sees a similar accumulated strain from the underlying layer. If the barriers are precisely strain-compensated even superlattices of considerable thickness can be grown.

This approach can also be used to incorporate higher strain into a SQW to increase the accessible wavelength range. Highly strained GaInAs/GaAs QWs as active region in diode lasers emitting at 1060 nm have been fabricated using GaAsP barriers [87]. Strain compensation may also help in improving device reliability, since at reduced overall strain the driving force for defect generation or defect motion is reduced. Additionally, strained barriers can be instrumental in optimizing the band structure. For red diode lasers tensile-strained AlGaInP barriers around the QW have improved carrier confinement and reduced absorption at the laser facet [90]. Normally, a transition from biaxial to uniaxial strain can lead to a band gap reduction and thus enhanced absorption in the barriers. If the active region is strain-compensated this effect can be avoided.

6   Device Results

The ultimate criterion for optimization of growth procedures is the device performance. However, a valid comparison of data available from the literature, if at all possible, is difficult. Usually, the basic quality of the layer structure is assessed on bare chips using pulsed excitation. Typical values for the pulse width are 0.5 ms to 1 ms at 5 kHz repetition rate to prevent the devices from self-heating. If CW or quasi-CW data of mounted devices are compared the efficiency of heat removal from the device is an important factor. The heat dissipation is partly determined by the thermal conductivity of the waveguide and cladding layers and partly by the quality of the mounting procedures and the thermal conductivity of submounts, etc.. Thus measurements on mounted devices often give information only on the lower limit of the material quality.

The geometry of the devices used for the assessment of the layer quality is an important factor. A larger width of the emitting aperture allows for higher output power since the power density is reduced. Longer devices facilitate heat removal again allowing for higher output power. The same holds true for reduced measurement temperatures where carrier leakage is also reduced.

The output power from one facet is measured and it is important to know whether a coating has been applied to alter the reflectivity of the facets. If no coating is applied the device emits from both facets. Summing the output power from both sides reflects a different situation than obtaining the same power from one facet with a high-reflectivity coating on the back facet. The combination of front and backside facet reflectivity and cavity length determines the resonator Q factor which has a strong impact on the measurement results.

The vertical laser structure can be optimized with respect to different criteria. A small vertical far-field angle improves the coupling efficiency into any optics. Additionally it reduces the power density on the facet by a broadened near field thus reducing the risk of mirror damage. On the other hand a small vertical far field is accompanied by a reduction of not only the optical but also the electrical confinement. This can lead to enhanced leakage currents and a reduction in efficiency.

There are no standards for aperture width, resonator length or facet reflectivity making the comparison of results obtained in different laboratories difficult. These difficulties are further enhanced by the different vertical lay-out as discussed above.

Even if a standard lateral design would be agreed upon a comparison of the data is only possible within the measurement accuracy. Integrating spheres used for the measurement of the output power usually have an accu-racy of ±10%. An improvement of efficiency by 10% (measured by the same equipment) obtained for example by optimizing the doping profile is already considered significant for high-power applications.

Due to the above reasons a valid comparison of the data found in the lit-erature with respect to the properties of the epitaxial layer structure is nearly impossible. Proprietary facet-passivation schemes and aggressive heat-sinking mechanisms, such as using diamond submounts or micro-channel coolers, are often employed in order to boost the power that can be extracted from the device. Although devices with 100 μm aperture have become standard for comparing the maximum power in a variety of material systems over a wide range of wavelengths, in particular, the record values for output power have to be carefully checked for device layout and measurement conditions. Addi-tionally, these values usually are limited by heat dissipation and the quality of the facet coating. We thus restrict ourselves to an overview on the mate-rial combinations used in different wavelength ranges. An overview of these combinations is given in Table 1 together with the respective references.

Table 1. Combinations of active regions and waveguide structures used for HPDLs at different wavelengths in the range 1060 to 600 nm. The entries in the table indicate reference numbers

QW	Waveguide/Cladding	λ (nm)									
		650	700	750	800	850	900	950	1000	1050	1100
GaInAs	AlGaAs/AlGaAs						91		92,	87	
	GaInP/AlGaAs								94,95		
	GaInP/GaInAsP								15,28		
	GaInP/AlGaInP										
GaAs	AlGaAs/AlGaAs				98,99,100	13					
GaInAsP	AlGaAs/AlGaAs				14,102						
	GaInAsP/GaInP				103						
	GaInP/ AlGaInP			104	16						
GaAsP	AlGaAs/AlGaAs			17	106						
AlGa(In)As	AlGaAs/AlGaAs			97		91,100					
GaInP	AlGaInP/ AlGaInP	108	107								
	AlGaInP/AlGaAs	109									

Strained InGaAs QWs are used as active regions in the wavelength range 880 nm to 1060 nm. Usually they are embedded in AlGaAs/AlGaAs wave-guide structures [91,92]. Especially for the longest wavelengths above 1000 nm, strain-compensating GaAsP spacer layers have been successfully utilized [87,93]. For wavelengths longer than 940 nm the successful combination of In-GaAs QWs with InGaP/InGaAsP waveguides has been demonstrated

[15,28,62] and is commercially available. The combination of GaInP wave-guiding layers and AlGaAs cladding layers [94,95] is also being worked on but has not found widespread use up to now. Similarly, there are only a few reports on AlGaInP/GaInP waveguide structures in the longer-wavelength region.

Shorter wavelengths can be obtained by adding Al to the active region. Such strained AlGaInAs QWs are used together with AlGaAs waveguides predominantly for the important 808 nm wavelength [91,96] (pumping of Nd:YAG). The presence of indium in the active region is thought to increase resistance to dark-line defect propagation [60]. By increasing the Al content, this material system has also been used for shorter wavelengths down to 730 nm [97].

Unstrained GaAs QWs in AlGaAs waveguides find limited application in the range 850 nm to 800 nm. Though high-power 808 nm devices have been demonstrated [98,99], achieving such wavelengths requires extremely thin ($\approx$ 3 nm) QWs which are very difficult to control. Consequently the application of GaAs QWs concentrates on the range 850 nm to 870 nm [13]. Again, adding Al to the QW shifts the wavelength to the shorter side, with AlGaAs-active devices being used at 808 nm [100] and high-power operation down to 715 nm being reported [101]. However, as even small amounts of Al in the active region drastically reduce the maximum internal power density at the catastrophic-optical-mirror-damage (COMD) level, AlGaAs-active devices in this wavelength range offer limited reliability [59].

Quaternary GaInAsP-active regions have been extensively investigated for emission between 875 nm and 730 nm. High-power 808 nm-emitting com-pletely Al-free lasers using such QWs embedded in InGaP/InGaAsP wave-guides have been demonstrated [102,103]. AlGaInP/GaInP waveguide struc-tures provide higher barriers and have been utilized at 808 nm [16] and also down to 730 nm [104]. A combination of unstrained and strained GaInAsP QWs with AlGaAs waveguide structures has been intensively studied in our own laboratory [105]. Despite the difficulties in combining these materials, high reliability at high output powers has been found in these studies.

Tensile-strained GaAsP is also a contender for the wavelength range 810 nm to 715 nm. High efficiencies and good reliability are reported for such QWs in AlGaAs waveguide structures [17,106].

For even shorter wavelengths GaInP QWs have to be used [107,108,109]. Between 650 nm and 690 nm compressively strained GaInP QWs are usually employed, while tensile strain is introduced for shorter wavelengths around 630 nm. In all cases the waveguiding layer is made from AlGaInP. The cladding material is predominantly Al(Ga)InP. Although in principle pos-sible, as in the case of red VCSEL structures (where the Bragg mirrors are made from AlGaAs), there are very few reports [109] on the use of AlGaAs cladding layers for the short-wavelength region.

In the short-wavelength region the conduction-band barrier for the electrons becomes particularly small. Thus, the efficiency decreases with decreasing wavelength, making high-power operation increasingly difficult. Improvements have been brought about by the use of superlattices in the upper waveguide structure that act as Bragg reflectors for the electrons. However, the additional heterointerfaces also lead to increased series resistances. The series resistance is especially a problem in red diode lasers due to the limitations in p-doping levels that can be obtained in Al(Ga)InP as discussed in Sect. 3.2. The shortest wavelength reported for GaAs-based lasers up to now is 595 nm, but only low output powers can be achieved for such short wavelengths.

As already mentioned, the rapid progress made in the fabrication of GaN-based laser structures holds promises for the extension of the wavelength range of high-power diode lasers to even shorter wavelengths into the blue in some years from now.

List of Acronyms

CBE	Chemical Beam Epitaxy (also: MOMBE)
ClVPE	Chloride Vapor-Phase Epitaxy
CW	Continuous Wave
DH	Double Heterostructure
DMZn	Dimethyl Zinc $Zn(CH_3)_2$
ELOG	Epitaxial Lateral OverGrowth
GSMBE	Gas-Source Molecular-Beam Epitaxy
HPDL	High-Power Diode Laser
HVPE	Hydride Vapor-Phase Epitaxy
LED	Light-Emitting Diode
LEED	Low-Energy Electron Diffraction
LPE	Liquid-Phase Epitaxy
MBE	Molecular-Beam Epitaxy
MOCVD	Metalorganic Chemical Vapor Deposition (also: MOVPE)
MOMBE	Metalorganic Molecular-Beam Epitaxy
MOVPE	Metalorganic Vapor-Phase Epitaxy (also: MOCVD)
PHEMT	Pseudomorphic High Electron Mobility Transistor
QW	Quantum Well
RHEED	Reflection High-Energy Electron Diffraction
SCH	Separate Confinement Heterostructure
TE	Transverse Electric
TEM	Transmission Electron Microscopy
TBP	TertiaryButyl Phosphine $C_4H_9PH_2$

TMA   Trimethyl Aluminum
TMG   Trimethyl Gallium
TMI   Trimethyl Indium
TM   Transverse Magnetic
UHV   Ultra-High Vacuum

References

1. R. D. Dupuis: An introduction to the development of the semiconductor laser, IEEE J. QE **23**, 651– 657 (1987)
2. A. Ishida, H. Fujiyasu: Recent progress in lead salt lasers, in: *Laser Diodes and Applications II*, SPIE Proc. **2682**, 206–215 (1996)
3. T. C. Hasenberg, R. H. Miles, A. R. Kost, L. West: Recent advances in Sb-based midwave-infrared lasers, IEEE J. QE **33**, 1403–1406 (1997)
4. J. I. Malin, C. L. Felix, J. R. Meyer, C. A. Hoffmann, J. F. Pinto, C.-H. Lin, P. C. Chang, S. J. Murry, S.-S. Pei: Type II mid IR lasers operating above room-temperature, Electron. Lett. **32**, 1593–1595 (1996)
5. J. Faist, A. Tredicucci, F. Capasso, C. Sirtori, D. Sivco, J. N. Baillargeon, A. Hutchinson, A. Y. Cho: High power continuous-wave quantum cascade lasers, IEEE J. QE **34**, 336–341 (1998)
6. A. Tredicucci, F. Capasso, C. Gmachl, A. Hutchinson, D. Sivco, A. Y. Cho: High performance interminiband quantum cascade lasers with graded superlattices, Appl. Phys. Lett. **73**, 2101–2103 (1998)
7. C. Gmachl, A. Tredicucci, F. Capasso, A. Hutchinson, D. Sivco, J. N. Baillargeon, A. Y. Cho: High power $\lambda = 8\,\mu\text{m}$ quantum cascade laser with near optimum performance, Appl. Phys. Lett. **72**, 3130–3132 (1998)
8. G. P. Agrawal, N. K. Dutta: *Long Wavelength Semiconductor Lasers* (Van Nostrand Reinhold, New York 1986)
9. P. J. A. Thijs, L. F. Tiemeijer, J. J. M. Binsma, T. van Dogen: Progress in long wavelength strained layer InGaAs(P) quantum well semiconductor lasers and amplifiers, IEEE J. QE **30**, 477 (1994)
10. A. Ishibashi: II-VI blue-green laser diodes, IEEE J. Select. Topics Quant. Electron. **1**, 741–748 (1995)
11. S. Nakamura, G. Fasol: *The Blue Laser Diode* (Springer, Berlin Heidelberg 1997)
12. R. Steele: Review and forecasts of Laser Markets 1998, Part II, Laser Focus World **34** (2), 72 (1998)
13. S. O'Brien, H. Zhang, R. J. Lang: High-power wide-aperture AlGaAs-based lasers at 870 nm, Electron. Lett. **34**, 184–185 (1998)
14. W. Pittroff, F. Bugge, G. Erbert, A. Knauer, J. Maege, J. Sebastian, R. Staske, H. Wenzel, G. Traenkle: Highly reliable tensely strained 810 nm QW laser diode operating at high temperatures, IEEE Proc. CLEO **1**, 278–279 (1998)
15. A. Al-Muhanna, L. J. Mawst, D. Botez, D. Z. Garbuzov, R. V. Martinelli, J. C. Connolly: High-power ($> 10\,\text{W}$) continuous wave operation from $100\,\mu\text{m}$ aperture $0.97\,\mu\text{m}$ emitting Al-free diode lasers, Appl. Phys. Lett. **73**, 1182–1184 (1998)
16. J. K. Wade, L. J. Mawst, D. Botez, J. A. Morris: 8.8 W CW power from broad-waveguide Al-free active-region ($\lambda = 805\,\text{nm}$) diode lasers, Electron. Lett. **34**, 1100–1101 (1998)

17. A. Knauer, G. Erbert, H. Wenzel, A. Bhattacharya, F. Bugge, J. Maege, W. Pittroff, J. Sebastian: 7 W CW power from tensile-strained GaAsP/AlGaAs ($\lambda = 735$ nm) QW diode lasers, Electron. Lett. **35**, 638–639 (1999)

18. D. Z. Garbuzov, R. J. Menna, R. U. Martinelli, L. DiMarco, M. G. Harvey, J. C. Connolly: Broadened-waveguide 1.5 μm SCH-MQW InGaAsP/InP laser diodes with CW output powers of 4.6 W, CLEO '96, Anaheim, CA (1996) Postdeadline paper CPD10

19. X. He, D. Xu, A. Ovtchinnikov, F. Malarayap, R. Supe, S. Wilson, R. Patel: High power efficient GaInAsP/InP (1.9 μm) laser diode arrays, Electron. Lett. **35**, 397–398 (1999)

20. M. Maiorov, R. Menna, V. Khalfin, H. Milgaso, A. Triano, D. Garbuzov, J. Connolly: 218 W quasi-CW operation of 1.83 μm two-dimensional laser diode array, Electron. Lett. **35**, 638–639 (1999)

21. S. Nakamura, M. Senoh, S. Nagahama, N. Iwasa, T. Matushita, T. Mukai: InGaN/GaN/AlGaN-based LEDs and laser diodes, MRS Internet J. Nitride Semicond. Res. **4S1**, G1.1 (1999)

22. N. Holonayak Jr., S. F. Bevacqua: Coherent (visible) light emission from Ga(As$_{1-x}$P$_x$) junctions, Appl. Phys. Lett. **1**, 82 (1962)

23. G. H. Olsen, C. J. Nuese, M. Ettenberg: Reliability of vapor-grown InGaAs and InGaAsP heterojunction laser structures, IEEE J. QE **15**, 688–693 (1979)

24. M. A. Herman, H. Sitter: *Molecular-Beam Epitaxy*, Springer Ser. Mater. Sci. **7** (Springer, Berlin Heidelberg 1989)

25. E. H. C. Parker (Ed.): *The Technology and Physics of Molecular-Beam Epitaxy* (Plenum, New York 1986)

26. H. Jaeckel, G.-L. Bona, P. Buchmann, H. P. Meier, P. Vettinger, W. J. Kozlovsky, W. Lenth: Very high-power (425 mW) AlGaAs SQW-GRINSCH ridge laser with frequency-doubled output, IEEE J. QE **27**, 1560–1567 (1991)

27. M. B. Panish: Molecular-beam epitaxy of GaAs and InP with gas sources for As and P, J. Electrochem. Soc. **127**, 2729 (1980)

28. G. Zhang, A. Ovtchinnikov, J. Näppi, H. Asonen, M. Pessa: Optimization and characteristics of Al-free strained-layer InGaAs/GaInAsP/GaInP SCH-QW lasers ($\lambda \approx 980$ nm) grown by gas-source MBE, IEEE J. QE **29**, 1943–1949 (1993)

29. A. Ovtchinnikov, J. Näppi, A. Jaan, S. Mohrdiek, H. Asonen: Highly efficient 808 nm-range Al-free lasers by Gas-Source MBE, Proc. SPIE **3004**, 34–42 (1997)

30. M. Toivonen, P. Savolainen, H. Asonen, M. Pessa: Solid-source MBE for growth of laser diode materials, J. Cryst. Growth **175/176**, 37–41 (1997)

31. E. Veuhoff: Potential of MOMBE/CBE for the production of photonic devices in comparison with MOVPE, J. Cryst. Growth **188**, 231–246 (1998)

32. G. B. Stringfellow: *Organometallic Vapor-Phase Epitaxy: Theory and Practice* (Academic, San Diego 1989)

33. M. Razeghi: *The MOCVD Challenge. 1: A Survey of GaInAsP - InP for Photonic and Electronic Applications* (Institute of Physics, Bristol 1989)

34. M. Razeghi: *The MOCVD Challenge 2: A Survey of GaInAsP - GaAs for Photonic and Electronic Device Applications* (Institute of Physics, Bristol 1995)

35. A. C. Jones, P. O'Brien: *CVD of Compound Semiconductors: Precursor Synthesis, Development and Applications* (VCH, Weinheim 1997)

36. J. S. Roberts, N. J. Mason: Factors influencing doping control and abrupt metallurgical transitions during atmospheric pressure MOVPE growth of AlGaAs and GaAs, J. Cryst. Growth **68**, 422–423 (1984)

37. S. D. Hersee, M. Krakowski, R. Blondeau, M. Baldy, B. De Créoux, J. P. Duchemin: Abrupt OMVPE grown GaAs/GaAlAs heterojunctions, J. Cryst. Growth **68**, 282–288 (1984)

38. S. J. Bass: Device quality epitaxial gallium arsenide grown by the metal alkylhydride technique, J. Cryst. Growth **31**, 172–178 (1975)

39. N. Puetz, G. Hiller, A. J. Springthorpe: The inverted horizontal reactor: Growth of uniform InP and GaInAs by LPMOCVD, J. Electron. Mater. **17**, 381 (1988)

40. R. D. Dupuis: GaAlAs-GaAs heterostructure lasers grown by metalorganic chemical vapor deposition, Jpn. J. Appl. Phys. **19**, Suppl. 1, 415–423 (1980)

41. K. Shiina, H. Tanaka, T. Ohori, K. Kasai, J. Komeno: Multi-wafer growth of 4-inch epitaxial layers by MOVPE for HEMT LSIs, Inst. Phys. Conf. Ser. **129**, 61–66 (1992)

42. W. Van der Stricht, I. Moerman, P. Demeester, J. A. Crawley, E. J. Thrush: Study of GaN and InGaN films grown by metalorganic chemical vapor deposition, J. Cryst. Growth **170**, 344–348 (1997)

43. A. I. Gurary, G. S. Tompa, A. G. Thompson, R. A. Stall, P. A. Zawadzki, N. E. Schumacher: Thermal and flow issues in the design of metalorganic chemical vapor deposition reactors, J. Cryst. Growth **145**, 642–649 (1994)

44. P. M. Frijlink, J. L. Nicholas, P. Suchet: Layer uniformity in a multiwafer MOVPE reactor for III - V compounds, J. Cryst. Growth **107**, 166–174 (1991)

45. K. F. Jensen, D. F. Fotiadis, T. J. Mountziaris: Detailed models of the MOVPE process, J. Cryst. Growth **107**, 1–11 (1991)

46. T. Bergunde, M. Dauelsberg, L. Kadinski, Y. N. Makarov, V. S. Yuferev, D. Schmitz, G. Strauch, H. Jürgensen: Process optimization of MOVPE growth by numerical modeling of transport phenomena including thermal radiation, J. Cryst. Growth **180**, 660–669 (1997)

47. P. Andre, M. Boulou, A. Mircea-Roussel: Luminescence of $Al_xGa_{1-x}As$ grown by MOVPE, J. Cryst. Growth **55**, 192 (1981)

48. M. J. Tsai, M. M. Tashima, R. L. Moon: The effect of the growth temperature on $Al_xGa_{1-x}As$ (0<x<0.37) LED materials grown by MOVPE, J. Electron. Mater. **13**, 437 (1984)

49. G. Beister, F. Bugge, G. Erbert, J. Maege, P. Ressel, J. Sebastian, A. Thies, H. Wenzel: Reliable high-power InGaAs/AlGaAs ridge waveguide laser diodes, CLEO/EUROPE '98, Tech. Dig. 88 (1998) Paper C Tul 36

50. O. Ueda: *Reliability and Degradation of III-V Optical Devices* (Artech House, Norwood, MA 1996)

51. M. Suzuki, M. Aoki, T. Tsuchiya, T. Taniwatari: 1.24 − 1.66 μm quantum energy tuning for simultaneously grown InGaAs/InP QWs by selective area MOVPE, J. Cryst. Growth **145**, 249–255 (1994)

52. S. Adachi (ed.): *Properties of Aluminium Gallium Arsenide*, EMIS Datarev. Ser. **7** (IEE INSPEC, Stevenage, Herts 1993)

53. S. Adachi: *Physical Properties of III-V Semiconductor Compounds : InP, InAs, GaAs, GaP, InGaAs, and InGaAsP* (Wiley, New York 1992)

54. I. Rechenberg, A. Oster, A. Knauer, U. Richter, J. Menninger, M. Weyers: Ordering in $Ga_xIn_{1-x}As_yP_{1-y}$ detected by diffraction methods, Mater. Res. Soc. Proc. **417**, 49 (1996)

55. T. Suzuki, A. Gomyo, S. Iijima: Strong ordering in GaInP alloy semiconductors: Formation mechanism for the ordered phase, J. Cryst. Growth **93**, 396 (1988)

56. J. W. Matthews, A. E. Blakeslee: Defects in epitaxial multilayers I. Misfit dislocations, J. Cryst. Growth **27**, 118–125 (1974)

57. T. F. Kuech, R. Potemski, F. Cardone, G. Scilla: Quantitative oxygen measurements in OMVPE $Al_xGa_{1-x}As$ grown by Methyl precursors, J. Electron. Mater. **21**, 341–346 (1992)

58. T. F. Kuech, E. Veuhoff, T. S. Kuan, V. Deline, R. Potemski: The influence of growth chemistry on the MOVPE growth of GaAs and $Al_xGa_{1-x}As$ layers and heterostructures, J. Cryst. Growth **77**, 257–271 (1986)

59. S. L. Yellen, A. H. Shepard, R. J. Dalby, A. Baumann, H. B. Serreze, T. S. Guido, R. Soltz, K. J. Bystrom, C. M. Harding, R. G. Waters: Reliability of GaAs-based semiconductor diode lasers: $0.6 - 1.1\,\mu m$, IEEE J. QE **29**, 2058–2067 (1993)

60. J. S. Roberts, J. P. R. David, L. Smith., P. L. Tihanyi: The influence of trimethylindium impurities on the performance of InAlGaAs single quantum well lasers, J. Cryst. Growth **195**, 668–675 (1998)

61. A. Gomyo, T. Suzuki, K. Kobayashi, S. Kawata, I. Hino: Evidence for the existence of an ordered state in $Ga_{0.5}In_{0.5}P$ grown by metalorganic vapor-phase epitaxy and its relation to band-gap energy, Appl. Phys. Lett. **50**, 673–675 (1987)

62. L. J. Mawst, A. Bhattacharya, J. Lopez, D. Botez, D. Z. Garbuzov, L. Di-Marco, J. C. Conolly, M. Jansen, F. Fang, R. Nabiev: 8 W CW front-facet power from broad-waveguide Al-free 980 nm diode lasers, Appl. Phys. Lett. **69**, 1532–1535 (1996)

63. A. Bhattacharya, L. J. Mawst, M. Nesnidal, J. Lopez, D. Botez: 0.4 W CW diffraction-limited beam Al-free $0.98\,\mu m$ triple core ARROW-type diode lasers, Electron. Lett. **32**, 657–658 (1996)

64. C. H. Chen, S. A. Stockman, M. Peanasky, C. P. Kuo, in G. B. Stringfellow, M. G. Crawford (Eds.): *High Brightness Light Emitting Diodes*, Semicond. Semimet **48** (Academic, San Diego 1997)

65. C. Geng, A. Moritz, S. Heppel, A. Mühe, J. Kuhn, P. Ernst, H. Schweizer, F. Phillipp, A. Hangleiter, F. Scholz: Influence of order-domain size on the optical gain of AlGaInP laser structures, J. Cryst. Growth **170**, 418–423 (1997)

66. H. Hamada, S. Honda, M. Shono, R. Hiroyama, Y. Yodoshi, T. Yamaguchi: AlGaInP visible laser diodes grown on misoriented substrates, IEEE J. QE **27**, 1483–1490 (1991)

67. A. Valster, C. T. H. F. Liedenbaum, M. N. Finke, A. L. G. Severens, M. J. B. Boermans, D. E. W. Vandenhout, C. W. T. Bulle-Lieuwma: High quality $Al_xGa_{1-x-y}In_yP$ alloys grown on (311)B GaAs substrates, J. Cryst. Growth **107,** 403 (1991)

68. K. Onabe: Calculation of miscibility gap in quaternary InGaAsP with strictly regular solution approximation, Jpn. J. Appl. Phys. **21**, 797 (1982)

69. A. Knauer, G. Oelgart, A. Oster, S. Gramlich, F. Bugge, M. Weyers: Ordering in GaInAsP grown on GaAs by metalorganic vapor-phase epitaxy, J. Cryst. Growth **195**, 694–699 (1998)

70. A. Knauer, G. Erbert, S. Gramlich, A. Oster, E. Richter, U. Zeimer, M. Weyers: MOVPE growth of GaInAsP/GaAs, J. Electron. Mater. **24**, 1655 (1995)

71. I. Rechenberg, A. Knauer, U. Zeimer, F. Bugge, U. Richter, A. Klein, M. Weyers: Composition fluctuations in (In,Ga)(As,P) single layers and laser structures based on GaAs, Inst. Phys. Conf. Ser. **149**, 109 (1996)
72. D. M. Follstaedt, R. P. Schneider, E. D. Jones: Microstructure of (InGa)P alloys grown on GaAs by metalorganic Vapor-Phase Epitaxy, J. Appl. Phys. **77**, 3077–3087 (1995)
73. I. Rechenberg, A. Knauer, F. Bugge, U. Richter, G. Erbert, K. Vogel, A. Klein, U. Zeimer, M. Weyers: Crystalline perfection in GaInAsP/GaAs laser structures with GaInP or AlGaAs cladding layers, Mater. Sci. Eng. B **44**, 368–372 (1997)
74. R. W. Glew, K. Scarrott, A. T. R. Briggs, A. D. Smith, V. A. Wilkinson, A. Zhou, M. Silver: Elimination of wavy layer growth phenomena in strain compensated GaInAsP/GaInAsP multi-quantum well stacks, J. Cryst. Growth **145**, 764–770 (1994)
75. R. S. McFadden, M. Skowronski, S. Mahajan: Influence of growth temperature on ordering in InGaAs grown on (001) InP using tertiarybutylarsine source MOCVD, Mater. Res. Soc. Symp. Proc. **326**, 287–292 (1994)
76. E. Veuhoff, T. F. Kuech, B. S. Meyerson: A study of silicon incorporation in GaAs MOCVD layers, J. Electrochem. Soc. **132**, 1958–1961 (1985)
77. T. Iwamoto, K. Mori, M. Mizuta, H. Kukimoto: Doped InGaP grown by MOVPE on GaAs, J. Cryst. Growth **68**, 27–31 (1984)
78. T. Kimura, T. Ishida, T. Sonoda, Y. Mihashi, S. Takamiya, S. Mitsui: Metalorganic Vapor-Phase Epitaxy growth of Be-doped InP using Bismethylcyclopentadienyl-Beryllium, Jpn. J. Appl. Phys. **34**, 1106–1107 (1995)
79. M. Kondo, C. Anayama, H. Sekiguchi, T. Tanahashi: Magnesium doping transients during MOVPE of GaAs and AlGaInP, J. Cryst. Growth **141**, 1–10 (1994)
80. K. Kojima, R. A. Morgan, T. Mulally, G. D. Guth, M. W. Focht, R. E. Leibenguth, M. T. Asom: Reduction of p-doped mirror electrical resistance of GaAs/AlGaAs VCSELs by delta doping, Electron. Lett. **29**, 1771–1772 (1993)
81. E. Yablonovitch, E. O. Kane: Reduction of lasing threshold current density by the lowering of valence band effective mass, IEEE J. Lightwave Technol. LT **4**, 504–506 (1986)
82. A. R. Adams: Band-structure engineering for low-threshold high-efficiency semiconductor lasers, Electron. Lett. **22**, 249 (1986)
83. K. Kim, Y. H. Lee: Temperature dependent critical-layer thickness for strained layer heterostructures, Appl. Phys. Lett. **67**, 2212–2214 (1995)
84. C. Köpf, H. Kosina, S. Selberherr: Physical models for strained and relaxed GaInAs alloys: Band structure and low-field transport, Solid State Electron. **41**, 1139 (1997)
85. U. Zeimer, F. Bugge, S. Gramlich, I. Urban, A. Oster, M. Weyers: High-resolution X-ray diffraction investigation of crystal perfection and relaxation of GaAs/InGaAs/GaAs quantum wells depending on MOVPE growth conditions, Il Nuovo Cimento D **19**, 369–376 (1997)
86. I. Rechenberg, F. Bugge, A. Höpner, A. Klein, U. Richter: Defects in $In_xGa_{1-x}As$/GaAs strained quantum wells, Inst. Phys. Conf. Ser. **135**, 327 (1993)

87. F. Bugge, U. Zeimer, M. Sato, M. Weyers, G. Tränkle: MOVPE growth of highly strained InGaAs/GaAs quantum wells, J. Cryst. Growth **183**, 511–518 (1998)

88. G. Erbert, F. Bugge, A. Knauer, J. Sebastian, A. Thies, H. Wenzel, M. Weyers, G. Tränkle: Tensile-strained GaAs$_{1-y}$P$_y$-AlGaAs quantum well diode lasers emitting between 715 nm and 790 nm, Proc. 16th IEEE International Semiconductor Laser Conf., Nara, Japan (1998) p. 49

89. B. I. Miller, U. Koren, M. G. Young, M. D. Chien: Strain-compensated strained layer superlattices for 1.5 μm wavelength lasers, Appl. Phys. Lett. **58**, 1952–1954 (1991)

90. A. Valster, A. T. Meney, J. R. Downes, D. A. Faux, A. R. Adams, A. A. Brouwer, A. J. Corbijn: Strain-overcompensated GaInP-AlGaInP quantum-well laser structures for improved reliability at high output powers, IEEE J. Select. Topics Quant. Electron. **3**, 180–187 (1997)

91. S. Gupta, A. Garcia, S. Srinivasan, X. He, S. Wilson, J. Harrison, R. Patel: High average power density (0.2–0.65 kW/cm^2) diode laser stacks for 808 nm, 915 nm and 940 nm, Proc. 16th IEEE International Semiconductor Laser Conf., Nara, Japan (1998) p. 53

92. T. Fujimoto, Y. Yamada, Y. Yamada, A. Okubo, Y. Oeda, K. Muro: High-power, InGaAs/AlGaAs laser diodes with decoupled confinement heterostructure, SPIE Proc. **3628**, 38–45 (1999)

93. H. Q. Hou, K. D. Choquette, K. M. Geib, B. E. Hammons: High-performance 1.06 μm selectively oxidized vertical-cavity surface-emitting lasers with InGaAs-GaAsP strain-compensated quantum wells, IEEE Photon. Technol. Lett. **9**, 1057–1059 (1997)

94. T. Hayakawa: High reliability in 0.8 μm high-power InGaAsP/InGaP/AlGaAs laser diodes with a broad waveguide, SPIE Proc. **3628**, 29–37 (1999)

95. F. Bugge, A. Knauer, S. Gramlich, I. Rechenberg, G. Beister, H. Wenzel, G. Erbert, M. Weyers: MOVPE growth of AlGaAs/GaInP diode lasers, J. Electron. Mater. (in press)

96. C. Hanke, L. Korte, B. D. Acklin, J. Luft, S. Grötsch, G. Herrmann, Z. Spika, M. Marciano, B. de Odorico, J. Wilhemi: Highly reliable 40 W CW InGaAlAs/GaAs 808 nm laser bars, SPIE Proc. **3628**, 64–70 (1999)

97. M. A. Emanuel, J. A. Skidmore, M. Jansen, R. Nabiev: High-power InAlGaAs-GaAs laser diode emitting near 731 nm, IEEE Photon. Technol. Lett. **9**, 1451–1453 (1997)

98. D. Z. Garbuzov, J. H. Abeles, N. A. Morris, P. D. Gardner, A. R. Triano, M. G. Harvey, D. B. Gilbert, J. C. Connolly: High power separate confinement heterostructure AlGaAs/GaAs laser diodes with broadened waveguide, SPIE Proc. **2682**, 20–26 (1996)

99. Y. Oeda, T. Fujimoto, Y. Yamada, H. Shibuya, K. Muro: High-power, 0.8 μm-band broad-area laser diodes with a decoupled confinement heterostructure, *CLEO'98*, OSA Tech. Dig. Ser. **10** (1998)

100. K. Shigihara, Y. Nagai, S. Karadida, A. Takami, Y. Kokubo, H. Matsubara, S. Kakimoto: High-power operation of broad-area laser diodes with GaAs and AlGaAs single quantum wells for Nd: YAG pumping, IEEE J. QE **27**, 1537 (1991)

101. P. L. Tihanyi, F. C. Jain, M. J. Robinson, J. E. Dixon, J. E. Williams, K. Meehan, M. S. O'Neill, L. S. Heath, D. M. Beyea: High power AlGaAs-GaAs visible lasers, IEEE Photon. Technol. Lett. **6**, 775–777 (1994)

102. F. Daiminger, S. Heinemann, J. Näppi, M. Toivonen, H. Asonen: 100 W CW Al-free 808 nm linear bar arrays, *CLEO'97*, OSA Tech. Dig. Ser. **9**, 482–483 (1997)

103. J. Diaz, H. J. Yi, M. Razeghi, G. T. Burnham: Long term reliability of Al-free InGaAsP/GaAs ($\lambda = 808$ nm) lasers at high power high temperature operation, Appl. Phys. Lett. **71**, 3042–3044 (1997)

104. A. Al-Muhanna, J. K. Wade, T. Earles, J. Lopez, L. J. Mawst: High-performance, reliable, 730 nm-emitting Al-free active region diode lasers, Appl. Phys. Lett. **73**, 2869–2871 (1998)

105. G. Erbert, F. Bugge, A. Knauer, J. Maege, A. Oster, J. Sebastian, R. Staske, A. Thies, H. Wenzel, M. Weyers, G. Traenkle: Diode lasers with Al-free quantum wells embedded in LOC AlGaAs waveguides between 715 nm and 840 nm, SPIE Proc. **3628**, 19–28 (1999)

106. G. Erbert, F. Bugge, A. Knauer, J. Sebastian, A. Thies, H. Wenzel, M. Weyers, G. Tränkle: High-power tensile-strained GaAsP-AlGaAs quantum well lasers emitting between 715 nm and 790 nm, IEEE. J. Select. Topics Quant. Electron. **5** (1999) (in press)

107. S. L. Orsila, M. Toivonen, P. Savolainen, V. Vilokkinen, P. Melanen, M. Pessa, M. J. Saarinen, P. Uusimaa, F. Fang, M. Jansen, R. Nabiev: High-power 600 nm-range lasers grown by solid-source molecular-beam epitaxy, SPIE Proc. **3628**, 203–208 (1999)

108. J. S. Osinski, B. Lu, H. Zhao, B. Schmitt: High-power continuous-wave operation of 630 nm-band laser diode arrays, Electron. Lett. **34**, 2336–2337 (1998)

109. H. Jäckel, G. L. Bona, H. Richard, P. Roentgen, P. Unger: Reliable 1.2 W CW red-emitting (Al)GaInP diode laser array with AlGaAs cladding layers, Electron. Lett. **29**, 101–102 (1993)

GaAs Substrates for High-Power Diode Lasers

Georg Müller[1], Patrick Berwian[1], Eberhard Buhrig[2], and Berndt Weinert[3]

[1] Crystal Growth Laboratory, Department of Materials Science (WW6),
University Erlangen-Nürnberg, Martensstraße 7, D-91058 Erlangen, Germany
georg.mueller@ww.uni-erlangen.de
[2] Institute of Nonferrous Metallurgy, Technical University Bergakademie Freiberg,
Leipziger Straße 23, D-09596 Freiberg, Germany
buhrig@inemet.tu-freiberg.de
[3] Freiberger Compound Materials GmbH (FCM),
Am Junger Löwe Schacht 5, D-09599 Freiberg, Germany
weinert@fcm-germany.com

Abstract. GaAs substrate crystals with low dislocation density (Etch-Pit Density (EPD) $< 500\,\mathrm{cm}^{-2}$) and Si-doping ($\approx 10^{18}\,\mathrm{cm}^{-3}$) are required for the epitaxial production of high-power diode-lasers. Large-size wafers ($\geq 3\,\mathrm{in}$) are needed for reducing the manufacturing costs. These requirements can be fulfilled by the Vertical Bridgman (VB) and Vertical Gradient Freeze (VGF) techniques. For that purpose we have developed proper VB/VGF furnaces and optimized the thermal as well as the physico-chemical process conditions. This was strongly supported by extensive numerical process simulation. The modeling of the VGF furnaces and processes was made by using a new computer code called CrysVUN++, which was recently developed in the Crystal Growth Laboratory in Erlangen.

GaAs crystals with diameters of 2 and 3 in were grown in pyrolytic Boron Nitride (pBN) crucibles having a small-diameter seed section and a conical part. Boric oxide was used to fully encapsulate the crystal and the melt. An initial silicon content in the GaAs melt of $c(\mathrm{Si}_{\mathrm{melt}}) = 3 \times 10^{19}\,\mathrm{cm}^{-3}$ has to be used in order to achieve a carrier concentration of $n = (0.8\text{–}2) \times 10^{18}\,\mathrm{cm}^{-3}$, which is the substrate specification of the device manufacturer of the diode-laser. The EPD could be reduced to values between $500\,\mathrm{cm}^{-2}$ and $50\,\mathrm{cm}^{-2}$ with a Si-doping level of 8×10^{17} to $1 \times 10^{18}\,\mathrm{cm}^{-3}$. Even the 3 in wafers have rather large dislocation-free areas. The lowest EPDs ($< 100\,\mathrm{cm}^{-2}$) are achieved for long seed wells of the crucible.

The fabrication of high-power diode-lasers requires GaAs substrates with special material and surface properties for the epitaxial processing. The most-important requirement is a low dislocation density expressed in terms of the Etch-pit density (EPD), with a value of EPD $< 500\,\mathrm{cm}^{-2}$. This low EPD is an indispensable prerequisite for diode-laser fabrication as dislocations are deleterious for the lifetime and performance of the devices. The n-type electrical conduction (achieved by Si doping) with a carrier concentration of about $10^{18}\,\mathrm{cm}^{-3}$ is necessary because the substrate forms one of the diode contacts. Other important properties of the substrate wafers like orientation, surface and mechanical properties are discussed in Sect. 4.

R. Diehl (Ed.): High-Power Diode Lasers, Topics Appl. Phys. **78**, 121–171 (2000)
© Springer-Verlag Berlin Heidelberg 2000

1   Selection of the Growth Method

The GaAs substrate wafers are cut from bulk single crystals which are grown from the melt by special techniques. In this section we discuss possible growth methods and come to the selection of the most appropriate one. In Sect. 2 the technical details of the growth setup and processing will be presented in detail, including the furnace concepts, design of growth containers, thermodynamics of the materials and means of process automation. A separate section (Sect. 3) is devoted to the numerical modeling of the crystal-growth process. For that purpose a special computer code (CrysVUN++) was developed for the optimization of both the growth setup and the growth process. Finally, in Sect. 4 the achieved results concerning crystal and wafer properties will be presented.

1.1   Important Features of GaAs Crystal-Growth Methods

Bulk GaAs crystals can be grown from the melt either by 'crystal pulling' (Czochralski methods) or by directional solidification in a crucible (Bridgman-, Stockbarger- or gradient freeze methods). An important issue of the growth of GaAs by these methods is the control of the composition of the melt, as GaAs has an equilibrium vapor pressure of about 2 bar (As_2, As_4) at the melting point $T_m = 1238°$ C (for details see Sect. 2). Two principles are in use, 'hot wall' (a) and 'liquid encapsulation' (b) depicted in Fig. 1, which will be briefly discussed in the following.

In the first method, the heated crucible containing the GaAs melt with a free surface is placed in a hermetically sealed envelope (Fig. 1a) together with a second crucible containing a certain portion of elemental solid As. Three temperature–pressure conditions have to be fulfilled to prevent dissociation of the GaAs melt:

(i) Melt–vapor equilibrium: the equilibrium partial vapor pressures of the decomposing GaAs melt, i.e. the vapor pressures $p(As_2), p(As_4)$ of the

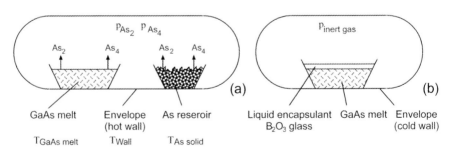

Fig. 1. Hot-wall principle (**a**) with separate vapor-pressure source (As reservoir with temperature $T_{As\ solid}$) and liquid-encapsulation principle (**b**) to prevent the dissociation of a decomposing melt (GaAs at melting temperature $T_{GaAs\ melt}$)

volatile component As (we can neglect the Ga pressure which is rather low) must be maintained in the whole volume of the envelope.

(ii) Vapor–solid equilibrium: the vapor pressures $p(As_2)$ and $p(As_4)$ of the solid reservoir (As) must exactly match the values defined in (i) for the melt at the given temperature $T_{GaAs\ melt}$, i.e. the melting temperature. Thus the temperature $T_{As\ solid}$ has to be controlled in the range 610–620° C to match the vapor pressures $p(As_2)$ and $p(As_4)$ of the GaAs melt. Any deviations would change the composition of the melt, which is undesirable.

(iii) Boundary condition for the envelope wall ('hot wall'): the temperature of the internal surface of the envelope wall must be higher than the temperature (610–620° C) of the solid As reservoir to avoid condensation of solid As at the wall, which otherwise would consume all the As from the solid As source and later from the melt. This condition $T_{wall} > T_{As\ solid}$ explains the name 'hot-wall principle'.

The important advantage of the hot-wall technique is the possibility to establish a well-defined melt composition by controlling the temperature of the As reservoir. The use of hot walls, however, can cause several technical problems as the material of the envelope has to sustain the chemical attack of the reactive As at $T \approx 610°$ C. This holds especially if a mechanical feed-through is necessary to introduce a pulling rod into the chamber, as in the case of the Czochralski method (Vapor-Controlled Czochralski (VCZ) and Hot-Wall Czochralski (HWCZ)).

The second method, the liquid-encapsulation method as sketched in Fig. 1b, overcomes the technical problems of a 'hot wall' elegantly but has the disadvantage that the melt composition cannot be controlled actively. The melt (GaAs) contained in the heated crucible is now covered by a chemically inert second fluid (boric oxide glass). This 'liquid encapsulant' floats on the top surface of the melt and also forms a thin layer between the crucible wall and the melt under proper preparation conditions.

The crucible with the encapsulated melt also has to be placed in a hermetically sealed envelope. But the condition for the gas atmosphere in the vessel is much simpler than in the hot-wall case.

As long as the pressure of the (typically inert) gas in the vessel is higher than the dissociation pressures $p(As_2)$ and $p(As_4)$ of the melt, no gas bubbles with components from the melt will penetrate through the encapsulation layer. The only loss of components from the melt is by diffusion through the encapsulant, which can normally be neglected. As the temperature of the wall of the growth vessel is not relevant for the melt condition, it can be kept at room temperature by water cooling and can be made from steel.

For cases of dissociation pressures $p \gg 1$ bar (e.g. for the in-situ synthesis of GaAs) the growth vessel has to be constructed for high-pressure use.

1.2   Liquid-Encapsulated Czochralski (LEC) and Vapor-Controlled Czochralski (VCZ) Techniques

The Czochralski method ('crystal pulling') [1] is the dominant principle for the production of bulk single crystals of a large number of electronic and optical materials. Several variants were developed for the growth of GaAs crystals (Table 1), some key items being common to all of these variants (Fig. 2a,b). The polycrystalline GaAs material is melted in a cylindrically shaped crucible. The GaAs melt is covered with a liquid encapsulant (B_2O_3). The seed crystal which is fixed at the pulling rod is dipped through the B_2O_3 layer into the melt ('seeding process'). It is important to adjust the power of the heaters carefully so that a certain portion of the dipped seed is remelted and a melt meniscus is formed. The pulling rod, i.e. the seed, is now slowly lifted (often under rotation) and the melt crystallizes at the interface of the seed by forming a new crystal portion. The shape of this crystal – especially the diameter – is controlled by weighing the crystal and properly adjusting the heating power.

Table 1. Variants of the Czochralski (Cz) technique and characteristic features

Cz. variant	Crucible	Heating system	Atmos.	Pressure (bar)	Melt surface	Crystal surface	Extras
LEC	pBN SiO$_2$	res. graph. RF	Ar, N$_2$	3–40	covered by B$_2$O$_3$	covered by B$_2$O$_3$	
VLEC VCZ	pBN	res. graph.	Ar, N$_2$ — As	3–40 several mbar	covered by B$_2$O$_3$	free	heated inner chamber with As source
HWCZ	pBN	res. graph.	As	~ 2	free	free	As source hot wall
MLEC	pBN	res. graph.	Ar, N$_2$	3–40	covered by B$_2$O$_3$	covered by B$_2$O$_3$	magnet system

LEC = Liquid-Encapsulated Cz, RF = radio frequency, atmos. = atmosphere VCZ (VLEC) = Vapor-pressure-controlled Cz (LEC), HWCZ = Hot-Wall Cz, MLEC = Magnetic LEC, res. graph = resistance heating system made from graphite.

1.3   Thermal Stress and Dislocation Density

It is common knowledge that a direct correlation exists between the thermal stress during crystal growth of GaAs and the resulting dislocation density, at least for an Etch-Pit Density (EPD) > 500 cm^{-2}. The levels of thermal stress can be calculated by using measured or simulated temperature distributions in the crystal and related quantitatively to experimentally determined EPDs (for a profound review see [2]).

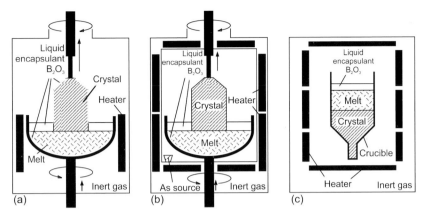

Fig. 2. Scheme of the Liquid-Encapsulated Czochralski (LEC) **(a)**, the Vapor-Controlled Czochralski (VLEC/VCZ) **(b)** and the liquid-encapsulated Vertical Gradient Freeze (VGF) **(c)** methods

From this kind of analysis the well-known conclusion is derived that the LEC technique is a 'high-stress' growth technology due to the extremely nonuniform heat transfer conditions in the region where the crystal leaves the B_2O_3 layer on top of the melt surface. The strong variations in the radial heat flow (bending of isotherms) combined with the relatively high axial temperature gradients (Fig. 3) lead to high values of the second derivatives of the temperature distribution within the elasto-plastic range of the crystal (from the melting point down to $\approx 600°$ C).

Thus, the EPD of LEC-grown GaAs exceeds values in the range of some 10^4 cm^{-2} for 4 in and even higher for 6 in-diameter wafers. Attempts to overcome this problem by hardening the GaAs lattice by alloying a few percent of the isoelectronic dopant In were in principle successful. But unfortunately this measure had a deleterious impact on the mechanical properties of the wafers (increased brittleness and change of lattice parameter) and was therefore dropped. Attempts were also undertaken to reduce the thermal stress by

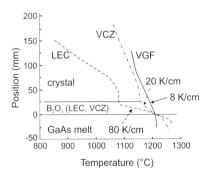

Fig. 3. Axial temperature profiles in GaAs crystals during growth by different techniques: LEC, VCZ, and VGF

using computer simulation to improve the thermal boundary conditions. But the options for the LEC process are rather limited [2]. Lower axial temperature gradients as shown in Fig. 3 lead to a decomposition of the crystal surface by arsenic evaporation and the formation of drops of liquid gallium at elevated temperatures. To overcome this problem the vapor-pressure-controlled Czochralski method (VLEC, VCZ) was developed.

The principle of the Vapor Pressure Controlled LEC (VLEC or VCZ) method is shown in Fig. 2b. It is a special variant of the LEC method (Fig. 2a) introduced in 1983 by *Tatsumi* et al. [3] with the objective to reduce the thermal stress in the growing crystal by an active in-situ after-heating of the grown crystal. The after-heating, however, has the consequence that the crystal portion outside the boric oxide encapsulation layer has such an elevated temperature (Fig. 3) that the grown crystal decomposes at the surface by dissociation and is damaged severely. To prevent this dissociation it is necessary to use an inner envelope with hot walls and a source for vapor-pressure control, as described in Sect. 1.2.

The condition for preventing the undesired decomposition of the GaAs crystal is that a solid supply of As provides a partial pressure which is higher than the equilibrium vapor pressure (in the millibar range) of the decomposing GaAs crystal. Even though this vapor pressure is much smaller than the vapor pressure of 2 bar which would be necessary to avoid a decomposition of the melt, the VCZ method is a complex and expensive technique. Several patents describe technical solutions of the complex double container with feed-through of the pulling rod and a separately heated vapor-pressure source.

Whereas in the case of conventional LEC the pulling rate is about 10 mm/h, with VCZ it has to be reduced to 3–8 mm/h due to the reduced axial temperature gradient removing the heat. More details on the use of the VCZ method for the growth of GaAs are given in a review article [4].

A considerable reduction of the thermal stress during growth and hence the danger of plastic deformation, i.e. dislocation formation, is possible by the directional solidification methods (e.g. Vertical Gradient Freeze, Fig. 2c) which are described in the next section.

1.4   Methods of Directional Solidification: Gradient Freeze and Bridgman Variants

Directional solidification is the basic principle of controlled solidification with greatest importance in the field of metallurgical production. The use of this principle for growing single crystals by melting a charge in a crucible and cooling the melt until it starts to crystallize from one end of the crucible to the other dates back to 1914 [5]. This method can be carried out by moving the growth interface in horizontal or in vertical direction (Fig. 4). The crystal-growth configuration typically consists of a tube furnace which provides a temperature profile with negative gradients parallel to the growth

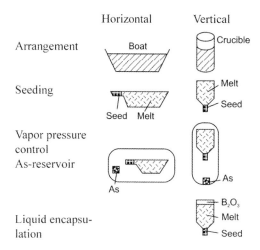

Fig. 4. Various principles of directional solidification methods (Gradient Freeze, Bridgman) and important features

direction, at least in the vicinity of the growth interface but mainly in the major portion of the melt and the crystal. A single crystal seed is used in one end of the boat or the lower end of the vertical crucible (Fig. 4).

Crystal growth is carried out by a controlled shifting of the temperature profile relatively to the boat or crucible as sketched in Fig. 5 exemplifying the vertical configuration. This can be achieved by three variants: mechanical movement of the crucible relatively to the fixed furnace, first introduced by *Bridgman* in 1925 [6], mechanical movement of the furnace relatively to the fixed crucible (*Stockbarger* in 1925 [7]) and without any mechanical movement. In the latter case, shifting of the temperature profile is achieved only by power control of the furnace heater (gradient freeze).

Table 2 compiles the Gradient Freeze and Bridgman–Stockbarger variants together with some important features which are presently in use for the

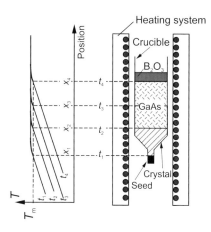

Fig. 5. Sketch of the furnace-temperature profiles for different time steps (t_1, t_2, t_3 and t_4) with positions x_1, x_2, x_3 and x_4 of the crystal–melt interface in a Vertical Bridgman (VB) or Gradient Freeze (VGF or Stockbarger) configuration

growth of GaAs crystals. The control of the melt composition with respect to the loss of As is again an important issue.

Table 2. VGF and Bridgman variants and characteristic features for the growth of GaAs crystals

Variant	Axis	Movement	Crucible	B$_2$O$_3$ encap.[c]	Silica ampoule	As source	Heating system	Growth direction
HB	hor.[a]	furnace	SiO$_2$	no	yes	yes	tube furnace	111[f]
HGF	hor.	no	SiO$_2$	no	yes	yes	tube furnace	111[f]
VGF	vert.[b]	no	pBN	yes	yes	no	tube furnace	100
			pBN	no[d]	yes	yes	tube furnace	100
	vert.	no	pBN / SiO$_2$[e]	yes	no	no	graphite heater	100
VB	vert.	furnace	pBN	yes	yes	no/yes	tube furnace	100
	vert.	crucible	pBN	yes	no	no	graphite heater	100

[a] hor. = horizontal, [b] ver. = vertical, [c] encap. = encapsulant,
[d] is possible but with low yield, [e] special technique for removal of SiO$_2$ crucible,
[f] low yield for < 100 > growth direction and crystals having a D-shaped cross section

The Horizontal Bridgman (HB) technique is the oldest industrial technique for the growth of GaAs single crystals and is still in use today for industrial production of Si-doped GaAs. However, its use is decreasing due to the fact that only <111>-oriented crystals can be grown with a high yield and the noncircular (D-shaped) cross section of the crystals gives a bad yield in producing round wafers.

The use of the VGF and VB methods for GaAs-crystal growth goes back to the 1960s but the real break-through came only with the pioneering work of *Gault* et al. [8] in 1986. A special article by *Monberg* [9] in the *Handbook of Crystal Growth* is devoted to this development. The increasing interest in the use of VGF results from the fact that it provides several advantages compared to the other methods. It is the simplest principle of melt growth, has stable hydrodynamic conditions (compare [10]), is well suitable for computer modeling (Sect. 3) and for automatic process control, has low thermal-stress conditions for the growing crystal (Fig. 3), and gives round-shaped crystals with easy diameter control and less-expensive equipment. The control of dissociation and vapor pressure is possible by the principles which were introduced in Sect. 1.4 and Fig. 1. Its technical realization is discussed in detail in Sect. 2.

However, the VGF method has one severe disadvantage compared to the Czochralski method. It does not allow for a direct observation of the growing crystal. Any disturbance of growth, e.g. by crystal–crucible interaction, cannot be 'repaired' by melting back but is detected only when the crystal is removed from the crucible after growth. Earlier, twin formation was a problem in the VGF growth of GaAs and still is for InP, but now it can be overcome for GaAs by proper preparation and shape of the crucible (Sect. 2.4).

As a consequence of the above discussion we have concluded that the VGF method seems to be best suited to develop a highly efficient industrial process for the growth of low-EPD Si-doped GaAs single crystals to be used for the fabrication of substrates for high-power diode-lasers.

In the next section our VGF process technology (which was developed in the framework of the German LASER 2000 project) will be described in detail.

2 Physico-Chemical Features of the VGF Technology for the Growth of Si-Doped Low-EPD GaAs Single Crystals

The selection of the proper VGF variants and the optimization of the growth setup (including materials preparation) are strongly related to the details of the physico-chemical processes and their kinetics involved in the system considered, including all participating species and materials. GaAs and the As species which evaporate during heating are highly reactive. A direct contact to hot metallic parts (including heaters) of the growth setup has to be avoided. Present VGF technology uses, therefore, highest-purity non-reactive materials like boric oxide (B_2O_3), pyrolytic boron nitride (pBN), quartz (SiO_2) and eventually graphite inside the growth setup.

In addition to the chemical aspects other factors like shape of the crucible or type of furnace have also to be considered.

2.1 Discussion of VGF Variants

We have selected a crystal-growth process without any mechanical movement of furnace or crucible, called 'VGF' (Sect. 1 and Table 2). In this case disturbance by mechanical vibrations is reduced and the growth setup is simplified.

Furthermore, in this article we are concentrating on pBN as a crucible material, although GaAs crystals were also grown successfully by using fused-silica VGF crucibles [11]. Silica has the advantage of a better match to the thermal conductivity of GaAs as compared to pBN, but it has some inherent disadvantages like strong reaction with boric oxide and softening at the growth temperature (1250–1280° C). pBN is chemically resistant and thermally stable but the strongly anisotropic heat conductivity causes some problems by bending the isotherms at the periphery (for details see Sect. 3).

2.1.1   The Hot-Wall VGF Process

In the hot-wall VGF process (Sect. 1.2) the crucible with the GaAs is placed in a sealed silica ampoule as sketched in Fig. 6b,c. In this case the internal space of the ampoule is a hermetically closed chemical system without any interaction with other constituents like heating, insulating or wall material. Both variants in Fig. 6b,c use a surplus of As supply as explained in Fig. 1a. In the first variant (Fig. 6b) the extra As will be totally evaporated. The necessary amount m of solid As to provide that As-vapor pressure which is equal to the equilibrium vapor pressure of the upper surface of the GaAs melt can be calculated to be

$$m = \frac{p_{\mathrm{GaAs},T} V r M_{\mathrm{As}}}{R T_{\mathrm{av}}} \text{ , with} \tag{1}$$

m = amount of the solid As (g)
$p_{\mathrm{GaAs},T}$ = As partial pressure of the GaAs melt at the corresponding (desired) composition and mean temperature T_{av} of the GaAs surface during growth (kPa)
V = free volume of the ampoule (dm^3)
r = mean number of As atoms of the dominating As molecules
M = molecular weight of As (g)
R = gas constant
T_{av} = mean temperature of the relevant free volume of the ampoule during growth.

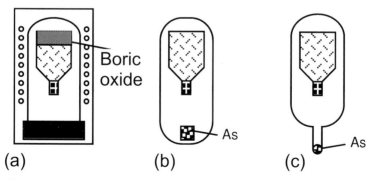

Fig. 6. VGF variants with liquid encapsulation (**a**) and closed ampoule without (**b**) and with (**c**) separately controlled temperature of the As-vapor source

In this method the As-vapor pressure cannot be adjusted during the entire growth process because T_{av} is changing. This can be overcome by the variant shown in Fig. 6c which uses an As source with a separately controlled temperature. Now the solid As is not totally evaporated as in the case shown in Fig. 6b but heated to yield a certain As-vapor pressure which is required to be in equilibrium with a certain As/Ga composition, eventually stoichiometry (As/Ga = 1).

Both variants (Fig. 6b,c) were studied with liquid encapsulation by boric oxide and without. Statistical evaluation of the results clearly showed that the yield in low-EPD single-crystalline material could be increased when boric oxide was used.

Although the variant with B_2O_3 (Fig. 6b,c) is not exactly the hot-wall principle as sketched in Fig. 1a, it exploits its advantages. The melt composition can still be controlled by the vapor pressure of the As source but with some kinetic slowdown due to the diffusive or convective transport of As through the boric oxide layer. On the other hand, this principle uses the advantages of the 'soft packaging' of the growing crystal by the thin liquid layer of boric oxide between pBN crucible and crystal.

2.1.2 The Cold-Wall VGF Process

In Sect. 1.2 we have seen that a decomposition of the GaAs melt can also be avoided by using inert gas overpressure *and* a liquid encapsulant boric oxide in *an open system* which is now an absolute necessity. In this case the sealed containment is a water-cooled steel autoclave, as sketched in Fig. 6a. In contrast to the hot wall variants (Fig. 6b,c) the Ga/As composition of the melt cannot be controlled. Some loss of As during heating ($T < 500°$ C) cannot be avoided as long as the boric oxide is too viscous to completely cover the polycrystalline GaAs charge. This evaporating As can react with the heaters and deposit on the steel walls. Thus graphite heaters have to be used to avoid damage.

Another solution of this problem is the use of a silica ampoule around the crucible which has a small hole in the lower cold end. In this case the silica ampoule contains the same inert-gas pressure as the autoclave but prevents the contamination of the heaters and the autoclave wall as the small amount of evaporating As deposits on the lower cold ampoule walls.

The transport of As through the boric oxide during growth is only by diffusion. This small loss of As is of no practical importance; it also deposits on the ampoule wall. In this case inexpensive metal wire heaters (e.g. made from KANTHAL) can be used.

2.2   Important Chemical Reactions in the VGF Growth of Si-Doped GaAs

The composition and doping of the GaAs crystals are strongly influenced by the thermodynamic equilibria and the kinetics of the chemical reactions which are involved.

2.2.1   GaAs Phase Diagram

The phase diagram of GaAs is shown in Fig. 7. The maximum of the liquidus–solidus line defines the melting point $T_m = 1238°$ C and does not exactly correspond to the ratio Ga/As = 1 (called stoichiometry). During crystallization the composition of $(Ga_x As_y)$ may change according to the solidus line in the phase diagram. The off-stoichiometry of GaAs crystals is related to certain concentrations of intrinsic and extrinsic point defects (see e.g. [12]).

A change in the composition of the GaAs melt is mainly caused by the loss of As due to evaporation. Another effect is the formation of arsenic and gallium oxides. Both oxides are dissolved in the boric oxide but with different solubilities [13]. The composition of the gas phase which is in equilibrium with GaAs is shown in Fig. 8. The dominating gas species which define the vapor pressure are As_4 and As_2 if the system is close to equilibrium.

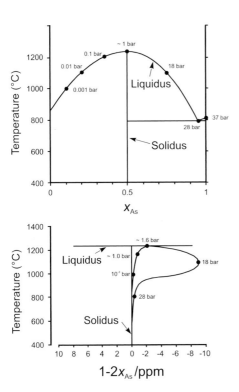

Fig. 7. Phase diagram and vapor pressure of GaAs (*upper figure*) with strongly enlarged solidus region (*lower figure*) [12]

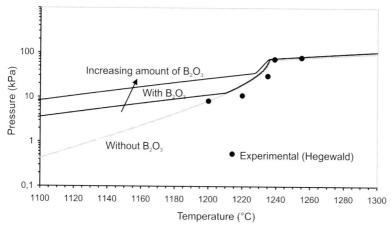

Fig. 8. Calculated arsenic partial pressure versus temperature above GaAs with/without B_2O_3 encapsulation [14]

2.2.2   Reactions in the System GaAs, Si and B_2O_3

For doping GaAs with Si, elementary Si or SiAs is added to the GaAs melt. Elementary Si might cause problems if an oxide layer prevents the dissolution.

Si is incorporated into the growing GaAs crystal according to the segregation coefficient k, defined by $k = c_s/c_l$ with the concentrations in the crystal (c_s) and the melt (c_l). Literature values of k are given in Table 3. As k is smaller than 1 the GaAs melt is enriched by Si during growth ('segregation') resulting in a continuously increasing Si concentration in the growing GaAs crystal (Fig. 9). The experimental data follow the Scheil equation [15] given in the figure.

Table 3. Segregation coefficients k of Si in GaAs obtained by different growth techniques as reported in the literature

Author	k		Growth method
[17]	0.11, 0.13		LEC
[18]	0.11		LEC
[19]	0.0185		LEC
[20]	0.025–0.028		LEC
[11]	0.18–0.31		VGF + SiO_2 crucible
[21]	0.14		HB/HGF
[22]	0.136		HB/HGF
[15]	0.11		VGF + pBN crucible
	$k(Si_{Ga})$	$k(Si_{As})$	
[23]	0.12	0.036	HB / HGF

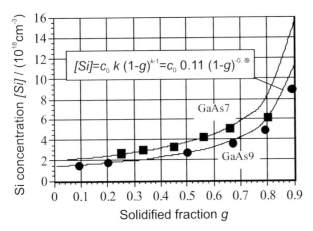

Fig. 9. Si concentration of a VGF-grown 2 in GaAs crystal versus solidified fraction g. *Symbols*: measured carrier concentration $n = $ [Si]$/0.67$ (explanation in Sect. 4). *Lines*: calculated according to [15] with $k = 0.11$ ([16])

The nonuniform Si doping along the growth axis shown in Fig. 9 gives a higher Si concentration in wafers grown at the tail end as compared to wafers from the seed end of a VGF crystal. This axial nonuniformity is only acceptable for wafer fabrication if the shape of the crystal–melt interface is nearly planar. Otherwise, as shown in Fig. 10a, the longitudinal section of a VGF crystal reveals a strong concave bending of the interface, resulting in a strong nonuniformity of the distribution of the Si in the wafers. Figure 10b shows such an example of a wafer with a ring-like nonuniform Si distribution (taken from the crystal of Fig. 10a). Such kind of nonuniformity has to be avoided by establishing proper thermal boundary conditions (see Sect. 3).

Results for crystals grown with different VGF variants show that the use of B_2O_3 can strongly reduce the doping level compared to concentrations expected from the segregation coefficient (Fig. 11). The Si concentration in the melt, however, should not exceed 4×10^{19} cm^{-3} to prevent deleterious precipitations at the surface or in the bulk of the GaAs crystals which will induce polycrystalline growth [15].

Furthermore, the GaAs melt is contaminated with B stemming from the B_2O_3 encapsulant. B competes in GaAs with the Si dopant in the reaction with oxygen according to

$$2B_2O_3 + 3Si \leftrightarrow 3SiO_2 + 4B \tag{2}$$

with the reaction constant $k_{B/Si} = |B|^{\frac{4}{3}}/Si$.

Si oxidation supports the observation that increased Si-doping concentrations foster the incorporation of B in GaAs [25] as the equilibrium (2) is shifted to the right. (In the case that silica is used as ampoule material the reaction of the SiO_2 with volatile components of the system also causes a

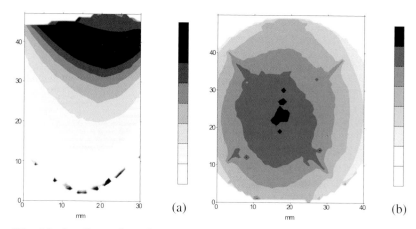

Fig. 10a,b. Examples of nonuniformity of VGF-grown GaAs crystals due to nonuniform Si distribution, caused by strong bending of the growth interface and unsteady growth conditions, as revealed by infrared-transmission topography [24]. Axial (**a**) and radial (**b**) crystal sections

contribution to the Si doping. But its level can be neglected compared to the Si-doping level of $10^{18}\,\mathrm{cm}^{-3}$.)

The reaction of the pBN crucible material with oxygen is used for crucible preparation to create a fresh B_2O_3 layer on the crucible wall. It has a thickness between $5\,\mu\mathrm{m}$ and $40\,\mu\mathrm{m}$ depending on the oxidation temperature around 1000° C. This procedure has to be optimized to provide a sufficiently thick B_2O_3 layer to avoid any contact of the GaAs melt or crystal with solid pBN. On the other hand, the oxidation process has to be as short as possible to save wall thickness for frequent use of the crucible.

The physico-chemical reactions mentioned in this section can be treated by numerical modeling with the appropriate software considering all relevant reactions of the elemental system components, as sketched in Fig. 12. An example of such a computer program is ChemSage [26], which was used to calculate the vapor pressures by the method of multi-component thermodynamics, as plotted in Fig. 8. These kinds of numerical models are very useful to study the qualitative influence of parameters to save experimental time.

An example of the usefulness of such thermodynamic calculations of equilibria among liquid GaAs and Si, B, O is found in [27]. A growth configuration with a SiO_2 (or pBN) crucible, B_2O_3 encapsulation and CO species in the surrounding gas phase are proposed from this kind of thermodynamic modeling. The corresponding results are summarized in Figs. 13 to 16.

Figure 13 shows the predominant areas of the various liquids as a function of the carbon activity and the oxygen partial pressure, taking into account the chemical potentials of the elemental constituents.

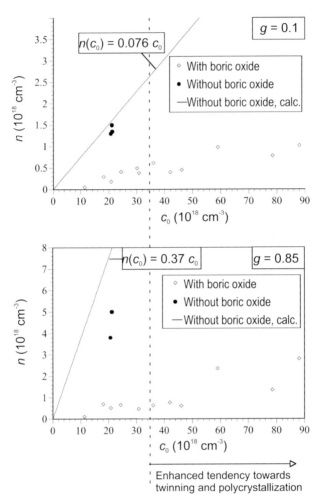

Fig. 11. Carrier concentration n versus Si concentration in GaAs grown with/without boric oxide encapsulation [15]

The shaded area in Fig. 13 indicates the experimental conditions under which GaAs crystals should be grown. This region is limited at low O potentials by the saturation of the GaAs melt with B and Si and the occurrence of carbides. For high carbon activity carbon is saturated. For high oxygen partial pressure, oxides occur (SiO_2, Ga_2O_3 and B_2O_3).

The results of the thermodynamical analysis of the system under consideration, as presented in the above diagrams, are obtained without taking into account a solid carbon source and the negligible impact of crucibles as dopant sources.

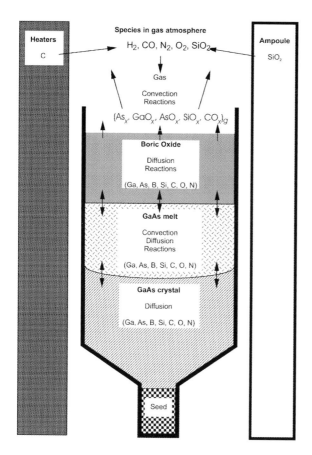

Fig. 12. Physico-chemical reactions in a system consisting of Si-doped GaAs solid and melt, boric oxide, carbon heaters, quartz ampoule and inert-gas atmosphere

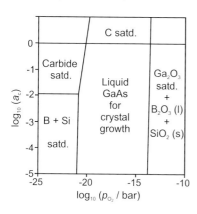

Fig. 13. Equilibrium phases depending on the oxygen partial pressure and the carbon activity in the chemical system illustrated in Fig. 12 [27]

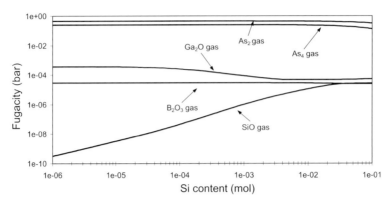

Fig. 14. Fugacity of species in the gas phase in equilibrium with boric oxide and GaAs melt versus variation of Si content [27]

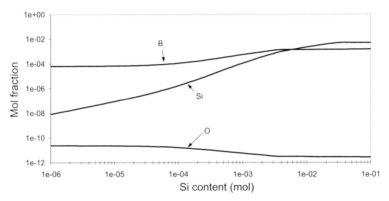

Fig. 15. Mol fraction of B, O and Si in the GaAs melt in equilibrium with liquid boric oxide versus variation of Si content [27]

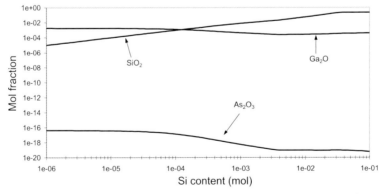

Fig. 16. Mol fraction of As, Ga and Si oxides in the liquid boric oxide in equilibrium with the GaAs melt versus variation of Si content [27]

The composition of the gas phase (in terms of the fugacities of the gaseous species) in equilibrium with liquid B_2O_3 and the GaAs melt as a function of the Si concentration (activity) is shown in Fig. 14. The dependence of the mol fractions of B, O and Si and of the mol fractions of SiO_2, Ga_2O_3 and B_2O_3 in liquid boric oxide and molten GaAs on the Si concentration in the system are shown in Figs. 15 and 16, respectively.

It can be derived from the analysis that the As content in liquid B_2O_3 is very low, which means that the B_2O_3 acts like a barrier for As, and the concentration of the As species in the gas phase may be neglected in practical crystal growth.

The boron concentration in the GaAs melt increases with increasing Si concentration in the melt, which is in accordance with experimental observations.

It is remarkable that the concentration of gallium oxide in the B_2O_3 melt is much larger than the As concentration. Hence, the composition of the GaAs melt is shifted towards a surplus of As as a consequence of using the B_2O_3 encapsulant.

In conclusion, knowledge of the system's thermochemistry as analysed by thermodynamic calculations is very useful to guide VGF growth experiments of Si-doped GaAs if the relevant boundary conditions such as starting composition and reaction temperatures are given.

2.3   VGF Furnace Concepts

The successful growth of Si-doped, low-EPD GaAs substrate crystals requires an adequate furnace concept to achieve the important prerequisites for the VGF technology (see also Sect. 3) such as

- adjustment of heating profiles with low axial gradients ($< 10\,\mathrm{K/cm}$)
- solid-liquid interface with slightly convex or planar shape
- defined control of growth rate
- axial symmetry
- mechanical and chemical stability under As contamination (Sect. 2.2).

These requirements can only be fulfilled by furnaces with several heating elements, three being regarded as the minimum. If a separately controlled As source is used an additional heater is necessary to provide an isothermal region with a temperature around $615°$ C. The optimum furnace concept should provide a nearly isothermal upper region for the melt in the range 1240 to $1250°$ C to avoid superheating, an adiabatic region where the growth interface is positioned and a region with a low axial temperature gradient ($\approx 5\,\mathrm{K/cm}$) in the crystal portion. The heat flux should be mainly axial, from top to bottom. Such heating profiles can best be established by multi-zone furnaces. Furnaces with 20–30 zones were successfully used for VGF growth [28]. The complex construction, however, and the expensive control systems of such multi-zone furnaces render this concept only useful for research and development but not for cost-effective production. Thus one important objective of the VGF furnace development is to reduce the number of heating zones to the necessary lower limit. We have used furnaces with about 10 zones down to 3 zones.

The desired temperature–time profiles of the heaters which result in low thermal stress and the corresponding growth rate of 2 to 4 mm/h are obtained by optimization strategies with both experiments and computer simulation (Sect. 3). Another important hardware item is the furnace control system which must operate in a very stable manner. High-precision solutions with $\Delta T < \pm 0.1\,\mathrm{K}$ were reported [29] and tested in the laboratory of one of the authors.

2.4 Preparation of Starting Materials and Procedure of VGF Growth

The procedure of a typical VGF-growth run including the preparation of the starting materials is presented in Fig. 17. The pre-synthesized polycrystalline GaAs has to be shaped cylindrically within the diameter of the crucible to achieve a high proportion of filling of the crucible volume. The pBN crucible can be pre-oxidized to achieve a better wetting of the boric oxide encapsulant. The oxidation procedure is optimized to give the best yield in single-crystal growth.

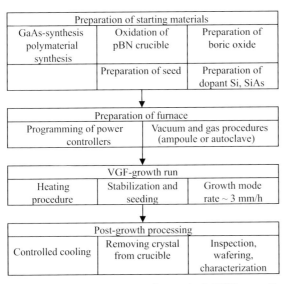

Preparation of starting materials		
GaAs-synthesis polymaterial synthesis	Oxidation of pBN crucible	Preparation of boric oxide
	Preparation of seed	Preparation of dopant Si, SiAs

Preparation of furnace	
Programming of power controllers	Vacuum and gas procedures (ampoule or autoclave)

VGF-growth run		
Heating procedure	Stabilization and seeding	Growth mode rate ~ 3 mm/h

Post-growth processing		
Controlled cooling	Removing crystal from crucible	Inspection, wafering, characterization

Fig. 17. Important steps of a typical VGF-growth process for the fabrication of Si-doped low-EPD GaAs wafers

A large cone angle ($> 140°$) of the crucible is favorable, as this angle causes a decreased length of the {111} edge facets in the conical part of the crystal [30] and hence a reduced tendency for twinning [31]. The seed crystal must fit very precisely into the seeding well to avoid a misorientation or wetting by the melt. The Si-doping source is placed into the crucible as a piece of solid Si or SiAs together with the feed material. The boric oxide is used with a certain content of water as specified by the supplier and placed on top of the feed material. The filled crucible is either placed in the silica ampoule or in a crucible holder in the furnace. In the hot-wall VGF (Fig. 6b in Sect. 2.2) the silica ampoule is evacuated and sealed.

The growth process starts by carefully heating the assembly close to the thermal profile suitable for seeding. The seeding procedure can be supported by the aid of thermocouples closely placed to the seed well. After seeding the growth mode is started by activating the computer-controlled power or temperature–time program for each of the furnace heaters. This program runs fully automatically by computer control without any interactions of an operator. The growth rate is in the range of 2–4 mm/h. When the growth mode is finished the program cools the crystal automatically under conditions which ensure low thermal stress in the GaAs crystal. All thermal program steps can be optimized by means of computer simulation prior to the growth run, which is described in detail in Sect. 3.

After having finished the cooling procedure, the crucible including the crystal is dismounted. The crystal can only be removed from the crucible after the boric oxide has been dissolved by an organic solvent, e.g. methanol. The optimization of this procedure is important for the reuse of the pBN crucible.

The yield of a VGF process strongly depends on the optimization of all steps in the VGF procedure. Of special importance are the preparation of the crucible and the proper thermal-process conditions which can be optimized by applying numerical process modeling as shown in the following section.

3   Numerical Modeling for VGF-Process Optimization

It is well acknowledged in semiconductor-crystal growth that the quality of the crystals depends strongly on the thermal boundary conditions [32]. This holds especially for the growth of low-EPD GaAs substrate crystals by the VGF crystal-growth technique. The main advantage of the gradient freeze methods is the possibility to establish well-defined temperature distributions with relatively small axial and radial temperature gradients (compare Fig. 3). These low temperature gradients result in low thermal stress fields which are necessary for the growth of low dislocation density GaAs crystals [2]. With the complexity of a VGF furnace setup of 10 or more separately controlled heating elements, an empirical optimization of a VGF-growth process is a rather time-consuming and inefficient procedure. On the other hand such a task can now be performed much better by the aid of numerical process modeling based on computer simulation [33]. Applying state-of-the-art computational techniques such calculations can be executed even on a personal computer within a few hours.

3.1 Principles and Strategy of the Numerical Modeling

The numerical modeling of crystal growth like VGF has two important objectives. Firstly, it supports the concept and design of a new furnace setup. The simulation yields quantitative results on temperature profiles and power consumption for a variety of differing details in geometry and materials data. The setup can be optimized before taking a piece of hardware in hand.

Secondly, when the equipment is already to hand, the simulation can be used to optimize the growth process itself. It is even possible to predict the crystal properties and optimize them, e.g. by a calculation of the thermal stress, as will be demonstrated in the following.

To tackle the task of optimizing the VGF growth of low-EPD GaAs crystals, the computer software must fulfil the following prerequisites:

- easy and efficient input of geometric details of the construction by reading CAD files
- treatment of anisotropic material parameters, especially the highly anisotropic heat conductivity of the pBN crucible and thermal insulating materials
- calculation of anisotropic thermal stress in the grown crystal
- calculation of thermal fields by considering radiative and conductive heat transfer in complex geometries (convection is less important in the VGF growth of GaAs)
- solving the Stefan problem (calculation of the shape of the growth interface under consideration of the latent heat at given growth rate)
- calculation of heater powers for given temperature conditions (to achieve low thermal stress in the growing crystals) by the so-called inverse modeling [34].

To the authors' knowledge the only commercially available computer program which presently fulfills all of these criteria is the software system CrysVUN++, which was developed at the Crystal Growth Laboratory in Erlangen for the needs of VGF modeling.

CrysVUN++ uses an unstructured grid and the finite-volume method for solving the set of coupled differential equations with given boundary conditions (for detailed explanations the reader is referred to [35]). The numerical simulation with CrysVUN++ is based on a so-called 'global heat-transfer model'. This means that the calculations include all details of the growth configuration like crystal, melt, crucible, heaters, insulation and the water–cooled pressure vessel. The important items of the software system are given in Fig. 18.

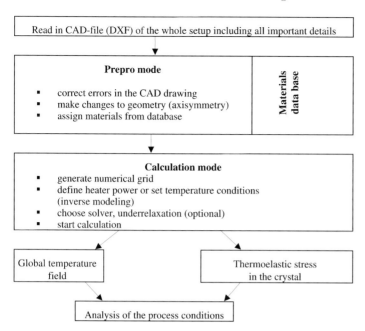

Fig. 18. Setup of the computer code CrysVUN++ [36]. CAD-DXF file means a Computer Aided Design drawing e.g. in the popular DXF format

After reading the geometry of the equipment (CAD file) the numerical grid is generated automatically even for complex geometries. Figure 19 (left-hand side) shows an example of such a numerical grid consisting of triangles. This grid is called 'unstructured' as the number of grid elements (triangles) surrounding a certain node is variable (in contrast to 'structured' grids). In regions of special interest, i.e. the crucible containing the semiconductor material, the grid can be refined without additional effort. The computer program can only treat axisymmetric geometries (so-called 2D model) which means that nonaxisymmetric parts have to be replaced by corresponding axisymmetric ones. Furthermore, the program considers heat conduction and radiation, but no fluid convection is taken into account. It has been shown that convective heat transfer in the VGF growth of GaAs is negligible, because of the relatively low gas pressure (up to 3 atm) and the high thermal conductivity of the GaAs melt [37]. Radiative heat exchange is implemented by the well-established view-factor method (for details see [38]) considering multiple reflections as diffuse gray-body radiation with emissivity $0 < \epsilon \le 1$. The assignment of material parameters is facilitated by a user-defined materials data base.

In the calculation mode of the code we have two options: if the power densities of the heaters are given, the program calculates temperatures in all parts of the growth setup as a unique solution. This is mathematically called

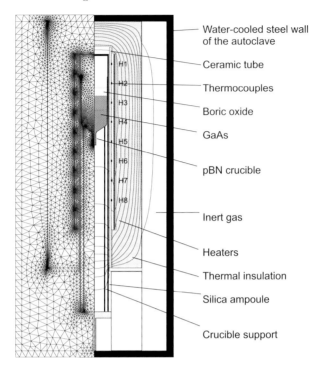

Water-cooled steel wall
of the autoclave

Ceramic tube

Thermocouples

Boric oxide

GaAs

pBN crucible

Inert gas

Heaters

Thermal insulation

Silica ampoule

Crucible support

Fig. 19. Numerical grid (*left*) for the setup shown on the *right-hand side* with all relevant parts for the growth of 2 in and 3 in GaAs single crystals by the vertical gradient freeze method. In the crystal – where stress has to be calculated – and in the heaters, a finer grid is used [39]

the 'normal' problem. However, usually of much higher interest in crystal growth is the adjustment of the heater power to achieve a desired temperature distribution. Thus, due to the objective of optimizing the growth process, the temperature field in the growing crystal has to fulfill several constraints, e.g. position and shape of the solid/melt interface or a certain temperature difference between seed and top of the melt. From the mathematical point of view it is much more difficult to solve this problem, which is called 'inverse modeling' or an 'inverse problem'. The program system CrysVUN++ contains a computation mode which allows us to calculate the set of heater powers in order to fulfill given thermal conditions in the crystal and melt with a minimum of deviation. The algorithm for this mathematical procedure is described elsewhere [33]. Its use for the VGF setup of Fig. 19 is demonstrated in Sect. 3.3. The analysis of thermoelastic stress in the crystal is important for predicting the quality of the grown GaAs crystal. A very useful scalar for the discussion of stress in solid bodies, especially for crystal growth, is the von-Mises stress σ_{vm}, which is computed from distinct components of the stress tensor $\sigma_{i,j}$ in cylindrical co-ordinates by

$$\sigma_{\mathrm{vm}} = \sqrt{\frac{(\sigma_{rr} - \sigma_{zz})^2 + (\sigma_{\varphi\varphi} - \sigma_{rr})^2 + (\sigma_{\varphi\varphi} - \sigma_{zz})^2 + 6\sigma_{rz}^2}{2}} \ . \tag{3}$$

The software CrysVUN++ determines the components of the stress tensor $\sigma_{i,j}$ from the calculated temperature field. The σ_{vm} stress is visualized in colored graphics and/or isolines of equal von-Mises stress. Further details and examples will be given in the next sections.

Even with the availability of good computer software it is necessary to develop a strategy to optimize the process model for a given task. For that purpose we use a concept which is illustrated in Fig. 20. In the first verification step the model is used to simulate an experimental setup where the crystal and melt are replaced by a solid graphite body (dummy) which contains sensors for temperature measurements. This step allows for an improvement

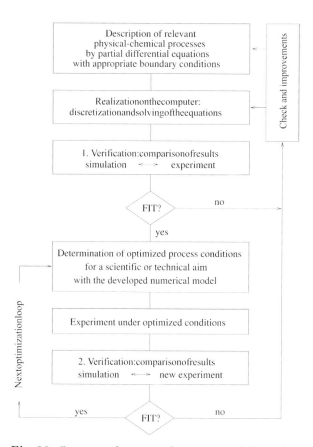

Fig. 20. Strategy of numerical process modeling: the starting point is the set of physical equations that determine the process to be simulated. After the first tests the model can be applied to improve crystal-growth processes as long as the physical model remains true and the discretization is well-suited to reflect its relevant properties on the computer

of the numerical model. After a second step the verification gives confidence for the computer simulation of the real VGF crystal-growth process.

In the following sections we will demonstrate how the numerical modeling was used to optimize VGF GaAs growth and VGF crystal-growth processes for GaAs.

3.2   Optimization of VGF-Growth Equipment

We have used the numerical model to develop VGF-growth equipment for GaAs crystals from diameters of 2 in up to 4 in. The 3 in setup was up-scaled from an earlier 2 in process [16]. The 4 in equipment is based on a totally new design mainly constructed from the results of numerical simulation [40]. The process model was used e.g. to analyse the effects of various insulating materials in order to estimate the required heater powers and to determine the geometry, i.e. the relative dimensions and positions of all important parts (heaters, crucibles, insulation, etc.).

In the following it is demonstrated by one example how a certain part of the furnace can be optimized by numerical modeling. We examine the region of the transition of the crystal shape from the conical to the cylindrical part. It is known that the thermal stress is strongly increased in this part of the crystal as the heat flux from the melt is redirected into the narrow seed channel.

One way to improve this situation may be the application of a certain material in the cone and channel region. If this issue were to be solved experimentally, the procedure would be very time consuming. Performing a real experiment would mean to select and to provide a specific material, to decide on its shaping, to install it in the furnace and to carry out growth runs. In the 'numerical experiment' one simply has to plot a block of space around the cone in the construction mode, assign an appropriate material to the new geometrical feature in the material mode and redo the calculation in the calculation mode. Figure 21 shows the results of such calculations where the influence of different kinds of material of the crucible support was studied by assigning four different thermal conductivities to the block of the material around the cone and the seed channel. The results for the von-Mises stress shown in Fig. 21 clearly demonstrate how efficiently the setup can be improved by computer simulation.

3.3   Optimization of Growth Runs by Inverse Modeling

One major disadvantage of VGF is that the semiconductor material is enclosed by the (nontransparent) crucible. As a result, the growth process cannot be directly observed. Furthermore, the temperature distribution in the crystal cannot be measured during growth. This is why numerical simulation is a powerful tool to analyse the crystal-growth process. The temperature

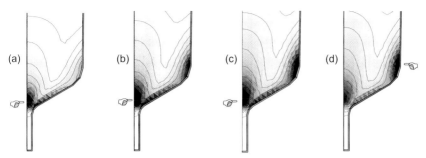

Fig. 21. Calculated distribution of thermal stress σ_{vm} (lines of equal von-Mises stress) in the conical part of the GaAs crystal for a VGF setup with different materials a–d (thermal conductivity) for the crucible support; σ_{max} is marked and has absolute values 3.57 MPa (**a**), 2.39 MPa (**b**), 1.89 MPa (**c**), 1.72 MPa (**d**), respectively. The distance between the lines of equal von-Mises stress is 0.32 MPa (**a**), 0.22 MPa (**b**), 0.17 MPa (**c**), 0.16 MPa (**d**), respectively

profile at each time step of the growth process which determines the movement of the solid–liquid interface and hence the thermal stress in the crystal, represents one of the most important sets of parameters to be controlled during the whole VGF process.

In this section we want to demonstrate that the inverse modeling can be used very well for finding an optimized thermal process for the VGF growth of GaAs crystals with the setup shown in Fig. 19. We use our numerical software system with the inverse modeling algorithm to maintain the controlling conditions during the VGF-growth run. The applied temperature conditions are given in Fig. 22 as follows: The position of the solid-liquid interface is fixed in the center of the crucible at a certain height by the condition $\vartheta_1 = 1511$ K. The condition of a fixed temperature gradient 2 K/cm in the melt in order to avoid instabilities of the interface is achieved by $\vartheta_2 = \vartheta_1 + 2$ K with an axial distance of 1 cm above the interface. The condition of a planar crystal–melt interface is simulated by setting the temperature $\vartheta_3 = \vartheta_1 = 1511$ K in a mesh point at the same height but on the periphery. Furthermore, we choose a condition to prevent the overheating of the GaAs melt. For that purpose we use a temperature condition ϑ_4 in the axial position at the top of the melt. We want to avoid that the temperature T at the top exceeds 1526 K, that is 15 K above the melting temperature of GaAs in order to avoid a great loss of arsenic. This condition is defined by $\vartheta_4 = \vartheta_1 + \min\{2 \text{K/cm } d_{14}, 15 \text{ K}\}$, where d_{14} denotes the distance between the two controlling nodes for ϑ_1 and ϑ_4. This condition means that for small distances d_{14} the melt overheating fits the thermal gradient fixed by ϑ_1 and ϑ_2, while for larger distances the melt overheating does not exceed 15 K.

The strategy of 'shifting' the control vertices through the crucible gives the heating power in the 8 heaters for each simulated position of the interface. This distribution of heating power is shown in Fig. 23. The symbols are the

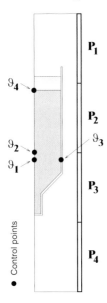

Fig. 22. Controlling conditions (fixed temperatures $\vartheta_1 - \vartheta_4$) for an optimized VGF-growth process of GaAs in the setup in Fig. 21. For details see text

calculated results. In the growth experiment the interpolated lines are used as time-dependent heater powers. In the simulation of the seeding and the final state, the growth velocity was set to $v_{\mathrm{growth}} = 0$. In all other states $v_{\mathrm{growth}} = 2.5\,\mathrm{mm/h}$ was applied. So the liberation of latent heat is taken into account in the quasi-steady-state approximation.

The calculations were performed with CrysVUN++ on a mesh with 5427 nodes covering the whole VGF setup. One solution of the inverse problem took less than three hours on a Pentium Pro 200 machine with 128 MB of RAM. Thus the whole simulation time was less than 50 h.

For a discussion of the results we will extract the temperature distributions in the GaAs (crystal and melt) which are obtained as a part of the inverse-modeling procedure (compare [33]). These temperature distributions are in fact identical to the results of a direct modeling with the heater powers of Fig. 23 as input data for a simulation of the VGF process. The direct simulation corresponds to a real crystal-growth experiment where the heater powers are given as input data.

First we extract the axial temperature profile along the center of the crucible from the calculated global temperature field. The result is plotted in Fig. 24 for the corresponding 15 investigated time steps of the growth run. These time steps correspond to certain equidistant positions of the growth interface on the axis of the crucible as the growth rate $v_{\mathrm{growth}} = 2.5\,\mathrm{mm/h}$ is constant. The different graphs represent the growth states which belong to the corresponding set of the heater powers $P_1 - P_8$ of the eight heaters $H_1 - H_8$ (Fig. 19), represented in Fig. 23 by one set of the 8 symbols for one fixed position of the solid–liquid interface at the furnace axis. Only for

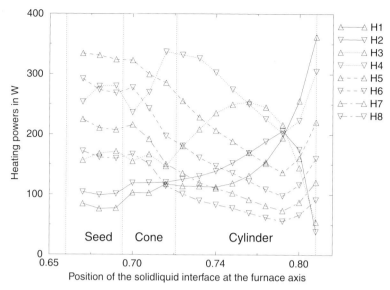

Fig. 23. Calculated heating powers $P_1 - P_8$ of the heaters $H_1 - H_8$ for 15 different positions of the growth interface under the conditions defined in Fig. 22. The controlling vertices were shifted by 1 cm per calculation. The resulting powers are plotted versus the position of the solid–liquid interface

the position of the seeding ($z = 0.665$ m) and for the tail end of the crystal ($z = 0.795$ m) we have set the growth velocity to $v_{\text{growth}} = 0$ mm/h.

One can clearly recognize the kink of the temperature profile at the position of the crystal–melt interface $T_{\text{melt}} = 1511$ K. The break of the slopes in the axial profiles on the interface corresponds to the Stefan condition

$$-\lambda_{\text{liquid}} \left(\frac{\partial T}{\partial z}\right)_{\text{liquid}} + \lambda_{\text{solid}} \left(\frac{\partial T}{\partial z}\right)_{\text{solid}} = \rho l v_{\text{growth}} \, , \qquad (4)$$

where λ denotes the different heat conductivities in the solid and the liquid whereas $\partial T/\partial z$ are the axial temperature gradients on the interface; ρ stands for the density and l is the latent heat.

The curve for the crystal tail end ($z = 0.795$ m) shows the end of the growth mode where we want to reduce the temperature gradients in the as-grown crystal to reduce the thermal stress during a following cooling procedure.

The results show clearly that important requirements for a good VGF process are fulfilled:

- The axial temperature gradients $\partial T/\partial z$ are nearly constant in the crystal in the region of the growth interface, which should provide conditions of low thermal stress (see later, Fig. 27).

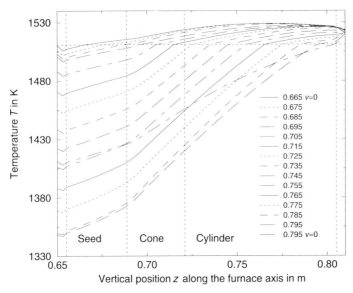

Fig. 24. Calculated axial temperature profiles in the center of the autoclave for different growth states, i.e. positions of the solid–liquid interface. The legend shows the positions of the interface at the crucible axis. The growth velocity is $v = 2.5 \, \text{mm/h}$, except for the seeding (position $z = 0.665 \, \text{m}$) and the end of the growth (position $z = 0.795 \, \text{m}$) where $v = 0 \, \text{mm/h}$ is applied. At the melting temperature of 1511 K we observe a kink in the temperature profile due to the different heat conductivities λ in the solid and liquid phases and due to the liberation of latent heat

- The temperature in the GaAs melt does not exceed the given limit of $T \leq 1526 \, \text{K}$.

Next, we analyse the shapes of the crystal–melt interfaces. They are plotted in Fig. 25 again for the different time steps considered corresponding to the set of heating powers depicted in Fig. 23.

Most of them have a planar shape, but deviations in the vicinity of the periphery of the crystal become apparent. This deviation can be explained by the strongly anisotropic conductivity λ of the pBN crucible

$$\lambda_{\parallel}^{\text{pBN}} = 62.7 \, \text{W/mK},$$

$$\lambda_{\perp}^{\text{pBN}} = 2.0 \, \text{W/mK},$$

$$\lambda_{\text{solid}}^{\text{GaAs}} = 7.1 \, \text{W/mK},$$

$$\lambda_{\text{liquid}}^{\text{GaAs}} = 17.8 \, \text{W/mK}.$$

This bad fitting of heat conductivities of pBN and GaAs always causes a strong bending of the heat flux in the vicinity of the crucible wall and hence a locally concave bending of the interface. This can clearly be seen from the enlarged temperature profile of this region depicted in Fig. 26. The strong

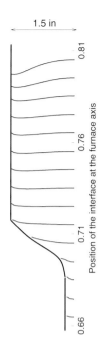

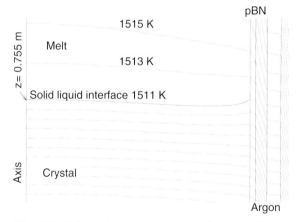

Fig. 25. Shapes of the solid–liquid interface for various positions during crystal growth calculated with the heating powers shown in Fig. 23

Fig. 26. Enlarged temperature profile around the solid–liquid interface and the crucible, for the position $z = 0.755\,\mathrm{m}$ of the interface. The isotherms differ by $u\Delta T = 2\,\mathrm{K}$

convex bowing of the solid–liquid interface at the tail end of the crystal (top of Fig. 25) can be explained by the fact that the top portion of the melt is heated from the side by the top heaters $(H_1 - H_3)$.

These deviations from the controlling conditions which we have prescribed by the temperature condition $\vartheta_1 = \vartheta_2$ (compare Fig. 22) show that the calculation cannot fulfill physically unreasonable conditions. But it gives a physically possible solution that comes closest to it – in the meaning of a minimum criterion (for details, see [33]).

Nevertheless one can demonstrate with a calculation of the thermal stress in the GaAs crystal during the VGF-growth run that the inverse modeling provides optimized process conditions. In Fig. 27 we show the distribution of the thermal stress in terms of the von-Mises criterion σ_{vm} (see Sect. 3.1), for three different growth stages. The values of $\sigma_{vm} = 1.2\,\text{MPa}$ are the best ones that have been achieved after lots of calculations and empirical optimizations. Especially the value of $\sigma_{vm} = 0.8\,\text{MPa}$ for the solid–liquid interface in the conical region ($z = 0.705\,\text{m}$) is quite remarkable.

Now we will demonstrate by two examples of crystal growth how such kinds of process optimizations can reduce the dislocation density of 3 in GaAs crystals considerably and improve VGF process efficiency (reduction of process time). In Table 4 the mean Etch-Pit Density (EPD) of investigated wafers cut at two positions of three VGF-grown crystals are shown. A reduction of the etch-pit densities by an order of magnitude was achieved by optimizing the thermal process.

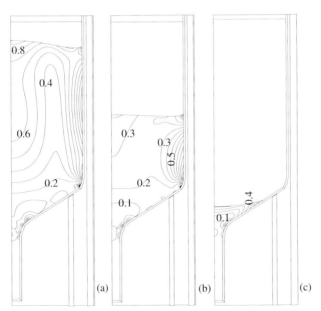

Fig. 27. Lines of equal von-Mises stress in MPa for three representative growth stages with the interface at the axial positions of $0.705\,\text{m}$ **(a)**, $0.775\,\text{m}$ **(b)** and $0.795\,\text{m}$ **(c)**, respectively. See also Fig. 25 corresponding to the calculated temperature distributions of Fig. 24

Table 4. Reduction of EPDs in VGF growth of 3 in GaAs crystals by means of numerical process simulation

	Before optimization Crystal 1	After optimization Crystal 2	Crystal 3
Seed end	2400	2000	570
Tail	4200	430	240

In Fig. 28 we compare for demonstration the EPDs of two wafers grown in the same VGF furnace but with two different thermal processes before and after optimization (for more details on the dislocation density see Sect. 4.3).

The computer modeling can also be used successfully to achieve a constant growth rate at a desired value of e.g. 2.5 mm/h. In this case the validity of the computer calculations was cross-checked by a special growth experiment using interface marking by controlled crucible rotation at certain intervals during the VGF-growth run. To detect this marking the grown crystal was sliced along the growth axis. Figure 29 shows near-infrared transmission topographs of longitudinally cut (110)-oriented wafers of one half of this crystal. The measured rotational striations are visualized by dashed lines. For comparison, the calculated interface shapes and positions are depicted by solid lines. As can be seen, the bending of the striations corresponds very well to the calculated shapes of the solid–liquid interfaces at equal axial positions. A nearly flat interface in the crystal center and a slightly concave one at the rim were achieved. This bending is caused by the thermal conductivity behavior of the pBN crucible.

The axial positions of the interface for equal temperature profiles at the heaters differ prominently, especially in the seed and conical regions: the simulated position '1' (solid line 1) is still in the seed well, whereas the experimentally observed corresponding position '1' (dashed line 1) is in the

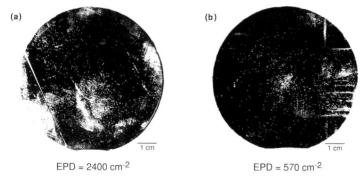

EPD = 2400 cm⁻²   EPD = 570 cm⁻²

Fig. 28. KOH etched {001} wafers of VGF-grown 3 in GaAs crystals and averaged Etch-Pit Densities (EPD) obtained before **(a)** and after **(b)** optimization by numerical modeling

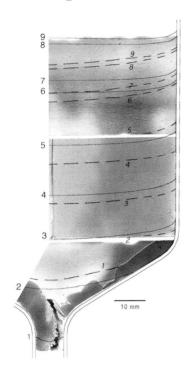

Fig. 29. Infrared-transmission topographs of longitudinally cut, (110)-
oriented wafers of one half of a 3 in GaAs
crystal with induced rotational striations, numbered 1 to 9 (*dashed lines*).
For comparison, the corresponding
calculated solid–liquid interface shapes
(*solid lines*) for the same temperatures
at the furnace heaters are also given
(numbered 1 to 9, respectively)

cone. The difference between these simulated and experimental solid–liquid
interface positions is about 2 cm. For interface position '2', the difference
is about 1.5 cm. This difference decreases systematically and disappears between positions 5 and 6. At the seed end the positions differ again slightly
but in the opposite direction. The comparison between calculated and experimentally observed solid–liquid interface positions (taken at the crystal
center) is given in Fig. 30. It is remarkable that, in spite of the slight differences to the numerical results especially in the seed and cone regions, a
constant growth rate of 1.8 mm/h is achieved in the experiment, which is
somewhat smaller than the calculated one (2.5 mm/h). The correspondence
of the growth speed predicted by the numerical model and the experimental
growth rate is not perfect. However, the simulations yield reliable results for
the expected thermal stress in the GaAs crystal and enable the crystal grower
to make considerable progress towards improving the quality of the crystals,
especially by significantly reducing the dislocation density of the wafers as
desired for epitaxial applications.

4   Crystal and Wafer Properties

The GaAs crystals (Fig. 31), which were grown by a VGF process as described in the previous Sects. 2 and 3, were cut and polished at Freiberger

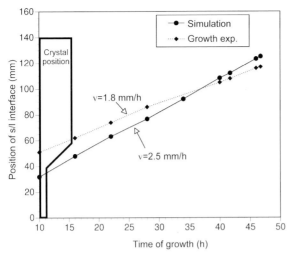

Fig. 30. Simulated and measured axial positions of the solid–liquid interface during growth of a 3 in Si-doped GaAs crystal by the VGF technique in an 8-zone furnace. The growth rates v are evaluated from the constant slopes

Fig. 31. Typical VGF-grown Si-doped GaAs crystals with diameters of 3 in (3.1 kg) and 2 in (1 kg), EPD $< 500\,\mathrm{cm}^{-2}$

Compound Materials (FCM). The wafering technology had to be especially optimized for the Si-doped low-EPD VGF material which behaves differently from undoped semi-insulating LEC-grown GaAs. Furthermore, the surface properties of the wafers had to be adjusted to the requirements of epitaxial processing technologies for high-power diode-lasers. A special 'epi-ready' surface finish was developed, so far for wafers with 2 in diameter.

4.1   Wafering

Important properties of the wafers are compiled in the list of specifications
(given in Table 5) which is presently in use as a standard for 2 in wafers. The
use of substrates with 3 in wafers for diode-laser fabrication is presently introduced; the corresponding list of standard specifications will be elaborated
in the near future.

Table 5. Specifications for 2 in GaAs substrates to be used for epitaxial processing
of diode-lasers

Dopant/Conduction Type	Si/n-Type
Carrier concentration	$(0.8 - 2.0)10^{18}/cm^3$
EPD	$\leq 500/cm^2$ (guaranteed area $\geq 85\%$)
Surface orientation	$(100) \pm 0.5°$ or
	$(100)\, 2°$ FO $< 101 >$, $\alpha = 135°$
Thickness	$375\,\mu m \pm 20\,\mu m$
Diameter	$50.8\,mm \pm 0.3\,mm$
Flats	according to SEMI standard M 9.3-89
Primary flat	$(0\overline{1}\overline{1}) \pm 0.5°$
Flat length	$15.9\,mm \pm 1.6\,mm$
Secondary flat	$(0\overline{1}1) \pm 5°$ (90° clockwise
	from primary flat)
Flat length	$8.0\,mm \pm 1.6\,mm$
Surface finish	front side: polished, epi-ready
	back side: as cut/etched
Warp	$\leq 10\,\mu m$
TTV	$\leq 8\,\mu m$
Packaging	individual fluoroware container
Remarks	wafer free of any mechanical stress
	and subsurface damage

Important crystal and wafer properties like doping, uniformity and dislocation density will be discussed in the following sections.

4.2   Electrical Characterization and Silicon Doping

The wafers are electrically characterized by measuring the Hall effect and
the resistivity according to the van-der-Pauw method. The Si concentration is
determined by Secondary Ion Mass Spectroscopy (SIMS), Atomic Absorption
Spectroscopy (AAS) and Glow Discharge Mass Spectrometry (GDMS). The
specified resistivity or carrier concentration is determined by averaging the
data across the wafers. These data can slightly differ from the local data
due to the segregation coefficient of Si and the curved solid–liquid interface
(Sect. 2). Typical results are shown in Fig. 32.

The electron mobility versus the carrier concentration of various Si-doped
GaAs VGF crystals at room temperature is plotted in Fig. 33. The solid

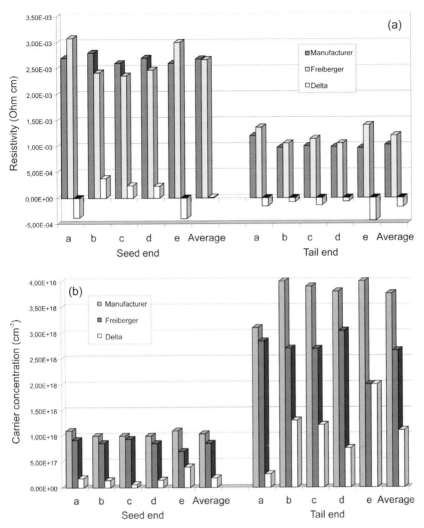

Fig. 32. Resistivity (**a**) and carrier-concentration (**b**) data detected by averaging across various wafers measured by the wafer manufacturer in comparison with data as specified by the crystal grower; delta means the difference

lines represent the theoretical mobilities for various compensation ratios $\Theta = N_A/N_D^+$ according to [41], with N_A^- and N_D^+ being the concentrations of ionized acceptors and donors. Si is an amphoteric impurity in GaAs and can substitute an As (Si_{As}) as well as a Ga (Si_{Ga}) atom. Si_{As} forms a shallow acceptor level with an activation energy $E_A = 25\,meV$ and Si_{Ga} a shallow donor level with $E_D = 5\,meV$ [42].

According to the literature [43,44] only 10–20% of the Si should be incorporated on As sites. This would lead to a compensation ratio of $\Theta \leq 0.25$.

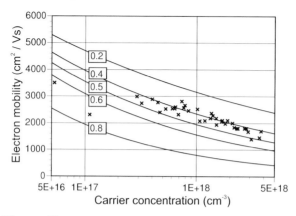

Fig. 33. Electron mobility and carrier concentration of various Si-doped GaAs VGF crystals at room temperature. The *solid lines* represent the theoretical mobilities for various compensation ratios $\Theta = N_A^-/N_D^+$ according to [41]

Figure 33 shows, however, that $0.35 < \Theta < 0.5$ is found for $n > 2 \times 10^{17}\,\mathrm{cm}^{-3}$. This can only be explained if we assume that one or more additional acceptors exist in addition to Si$_{As}$. Possible candidates could be Si complexes [45,46,47,48] and intrinsic anti-sites Ga$_{As}$ [49,50].

Comparing these results on VGF material with Horizontal Bridgman (HB) and Liquid-Encapsulated Czochralski (LEC) grown material yields the following: The boric oxide-free HB technique with As-vapor source provides material with the lowest $\Theta = 0.35$ [51]. VCZ-grown crystals reveal high mobilities and also a low compensation ratio $\Theta = 0.4$ [52]. If a SiO$_2$ crucible is used [45] in the VGF technique instead of boron nitride (pBN) the electron mobility is strongly reduced and $0.5 < \Theta < 0.6$ results. This is attributed to a more Ga-rich melt composition providing higher concentration of arsenic vacancies V_{As}. This effect can be reduced by incorporating an As-vapor source (Sect. 2).

In Fig. 34 it is shown that the electron concentration is about 0.63 times the Si concentration for both VCZ and VGF crystals grown from pBN crucibles, whereas SiO$_2$ crucibles give $n = 0.43$ [Si] for $n < 6 \times 10^{18}\,\mathrm{cm}^{-3}$. This means that about 60% of Si is incorporated as donor (Si$_{Ga}$), 10–20% as acceptor (Si$_{As}$) and the rest presumably as defect complexes.

4.3 Optical Characterization by Infrared and Photoluminescence Mapping

The homogeneity of the wafers is characterized by PhotoLuminescence (PL) topography at a near-band-gap wavelength. The PL intensity gives qualitative results on the nonuniformity of the Si doping, eventually caused by striations or by an asymmetry of the solid–liquid phase boundary and defect structures like dislocations and precipitations which are close to the wafer

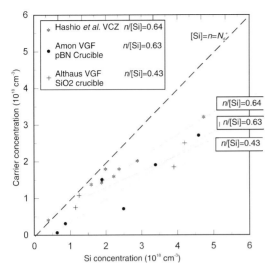

Fig. 34. Electron concentration n versus chemically determined (SIMS) Si concentration [Si] of GaAs crystals grown by VGF and VCZ for ratios $n/$[Si]; the dashed line $n =$ [Si]

surface. PL topograms can be obtained routinely by automated commercial equipment within a short time. Figure 35 shows a typical example with radial nonuniformity and increased PL intensity in the areas of increased dislocation density.

A quantitative evaluation of the PL intensity, however, with respect to electron concentration, Si concentration or dislocation density is not possible. Nevertheless the PL mapping is a fast nondestructive characterization method which is very useful, e.g., to evaluate the distribution of dislocations, as demonstrated in Fig. 36.

A quantitative analysis of the carrier distribution, i.e. Si distribution, is possible by infrared-transmission topography. In this method the IR absorption by the conduction electrons can be calibrated and used to show doping nonuniformity [24]. Results obtained by this method are exemplified in Sect. 2 (Fig. 10).

4.4   Residual Dislocations

4.4.1   Distribution of Dislocations Across Wafers

The residual dislocations of VGF-grown wafers can be characterized by wet chemical etching, which results in etch pits indicating the outcrop of dislocations on the wafer surface, by infrared microscopy and by X-ray topography. The first method is the standard technique for characterizing the quality

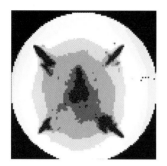

Fig. 35. Photoluminescence topogram of a 2 in GaAs:Si wafer with striations and increased PL intensity in the region of increased dislocation density

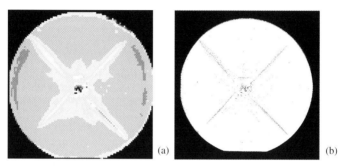

(a)   (b)

Fig. 36. PL mapping (**a**) in comparison with the distribution of the EPD (**b**) on the same wafer (2 in Si-doped VGF crystal from the seed end)

of wafers. For example GaAs LEC wafers are characterized by the ASTM F 1404-92 procedure which followed the earlier DIN 50454 standard of the year 1991. Both standards, however, are only useful for a uniform EPD $\geq$ $10\,000\,\mathrm{cm}^{-2}$ which is not the case in VGF wafers. Recently a revised standard (DIN 50454-1) was defined which considers the situation with low-EPD wafers.

In general the following procedure is used. The polished wafer surface is etched by dipping into a KOH melt at $400°\,\mathrm{C}$ for 4 to 10min. This etching creates pits with a diameter of about $70\,\mu\mathrm{m}$ around each dislocation line which penetrates the wafer surface. With the standard evaluation (like DIN 50454-1) the EPD is counted by using a representative spot checking on the wafer (see Fig. 37) and averaging the resulting EPD. The precision of this method increases with increasing ratio of characterized area to total area and dislocation density. For example, to achieve a precision of 10% by using testing areas around $1\,\mathrm{mm}^2$ it is necessary to count on 25% of the wafer surface for an EPD $< 1000\,\mathrm{cm}^{-2}$ and on only 10% of the wafer surface for an EPD $> 1000\,\mathrm{cm}^{-2}$.

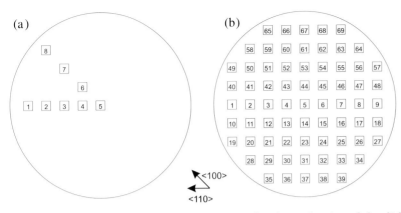

Fig. 37. Two examples of representative areas for the evaluation of the dislocation density; DIN 50454-1 **(a)** and a more precise method proposed by *Amon* [16] **(b)**

The optimum is, however, to count all etch pits, which is possible by an automated system developed at Freiberger Compound Materials. In this technique the wafer is illuminated in such a way that only one crystallographic plane of the etch pit reflects the light. This avoids the problem that any shallow pits or clusters give disturbing signals. The reflecting pits with their coordinates are detected by automated image analysis (Fig. 38).

Now the density and distribution of the dislocations can be evaluated by numerical programs, even for nonuniform distributions and very low dislocation densities (see e.g. Fig. 39).

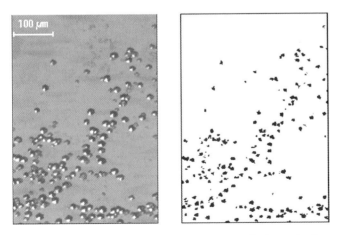

Fig. 38. Image of illuminated etch pits on a GaAs wafer by microscopy (*left*) and representation of the coordinates of the etch pits after image processing (*right*)

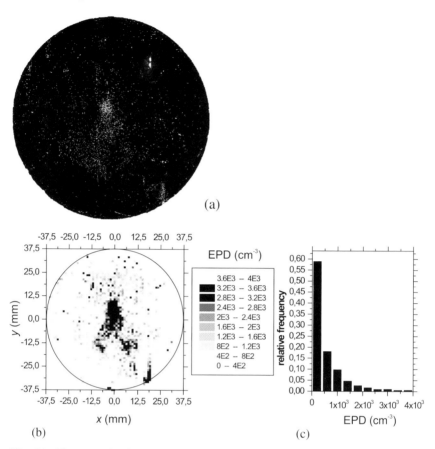

Fig. 39. Photograph of the etch pits on the illuminated wafer crystal (**a**), topogram of the etch pits (**b**) and frequency curve (**c**) of the EPD classes of the same wafer from a 3 in VGF crystal

4.4.2   Types of Residual Dislocations in VGF-Grown GaAs

It is known from the literature that Si doping can be used in LEC and Horizontal Gradient Freeze (HGF) growth to reduce the dislocation density by a lattice-hardening effect (Fig. 40). The VGF-grown crystals, however, show independently from the Si-doping level a low EPD ($< 500\,\mathrm{cm}^{-2}$) for the range $5 \times 10^{16}\,\mathrm{cm}^{-3} \leq n \leq 2 \times 10^{18}\,\mathrm{cm}^{-3}$. Obviously the thermal stress in an optimized VGF process is so small (Sect. 3) that the crystals can be grown with a small EPD without the necessity of lattice-hardening. The residual dislocations in VGF-grown GaAs are eventually not caused by multiplication according to the Haasen–Alexander model [53].

We have tried to analyse the character of these residual dislocations by the aid of DSL etching [56,57] and infrared microscopy. Furthermore, X-ray topography with synchrotron radiation (SRWBT) was carried out for several

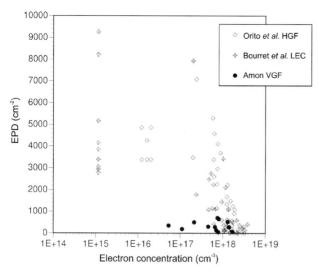

Fig. 40. Dislocation density (EPD) versus electron concentration of Si-doped GaAs (2 in) grown by HGF [54], by LEC [55] and by VGF [16]

crystal sections. By this method we could identify [58] three different types of straight dislocations which can be distinguished according to their line vector l and their Burgers vector b (see textbooks on X-ray topography, e.g. [59]), and a fourth type with a more wavy shape:

type (1): These dislocations have a line vector l which is nearly exactly parallel to the [001] growth direction. They predominantly occur in the center of the crystal. The type (1) is depicted in Fig. 41a at the left-hand side. This type is invisible for diffraction vector h parallel to l (004-reflection, Fig. 41). Consequently, the Burgers vector is perpendicular to l lying in the (001) plane. From that it can be concluded that type (1) is an edge dislocation. It was shown by near-infrared transmission microscopy that these dislocations can have a Burgers vector of <110> (more likely) or <100> type, which is consistent with the observed invisibility for h parallel to l.

type (2): This type consists of dislocations of the cross-shaped arrangement which is visible in Fig. 42. These dislocations appear on {001} wafers in [100] directions. The Burgers vector has not been determined but it can be shown by near-infrared transmission microscopy that these dislocations have a line vector which is only slightly tilted (6°–10°) against the [001] growth direction.

type (3): This type is also observed on (110) wafers (Fig. 41a,b) having line vectors $l = [\bar{1}12]$ and $[1\bar{1}2]$. For h parallel to l (in the $\bar{2}24$ or the $2\bar{2}4$ reflections, respectively) the dislocations are invisible (Fig. 41c). Therefore, this type is an edge dislocation as well, having a Burgers

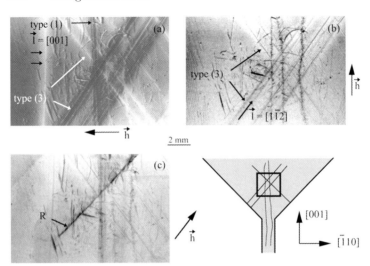

Fig. 41. White-beam topographs of a (110)-oriented wafer (thickness $t = 1.65$ mm), cut parallel to the growth direction of the conical region of a VGF-grown crystal (see sketch on the *lower-right side*) for different diffraction vectors $\boldsymbol{h}$. (a) $[\bar{2}20]$ reflection, $\lambda \approx 0.27$ Å($E \approx 45$ keV), $\mu t \approx 2.3$ (b) 004 reflection, $\lambda \approx 0.19$ Å($E \approx$ 65 keV), $\mu t \approx 0.9$ (c) $[\bar{2}24]$ reflection, $\lambda \approx 0.22$ Å($E \approx 56$ keV), $\mu t \approx 1.3$. The dark line denoted by 'R' in this image is caused by the crystal rim of an overlapping topograph.

vector of the <110> type. It originates at the rim of the crystal in the conical region at {111} facets.

type (4): A wavy network of dislocations was found which does not vanish for the used orientations of the diffraction vector $\boldsymbol{h}$ (symmetrical Laue cases) which can be seen for example in Fig. 43 in the VGF-grown crystal.

Another type of dislocation network, which we could not specify up to now, can be observed in the center of Fig. 41b: three frayed lines from the bottom to the top of the topograph.

The mean EPD of {001}-oriented wafers with 2 in diameter is found to depend on the length of the crystal portion which is grown in the seed well as illustrated by Fig. 42. Using a 'long' seed well (Fig. 42a) extremely low dislocation densities can be achieved. It is concluded that dislocations are suppressed in the small-diameter seed well in an effective way by growing out to the rim of the crystal. This argument does not hold for dislocations of type (1), because these dislocations grow nearly exactly in the growth direction. The observed cross-shaped arrangement of dislocations of type (2) can be suppressed by using a long seed well. As type (2) has a line vector which is slightly tilted against the [001] growth direction (by 6° to 10°) it also grows out to the rim of the crystal. Using a short seed well, these dislocations

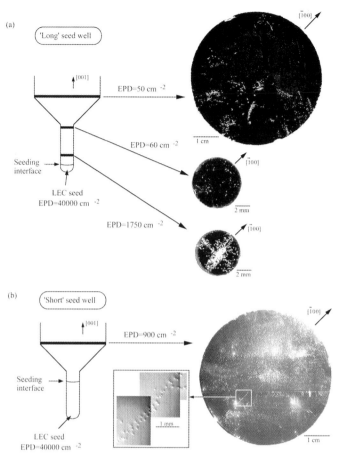

Fig. 42. KOH-etched {001} wafers cut perpendicular to the growth direction of VGF-grown crystals and corresponding EPDs for different positions of the seeding interface in the seed well: (a) 'long' seed well and (b) 'short' seed well

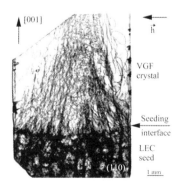

Fig. 43. White-beam topograph of a (110)-oriented section cut parallel to the growth direction intersecting the seeding interface of a VGF-grown crystal. Wafer thickness $t = 0.55$ mm, 220 reflection, wavelength $\lambda \approx 0.23$ Å(energy $E \approx 55$ keV), $\mu t \approx 1.4$ (μ–linear absorption coefficient)

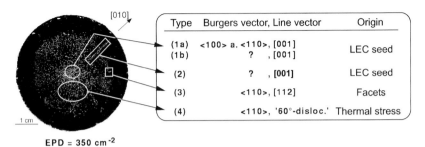

Type	Burgers vector, Line vector	Origin
(1a)	<100> a. <110>, [001]	LEC seed
(1b)	? , [001]	
(2)	? , [001]	LEC seed
(3)	<110>, [112]	Facets
(4)	<110>, '60°-disloc.'	Thermal stress

EPD = 350 cm⁻²

Fig. 44. Different types of dislocations in a 2 in VGF wafer (for the nomenclature see Fig. 35 and text)

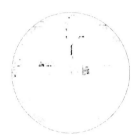

Fig. 45. KOH-etched wafer of a VGF-grown 4 in GaAs crystal with an averaged EPD of $138\,\mathrm{cm}^{-2}$

still appear on {001} wafers (insert in Fig. 42b). Therefore, it is concluded that these dislocations originate from the LEC-grown seed crystal, which was used in these growth experiments.

From the investigation of different crystals with 2 in diameter and dislocation densities between $50\,\mathrm{cm}^{-2}$ and $1000\,\mathrm{cm}^{-2}$ we can state the following (Fig. 44): for mean etch-pit densities of less than $200\,\mathrm{cm}^{-2}$ we found type (1) and type (2) to be dominant. For an increasing dislocation density, type (4) becomes more and more important because the absolute number of type (1) and (2) dislocations is nearly constant. Consequently, its relative contribution becomes smaller with increasing EPD. Type (4) dislocations are believed to be correlated to the thermal stress occurring during the crystal-growth process, because their formation can largely be avoided by minimizing thermal stress by the aid of numerical simulations (Sect. 3).

Finally it was even possible to grow GaAs crystals with a diameter of 4 in with very low EPD by using optimized VGF processing conditions (Fig. 45).

5   Conclusion

Numerical simulation is a valuable tool to develop and optimize crystal-growth equipment and a process like the vertical gradient freeze process for GaAs. It can be used to decrease the thermal stress during growth and hence the dislocation density to very low values. This allows the industrial production of low-EPD GaAs wafers, which is the prerequisite for a fabrication of high-power diode-lasers with a long lifetime.

The VGF processing itself provides an efficient and reproducible growth technique with relatively low investment for equipment. The potential of the VGF technique with respect to crystal diameter, length, uniformity and electronic properties, however, is far from being exhausted.

Acknowledgements

Important results were contributed to this article by Dr. Amon during his PhD work at the Erlangen Crystal Laboratory and by C. Hannig at the Institute of Nonferrous Metallurgy in Freiberg. The authors would also like to express their acknowledgement to Dr. Sonnenberg (Research Centre Jülich) for his valuable contributions to the research project and to D. Jockel (Materials Centre Göttingen). Further contributions came from Dr. Gärtner (TU Freiberg), B. Birkmann, Dr. M. Kurz, L. Kowalski, A. Pusztai, J. Stenzenberger (University Erlangen), J. Härtwig (ESRF), T. Bünger, Dr. M. Jurisch, A. Köhler, Dr. A. Kleinwechter, and U. Kretzer (Freiberger Compound Materials).

List of Symbols

c	Concentration
n	Carrier concentration in a crystal or melt
T	Temperature
T_m	Melting temperature
$p_{GaAs,T}$	As partial pressure of a GaAs melt at temperature T
m	Mass of solid As
V	Free volume of an ampoule
r	Mean number of As atoms of the dominating As molecules
M	Molecular weight of As
T_{av}	Mean temperature of the relevant free volume of the ampoule during growth
g	Solidified fraction of a GaAs ingot
c_s	Dopant concentration in solid GaAs
c_l	Dopant concentration in liquid GaAs
c_0	Starting concentration of a dopant in the GaAs melt
Si_{Ga}	Si atom on a Ga site in the GaAs lattice
Si_{As}	Si atom on an As site in the GaAs lattice
k	Segregation coefficient
$k_{B/Si}$	Reaction constant
a_c	Carbon activity in the chemical system
σ_{vm}	von-Mises stress
ϑ	Temperature

$\sigma_{i,j}$	Stress tensor
Θ	Compensation ratio of carriers
P	Heater power
v_{growth}	Growth rate of a crystal
λ	Heat conductivity
ϱ	Density
$\lambda_\parallel^{pBN}$	Heat conductivity of pBN parallel to its surface
$\lambda_\perp^{pBN}$	Heat conductivity of pBN perpendicular to its surface
$\boldsymbol{l}$	Line vector of a dislocation
$\boldsymbol{b}$	Burgers vector of a dislocation
$\boldsymbol{h}$	X-ray diffraction vector
μ	Absorption coefficient of X-rays
t	Thickness of a wafer
λ	wavelength of X-rays
N_A^-	Concentration of ionized acceptors
N_D^+	Concentration of ionized donors
E_A	Ionization energy of acceptors
E_D	Ionization energy of donors
V_{As}	Vacancy on an As site in the GaAs lattice

List of Acronyms

EPD	Etch-Pit Density
VB	Vertical Bridgman
VGF	Vertical Gradient Freeze
CrysVUN++	Crystal Growth Simulation by the Finite-Volume Technique on Unstructured Grids
LEC	Liquid-Encapsulated Czochralski
VLEC	Vapor-Pressure-Controlled Czochralski
VCZ	Vapor-Pressure-Controlled Czochralski
HWCZ	Hot-Wall Czochralski
MLEC	Magnetic LEC
RF	Radio Frequency
HB	Horizontal Bridgman
HGF	Horizontal Gradient Freeze
pBN	Pyrolytic Boron Nitride
FE-VB	Fully Encapsulated Vertical Bridgman
CAD	Computer-Aided Design
DXF	Drawing Exchange Format
SEMI	Semiconductor Equipment and Materials International
SIMS	Secondary Ion Mass Spectroscopy
AAS	Atomic Absorption Spectroscopy
GDMS	Glow Discharge Mass Spectroscopy

ASTM	American Society for Testing and Materials
SRWBT	Synchrotron Radiation White-Beam Topography
MB	MegaByte
RAM	Random access memory
PL	Photo luminescence
IR	Infrared

References

1. D. T. J. Hurle: *Crystal Pulling from the Melt* (Springer, Berlin, Heidelberg 1993)
2. J. Völkl: "Stress in the Cooling Crystal" in D. T. J. Hurle (ed.): *Handbook of Crystal Growth*, Vol 2 (Elsevier Science, Amsterdam 1996) p. 821
3. M. Tatsumi, T. Kawase, Y. Iguchi, K. Fujita, M. Yamada: in Proc. 8th Conf. Semi-insulating III–V Materials, Warsaw, PL, Tech. Dig. (1994) p. 11
4. P. Rudolph: *Profilzüchtung von Einkristallen* (Akademie-Verlag, Berlin 1982)
5. G. Tamman: *Lehrbuch der Metallographie* (Voss, Leipzig 1914)
6. P. W. Bridgman: Proc. Am. Acad. Sci. **60**, 305 (1925)
7. D. C. Stockbarger: Proc. Am. Acad. Sci. **60**, 133 (1925)
8. W. A. Gault, E. M. Monberg, J. E. Clemans: J. Cryst. Growth **74**, 149 (1986)
9. E. Monberg: "Bridgman and Related Growth Techniques" in D. T. J. Hurle (ed.): *Handbook of Crystal Growth*, Vol 2 (Elsevier Science, Amsterdam 1996) p. 51
10. G. Müller: in *Crystal Growth from the Melt*, Crystals **12** (Springer, Berlin, Heidelberg 1988)
11. M. Althaus: Forschungsbericht Jül-3252 (Forschungszentrum Jülich, Jülich 1996)
12. H. Wenzl, W. A. Oates, K. Mika: "Defect Thermodynamics and Phase Diagrams in Compound Crystal Growth Processes" in D. T. J. Hurle (ed.): *Handbook of Crystal Growth*, Vol 1A (Elsevier Science, Amsterdam 1993) p. 103
13. R. Karl: Chemisch-thermodynamische Untersuchung des Systems GaAs/B$_2$O$_3$/Gasphase; ein Beitrag zur Prozeßchemie des LEC-Verfahrens, Dissertation RWTH Aachen (1996)
14. S. Hegewald, K. Klein, C. Frank, M. John, E. Buhrig: Cryst. Res. Technol. **4**, 567 (1994)
15. W. G. Pfann: J. Metals **4**, 747 (1952)
16. J. Amon: Züchtung von versetzungsarmem, Silicium-dotiertem GaAs mit dem vertikalen Gradient-Freeze-Verfahren, Dissertation, University Erlangen-Nuremberg, Germany (1998)
17. A. G. Elliot, C. L. Wei, R. Farraro, G. Woolhouse, M. Scott, R. Hiskes: J. Cryst. Growth **70**, 169–178 (1984)
18. A. Flat: J. Cryst. Growth **109**, 224 (1991)
19. J. B. Mullin, A. Royle, S. Benn: J. Cryst. Growth **50**, 625–637 (1980)
20. R. Fornari: J. Cryst. Growth **94**, 433 (1989)
21. R. K. Willardson, W. P. Allred: in J. Franks (ed.): Inst. Phys. Conf. Ser. **3** (Institute of Physics, London 1967) p. 35
22. O. V. Pelevin, I. N. Shershakova, F. A. Gimel'farb, M. G. Mil'vidskii, T. A. Ukhorskaya: Soviet-Phys. Cryst. **16**, 528 (1971)

23. P. D. Greene: J. Cryst. Growth **50**, 612 (1980)
24. G. Gärtner, C. Hannig, G. Schwichtenberg, E. Buhrig: Züchtung und Charakterisierung von Si-dotierten GaAs-Einkristallen nach dem VGF-Verfahren, 26. DGKK Jahrestagung, March 1996, Cologne, Germany
25. G. Frigerio, C. Mucchino: J. Cryst. Growth **99**, 685 (1990)
26. ChemSage Handbook (GTT Technologies, Herzogenrath 1998)
27. W. A. Oates, H. Wenzl: J. Cryst. Growth **191**, 303 (1998)
28. D. Hofmann,. T. Jung, G. Müller: J. Cryst. Growth **128**, 213 (1993)
29. H. Sauermann, P. Deus, H. Krause, P. Metzing: Model calculation of tubular multizone furnace, Syst. Anal. Model. Sim. J. **13**, 275–300 (1993)
30. J. Amon, F. Dumke, G. Müller: J. Cryst. Growth **187**, 1 (1998)
31. D. T. J. Hurle: J. Cryst. Growth **147**, 239 (1995)
32. G. Müller: "Melt Growth of Semiconductors" in R. Fornari, C. Paorici (ed.): Proc. 10th Int. Summer School on Crystal Growth (Trans Tech, Aedermannsdorf, Switzerland 1998)
33. G. Müller: "Numerical Simulation of Crystal Growth Processes" in M. Griebel, C. Zenger (eds.): *Numerical Simulation in Science and Engineering* Notes Fluid Mech. **48**, 130–141 (Vieweg, Braunschweig 1994)
34. M. Kurz, G. Müller: J. Cryst. Growth **208**, 341 (2000)
35. M. Kurz, A. Pusztai, G. Müller: J. Cryst. Growth **198/199**, 101 (1998)
36. M. Kurz: Development of CrysVUN++, a Software System for Numerical Modeling and Control of Industrial Crystal Growth Processes, Dissertation, University Erlangen-Nuremberg, Germany (1998)
37. A. S. Jordan: J. Cryst. Growth **71**, 551 (1985)
38. F. Dupret, P. Nicodème, Y. Ryckmans, P. Wouters, M. J. Crochet: Int. J. Heat Mass Transfer **33**, 849–871 (1990)
39. J. Amon, P. Berwian, G. Müller: J. Cryst. Growth **198/199**, 361 (1998)
40. G. Müller, J. Stenzenberger: "Growth of GaAs-Substrates with low Dislocation Density by the Vertical Gradient Freeze Method" in P. Kiesel, S. Malzer, T. Marek (eds.): *Proc. Symp. Non-Stoichiometric III-V Compounds* (Verlag Inst. für Mikrocharakterisierung, University Erlangen-Nuremberg, Erlangen 1998) p. 127–134
41. W. Walukiewicz, L. Lagowski, L. Jastrzebski, M. Lichtensteiger, H. C. Gatos: J. Appl. Phys. **50**, 899–908 (1979)
42. M. Seifert, H. Weiß: in *Werkstoffe der Halbleitertechnik* (VEB Grundstoffindustrie, Leipzig 1990) pp. 284
43. K. Laithwaite, R. C. Newman, J. F. Angress, G. A. Gledhill: Inst. Phys. Conf. Ser. **33a**, 133–140 (1977)
44. M. R. Brozel, J. B. Clegg, R. C. Newman: J. Phys. D **11**, 1331–1339 (1978)
45. R. Fornari: Cryst. Res. Technol. **24**, 767–772 (1989)
46. J. K. Kung, W. G. Spitzer: J. Appl. Phys. **45**, 2254 (1974)
47. P. W. Yu, D. C. Reynolds: J. Appl. Phys. **53**, 1263 (1982)
48. C. Frank, K. Hein, C. Hannig, G. Gärtner: Cryst. Res. Technol. **31**, 753 (1996)
49. B. Pödör: J. Appl. Phys. **55**, 3603–3604 (1984)
50. K. R. Elliot: Appl. Phys. Lett. **42**, 274–276 (1983)
51. H. Boudriot, W. Siegel, K. Deus, E. Buhrig: Solid-State Commun. **89**, 889–891 (1994)
52. K. Hashio, S. Sawada, M. Tatsumi, K. Fujita, S. Akai: J. Cryst. Growth **173**, 33 (1997)

53. J. Völkl, G. Müller: J. Cryst. Growth **97,** 136 (1989)
54. F. Orito, H. Okada, M. Nakajima, T. Fukada, T. Kajimura: J. Electron. Mater. **15,** 87 (1986)
55. E. D. Bourret, M. G. Tabache, J. W. Beeman, A. G. Elliot, M. Scott: J. Cryst. Growth **85,** 275 (1987)
56. J. Weyher, J. Van de Ven: J. Cryst. Growth **63,** 285 (1983)
57. J. Weyher, W. J. P. Van Enckevort: J. Cryst. Growth **63,** 292 (1983)
58. J. Amon, J. Härtwig, W. Ludwig, G. Müller: J. Cryst. Growth **198/199,** 367 (1999)
59. B. K. Tanner: *X-Ray Diffraction Topography* (Pergamon, Oxford 1976)

High-Power Broad-Area Diode Lasers and Laser Bars

Götz Erbert[1], Arthur Bärwolff[2], Jürgen Sebastian[1], and Jens Tomm[2]

[1] Ferdinand-Braun-Institut für Höchstfrequenztechnik Berlin,
   Albert-Einstein-Straße 11, D-12489 Berlin, Germany
   erbert@fbh-berlin.de
[2] Max-Born-Institut für Nichtlineare Optik und Kurzzeitspektroskopie,
   Max-Born-Straße 2A, D-12489 Berlin, Germany
   baerwolff@mbi-berlin.de

Abstract. This review presents the basic ideas and some examples of the chip technology of high-power diode lasers ($\lambda = 650\,\mathrm{nm} - 1060\,\mathrm{nm}$) in connection with the achievements of mounted single-stripe emitters in recent years.

In the first section the optimization of the epitaxial layer structure for a low facet load and high conversion efficiency is discussed. The so-called broadened waveguide Large Optical Cavity (LOC) concept is described and also some advantages and disadvantages of Al-free material. The next section deals with the processing steps of epitaxial wafers to make single emitters and bars. Several possibilities to realize contact windows (implantation, insulators, and wet chemical oxidation) and laser mirrors are presented. The impact of heating in the CW regime and some aspects of reliability are the following topics. The calculation of thermal distributions in diode lasers, which shows the need for sophisticated mounting, will be given. In the last part the current state-of-the-art of single-stripe emitters will be reviewed.

High-power diode lasers are on the point of pushing laser technology forward to new states-of-the-art in modern industry. In the 1980s, low-power diode lasers were developed for communication and data-storage purposes. The next step in exploring semiconductor laser technology was the development of high-power diode lasers as key devices for laser systems used e.g. in materials processing and in medical applications. The breakthrough came in the early 1990s when the power and lifetime of $1\,\mathrm{cm} \times 0.06\,\mathrm{cm}$ laser bars reached more than 20 W and 10 000 h, respectively, opening up potential for volume application. The progress was achieved mainly by substantial improvements of the GaAs crystal-growth technology, supplying substrates of low defect density as well as in the epitaxial growth process of AlGaInAs and GaInAsP layers on GaAs. The advantage of laser systems based on high-power diode lasers as compared to lamp pumping or gas discharge for generating coherent light, are the smaller size, the higher electrical-to-optical conversion efficiency, the simpler power and cooling supply, and the higher reliability. First, high-power diode lasers are likely to replace the lamps in solid-state laser systems. Second, for applications, which do not require high optical power densities, many high-power diode lasers can be easily assembled to get the desired

R. Diehl (Ed.): High-Power Diode Lasers, Topics Appl. Phys. **78**, 173–223 (2000)
© Springer-Verlag Berlin Heidelberg 2000

output power for direct-diode application. Third, in applications demanding higher beam quality, replacement of solid-state laser systems by sophisticated high-power diode-laser stacks is envisaged. Prerequisites for the latter are lasers with a much higher spatial beam quality, eventually using wavelength multiplexing for combining many diode lasers to form a single beam. In this review we describe the main aspects of physics and technology of high-power diode-laser realization preferably for the first and second modes of application cited above.

Opportunities for enhanced high-power diode-laser employment are deduced from the following discussion :

- The electro-optical conversion efficiency of lamps is in the range around 50% [1] providing a first benchmark for high-power diode lasers. Currently, the conversion efficiency of HPDLs surpasses 60% [2].
- The typical lifetime of lamps is around 1000 h. Optical power from lamps is relatively cheap. To become competitive the lifetime of high-power diode lasers should be at least an order of magnitude longer.
- The optical output power of lamps for high-power diode-laser systems is in the range around several kilowatts. To achieve a similar power with diode lasers, about 100 high-power bars have to be provided. Using the rack-and-stack-technology such diode laser 'lamps' have been realized successfully [3]. Further research is directed towards increasing the output power per bar, thus reducing the effort in mounting technology and hence the cost per watt of optical diode-laser power. The near-future objective is to achieve a reliable 100 W from one diode-laser bar.
- For pumping the laser crystal of a solid-state laser system the wavelength of the diode lasers has to meet exactly the absorption wavelength of the crystal to be excited (e.g., for a Nd:YAG crystal $\Delta\lambda \leq 3\,\mathrm{nm}$ around 808 nm is typical). Wavelengths emitted by lamps are determined by the electronic transitions of the excited atoms. These wavelengths depend only on the gas filling and have, by nature, fixed values which are very reproducible. Therefore, their match to the optical transitions in the laser crystal are either good or bad. In practical systems only part of the broad lamp spectrum is absorbed for pumping, the rest heats up the crystal and degrades the solid-state performance. The wavelength of a diode laser is determined essentially by the thickness and the composition of the Quantum Well(s) (QWs) forming the active zone. Apart from a slight wavelength shift which can achieved by temperature and resonator quality, the requirements for reproducible epilayer growth are very stringent.
- A competitive issue is the price of the whole laser system in operation. For systems based on high-power diode lasers this is mainly determined by the costs of the diode lasers. Hence, efforts have to be focused on a reproducible semiconductor technology which should be as simple as possible.

Here, we will discuss the current developments of specific epitaxial waveguide structures which are used for high-power diode lasers. Thereafter the typical processing technology of Broad-Area (BA) stripe-diode lasers and laser bars will be presented. This is followed by treating the issues of operation in the CW regime and the thermal behavior, especially for single-stripe diode lasers which recently became commercially available.

1 Epitaxial Waveguide Structures for High-Power Diode Lasers

1.1 The Large Optical Cavity (LOC) Concept

The layer structure and the dimensions of a typical BA edge-emitting diode laser conceived for a high optical output power, together with the shape of the emitted beam, are shown in Fig. 1.

The vertical waveguide consists of a core having a refractive index higher than that of the cladding layers. In the core forming the active region, typically a strained quantum well is embedded. An optimized structure has to deal essentially with the following, partially conflicting, requirements:

- a high confinement factor to reduce the threshold current,
- a large near-field width to reduce the facet load,
- a small near-field width to reduce the overall thickness of the epitaxial layers as well as the thermal and series resistances,
- a large near-field width to achieve a small vertical divergence,
- low scattering losses,
- low losses of free-carrier absorption,
- high doping level to reduce the series resistance,
- high barriers for the carriers to improve the electrical confinement and therefore the internal efficiency and the thermal stability,
- small barriers between the different layers to achieve a low voltage across the diode.

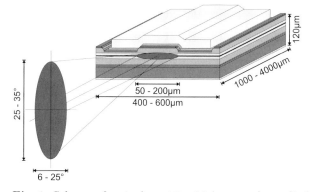

Fig. 1. Scheme of a single-emitter high-power laser diode with typical dimensions of the chip and the aperture

As can be deduced from the conflicting demands, a careful design of the entire diode structure is required with emphasis on the width of the optical mode and the doping profile. Trade-offs are unavoidable and depend, above all, on the desired output power per stripe width. Figure 2 shows different types of waveguides. A typical structure as drawn in the center yields a minimal threshold-current density. It has a relatively large refractive-index difference between cladding and core layers. The thickness of the core is optimized for a large confinement factor. However, the divergence is large and the intensity distribution shown exhibits a high peak resulting in a high facet load which lowers the maximal output power. As reduction of the threshold current of high-power diode lasers is not a predominant issue, a search for alternative waveguide structures holds some promise. Two examples are shown on the left-hand and on the right-hand sides of Fig. 2. Both structures represents a principal approach to get a larger width of the optical mode. One is to make the waveguide laser very thin by which the optical confinement stays relatively high but the optical field penetrates deeply into the cladding layers. The other is to increase the core width and to reduce the refractive-index difference between core and cladding.

The last approach discussed here, the so-called Large Optical Cavity (LOC) with broadened waveguide layers, allows a simple and good approximation of the near-field distribution to a Gaussian one [4] and results in improved diode-laser performance. In this case the optical confinement is smallest which is likewise true for the facet load. On the other hand the field distribution is smoothed and the energy being transported in the cladding layers is very small. Therefore, the cladding layers can be highly doped and made relatively thin, both measures leading to small series and thermal resistances. The disadvantage of a lower modal gain is compensated by the very low loss of such waveguide structures. The low loss allows us to make long resonator lengths in the range around 2 mm maintaining a high external efficiency. The vertical divergence is determined by the difference of refractive indices between waveguide and cladding layer.

There are many possibilities to realize a waveguide depending on material composition and thickness of the epitaxial layers of the waveguide and of the cladding layers, as well. In the following, two examples for the broadened-waveguide concept will be given, one based on an AlGaAs structure used for the wavelength region around 800 nm, the other based on Al-free layer structures for diode lasers at longer wavelengths above 900 nm. Some other approaches, especially for the short-wavelength regions will be described in the final part of this section.

1.1.1   AlGaAs-Waveguide Structures for Wavelengths Around 800 nm

The history of high-power diode lasers starts with the challenge to pump Nd:YAG solid-state lasers. The most important absorption band in Nd:YAG

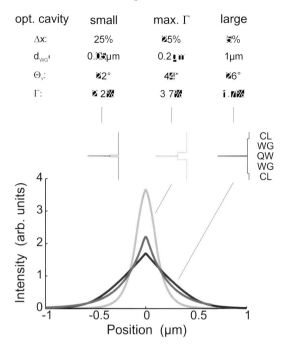

opt. cavity	small	max. Γ	large
Δx:	25%	25%	5%
d_{WG}:	0.05μm	0.2μm	1μm
Θ_v:	22°	45°	26°
Γ:	2.2%	3.7%	1.7%

Fig. 2. Vertical waveguide structures and calculated near-field intensity distribution for diode lasers emitting at $\lambda = 810$ nm. The calculation is done for $Al_{0.7}Ga_{0.3}As$ cladding layers and $Al_{0.45}Ga_{0.55}As$ waveguide layers in the *middle* and on the *left-hand side*; on the *right-hand side* the composition of the waveguide layer is $Al_{0.65}Ga_{0.35}As$

is at 808 nm which perfectly matches the technological possibilities to realize diode lasers with the established AlGaAs system. Until now the market for high-power diode lasers was biggest at this wavelength. Not only for high-power solid-state laser systems in materials processing but also for small low-power systems used in measurement applications, lots of diode lasers emitting around 808 nm are needed.

The first structures applying the LOC concept with broadened waveguides for this wavelength were presented by *Garbusov* et al. [5]. They used a 1.2 μm-thick waveguide core consisting of $Al_{0.4}Ga_{0.6}As$ and a cladding layer with a composition of $Al_{0.6}Ga_{0.4}As$. A version with a higher content of AlAs in the waveguide and cladding layers but a smaller difference in the AlAs content between core and cladding is shown in Fig. 3 [6]. This waveguide configuration leads to a broader near-field distribution which reduces the facet load. A further advantage is the smaller vertical divergence which is desired for fiber coupling and beam collimation.

In Fig. 4 the far-field intensity distribution is shown. The properties of the diode lasers are determined by the embedded quantum well.

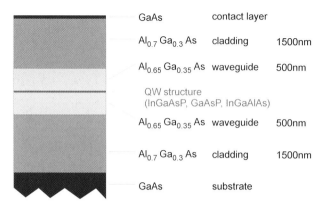

GaAs   contact layer

$Al_{0.7}Ga_{0.3}As$ cladding   1500nm

$Al_{0.65}Ga_{0.35}As$ waveguide   500nm

QW structure
(InGaAsP, GaAsP, InGaAlAs)

$Al_{0.65}Ga_{0.35}As$ waveguide   500nm

$Al_{0.7}Ga_{0.3}As$ cladding   1500nm

GaAs   substrate

Fig. 3. Scheme of the vertical layer sequence of a diode laser with an AlGaAs LOC broadened waveguide-structure

In Table 1 experimental results are listed for different types of quantum wells ranging from unstrained GaInAsP to tensile-strained GaAsP and compressively strained GaInAsP or AlGaInAs [7]. The gain coefficients, internal loss, internal efficiency and the transparency current density are determined by the length-dependence of the threshold-current density and the external differential efficiency. This method assuming a logarithmic dependence of the gain/current relation for a quantum-well active region was introduced by *McIlroy* et al. [8]. To demonstrate the method, the threshold-current density

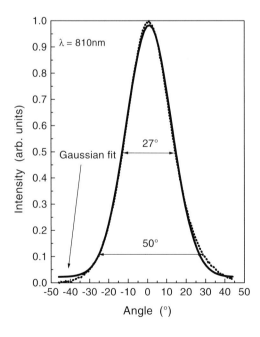

Fig. 4. Experimental far-field distribution of a AlGaAs diode laser with a LOC broadened-waveguide structure, Gaussian fit, $\lambda = 810\,\mathrm{nm}$

Table 1. Characteristic data for LOC-AlGaAs diode-laser structures (Fig. 3) with QWs for 810 nm (pulsed measurement, uncoated samples, 100 μm stripe width)

Comp.-QW [a]	N [b]	d-QW [c] (nm)	Strain	j_T [d] (A/cm²)	η_i [e] (%)	ΓG_0 [f] (cm⁻¹⁹)	I_{th} [g] (mA)	η_d [h] (%)	$\Theta_\perp$ [i] (°)
InGaAsP	1	18	-	222	80	23	362	73	25
InGaAsP	1	12	-	144	80	14	310	73	24
InGaAsP	1	12	comp	117	94	14	292	81	26
InGaAsP	1	6	comp	121	92	18	250	85	28
InGaAsP	2	4	comp	222	85	34	340	75	33
InAlGaAs	2	7	comp	225	90	45	290	85	29
GaAsP	1	9	tensile	163	91	19	290	90	25
GaAsP	1	15	tensile	140	90	18	260	88	27
GaAsP	1	20	tensile	153	92	22	254	88	27

[a] composition of the QW, [b] number of QWs, [c] QW thickness,
[d] transparency current density, [e] internal efficiency, [f] modal gain coefficient,
[g] threshold current ($L = 1000$ μm), [h] slope efficiency ($L = 1000$ μm),
[i] vertical far-field angle (FWHM)

versus the inverse resonator length is plotted in Fig. 5 for several quantum wells on a logarithmic scale.

In Fig. 6 the inverse external differential efficiency is given versus the resonator length. Both curves should be linearly fitted to get the values as above-mentioned.

For practical use, the threshold current and the differential efficiency at a resonator length of 1000 μm are also given. Such a resonator length results in a mirror loss of about 12 cm⁻¹ for a diode laser with uncoated facets, which corresponds to a mirror reflectivity of 30%. This mirror loss is typical for the

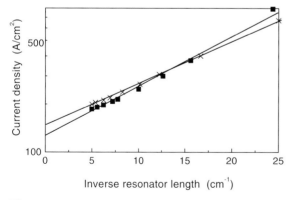

Fig. 5. Threshold-current density versus inverse resonator length of uncoated diode laser with an AlGaAs LOC broadened-waveguide structure; $\lambda = 810$ nm measured in pulsed regime (1μs; 5kHz); ■ compressively strained InGaAsP QW; × tensile-strained GaAsP QW

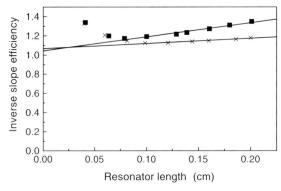

Fig. 6. Inverse slope efficiency versus resonator length of uncoated diode laser with an AlGaAs LOC broadened-waveguide structure; $\lambda = 810\,\text{nm}$ measured in pulsed regime $(1\,\mu\text{s}, 5\,\text{kHz})$; ■ compressively strained InGaAsP QW; × tensile-strained GaAsP QW

Anti-Reflection (AR) – High-Reflection (HR) facet-coated diode lasers used in real systems.

From Table 1 the progress by introducing strained quantum wells can be seen. Strained quantum wells account for lower threshold-current densities and higher efficiencies. The choice of the material composition of the quantum well as the active region depends on the epitaxial equipment, the experience and the application (TE or TM polarization, compressive- or tensile-strained QWs). The question whether one or two quantum wells should be better, is answered by both the desired output power per facet width and the resonator length. For medium-power applications a shorter resonator length below 1 mm is desirable, since material utilization for the diode laser is reduced in this case. Moreover, a Double Quantum Well (DQW) should be used. However, for higher output power, the resonator length is typically increased to 1.5 mm or longer. In this case a Single Quantum Well (SQW) would yield a lower threshold current.

The application of the waveguide structure shown in Fig. 3 is not limited to the wavelength range around 800 nm. It was also used successfully for shorter wavelengths down to 720 nm [9] and at wavelengths around 940 nm. For wavelengths around 800 nm an LOC broadened-waveguide structure was also realized with GaInP waveguide and AlGaInP cladding layers [10].

1.1.2   Al-Free Waveguides for Wavelengths from 940 nm to 980 nm

High-power diode lasers emitting in the 940 nm and 980 nm wavelength regions have recently aroused interest for pumping Yb:YAG solid-state lasers as well as Er-doped fiber lasers and amplifiers, respectively. In this wavelength range *Mawst* et al. [11] reported for the first time the successful realization of an Al-free LOC broadened-waveguide structure achieving high output power. Al-free epilayers on GaAs are not as common as AlGaAs and more difficult to grow, but they promise some benefits with respect to higher facet stability, reduced generation of defects and higher electrical conductivity. The layer structure explored is given schematically in Fig. 7.

The cladding layers consist of GaInP. A quaternary material GaInAsP lattice-matched to GaAs with a band gap of around 1.6 eV is used for the waveguide. The thickness of the waveguide is about 1 μm. The cladding layers can be made as thin as 700 nm, which reduces thermal and series resistances. The overall thickness of the epitaxial layers of only 2.5 μm clearly demonstrates one of the main benefits of the broadened-waveguide concept. On the other hand, a quaternary material with a higher band gap is more difficult to grow, and the high difference in refractive index between cladding and waveguide core is a critical issue of the structure. Hence, the divergence of these diode lasers is relatively high, viz. about 40° FWHM. However, the higher Catastrophic Optical Mirror Damage (COMD) level of the Al-free material completely compensates this drawback compared to AlGaAs structures. In Table 2 characteristic data of such structures with different GaInAs quantum wells and varying types of waveguides are presented.

Very low threshold-current densities have been achieved with a 1 μm-thick waveguide core. In Fig. 8 the threshold-current density versus inverse resonator length and in Fig. 9 the differential external efficiency versus resonator length are plotted for a typical waveguide structure with a single and a double quantum well.

Diode lasers with such a structure stand out for threshold-current densities below 200 A/cm² for 1000 μm resonator length and of only around 120 A/cm² for longer ones with differential external efficiencies of over 80%.

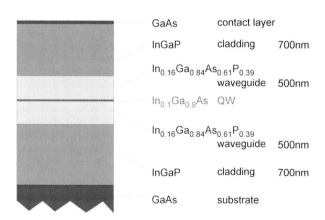

Fig. 7. Scheme of the vertical layer sequence of a diode laser with an Al-free LOC broadened-waveguide structure for emission wavelengths around $\lambda = 950$ nm

Table 2. Characteristic data for Al-free LOC diode-laser structures for wavelengths around 950 nm with InGaAs QWs (Fig. 3); pulsed measurement, uncoated samples, 100 μm stripe width

N [a]	d-QW [b]	j_T [c] (A/cm^2)	η_i [d] (%)	α_i [e] (cm^{-1})	I_{th}^f [f] (mA)	η_d [g] (%)	λ [h] (nm)	$\Theta_\perp$ [i] (°)	Ref.
1	5	82	97	1.5	175	87	935	43	[12]
2	5	143	95	2	220	86	948	42	[12]
1			95	2.3	140	80	970		[13]
2			97	2.5	170	80	970		[13]

[a] number of QWs, [b] QW thickness, [c] transparency current density,
[d] internal efficiency, [e] absorption losses, [f] threshold current ($L = 1000\,\mu m$),
[g] slope efficiency ($L = 1000\,\mu m$), [h] wavelength ($L = 1000\,\mu m$),
[i] vertical far-field angle (FWHM)

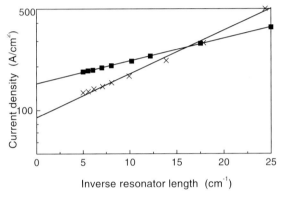

Fig. 8. Threshold-current density versus inverse resonator length of uncoated diode laser with an Al-free LOC broadened-waveguide structure; $\lambda = 935$ nm measured in pulsed regime (1μs, 5 kHz); × InGaAs SQW; ■ InGaAs DQW

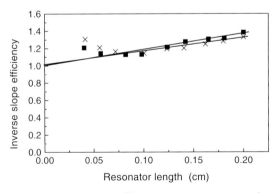

Fig. 9. Inverse slope efficiency versus resonator length of uncoated diode laser with an Al-free LOC broadened-waveguide structure; $\lambda = 935$ nm measured in pulsed regime (1μs, 5 kHz); × InGaAs SQW; ■ InGaAs DQW

Further broadening of the core leads to a lower divergence but also to higher threshold currents due to the smaller confinement factor.

1.1.3   Structures for Short-Wavelength Diode Lasers from 630 nm to 690 nm

The development of diode lasers emitting in the red spectral range was triggered from data-storage applications such as Digital Video Disk (DVD) systems. Until now red-emitting high-power diode lasers had a relatively small market. For example Cr-doped lithium strontium aluminum fluoride (Cr:LiSAF) solid-state lasers generating femtosecond pulses can be pumped by such diode lasers emitting between 650 nm and 740 nm [14]. In the future, however, a large market is anticipated in laser-projection technology when the red color (630 nm to 640 nm) will be supplied by high-power diode lasers. Another application with volume potential is foreseen in medical therapeutics. The difficulties of realizing the pertinent laser structures arise from the only possible small difference between the energy gap of the quantum-well material (typically GaInP) and that of the surrounding waveguide-layer material (AlGaInP). Therefore, the barrier height is quite low for the carriers in the active region. Quantum wells with tensile or compressive strain can be realized by using a certain Ga/In ratio. For the waveguide and cladding layers, material compositions with a band gap wider than 1.9 eV which corresponds to a wavelength of 650 nm are required. $(Al_xGa_{1-x})_{0.5}In_{0.5}P$ is used for the waveguide, and a composition with a higher Al value is used for the cladding layers, in many cases up to AlInP. In Fig. 10 a typical structure for a wavelength of 650 nm [15] is shown.

Due to the lower barrier height for the carriers in the quantum wells, a high confinement factor is more important than for diode lasers operating in

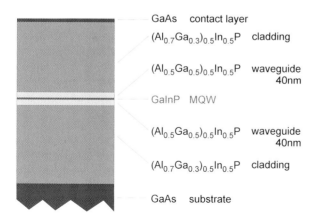

GaAs   contact layer

$(Al_{0.7}Ga_{0.3})_{0.5}In_{0.5}P$ cladding

$(Al_{0.5}Ga_{0.5})_{0.5}In_{0.5}P$ waveguide
                            40nm

GaInP   MQW

$(Al_{0.5}Ga_{0.5})_{0.5}In_{0.5}P$ waveguide
                            40nm

$(Al_{0.7}Ga_{0.3})_{0.5}In_{0.5}P$ cladding

GaAs   substrate

Fig. 10. Scheme of the vertical layer sequence of a diode for emission wavelengths around $\lambda = 650$ nm

the longer-wavelength range, and therefore a waveguide structure as shown on the left-hand side of Fig. 2 is used. The waveguide layers are relatively thin (30 nm – 100 nm), and the difference of the refractive index between wave-guide and cladding layer is large. However, because the wavelength is short, the divergence is 40° (FWHM) for the thicker and about 20° for 30 nm- to 40 nm-thick waveguide layers. Additionally, a reflector for electrons, consist-ing of thin layers with a periodically varying band gap, is grown to improve the carrier confinement and temperature stability. In Table 3 some published experimental results are listed for the wavelength range discussed here.

Table 3. Data of diode lasers in the red spectral range

N	d-QW	Strain	Structure	L	W	I_{th}	η_d	λ	$\Theta_\perp$	Ref.
				(μm)	(μm)	(mA)	(%)	(nm)	(°)	
3	6	comp	RW	200	2	10	72	690		[16]
1	5	comp	BR	650	5.5	60	60	685	21	[15]
		comp	RISA	500	3	18		660	22	[17]
1	5	comp	RW	800	5.5	60	91	656	40	[18]

2   Technology for Broad-Area Diode Lasers and Laser Bars

The objectives of device technology for power diode lasers are threefold. First, it is necessary to shape the optical resonator for lateral waveguiding. In the vertical dimension waveguiding is due to the epitaxial structure, in the lateral dimension waveguiding is usually achieved by mesa etching in the case of index-guided diode lasers, or more simply by a defined contact geometry in the case of gain-guided diode lasers. For a typically used stripe width beyond 50 μm the difference in performance is small.

Second, mirrors have to be generated to define the resonator. For high-power diode lasers this is usually done by cleaving the crystal and coating the mirror facets. This process is unique in semiconductor-device fabrication technology and a critical one for edge-emitting diode lasers.

Third, further processing has to ensure electrical contacts for the power supply and to allow mounting of the diode laser chip for handling and thermal heat sinking. The designed structure and contact metallization must allow for effective input of the electrical energy into the diode laser and to efficiently remove the waste heat. As diode lasers are rather small devices their thermal footprint is very high, reaching values on the order of kW/cm^2 although their conversion efficiency exceeds 50%.

Apart from facet coating, nearly all of these processing steps described in detail in [19] and [20] resemble those of integrated circuits. Because of the high current density and light intensity, all of the processing procedures must be designed to minimize the amount of stress and damage that the grown diode-laser epilayers and substrate experience during the processing steps in order to minimize the introduction and growth of defects. This is valid especially for high-power diode lasers with their larger active volume and high internal optical intensity $(1 - 10\,\text{MW/cm}^2)$.

All high-power diode-laser structures presented in the following are multimode lasers because of their optical confinement dimensions (for a typical broad-area diode laser see Fig. 1).

2.1   Processing of Contact Windows

The most-simple gain-guided diode-laser structure is based on laterally restricted current injection. This is realized by a contact window on the semiconductor surface. To prevent current spreading this surface is on top of the epitaxial layers. Via this contact window the current is injected into the active region using an appropriate metallization. Areas outside this window have to be electrically isolated.

Insulation is achieved either by strongly reducing the conductivity of the highly doped semiconductor contact layer caused by implantation or by a dielectric insulator placed between the semiconductor contact layer and the metallization layer. Diffusion of a dopant material, usually Zn, is a further approach to create a contact window. Selective diffusion requires a mask consisting of a stable dielectric insulator. For the processing of high-power diode lasers the implantation and the dielectric layer are mostly reported [21,22].

2.1.1   Implantation

The insulating effect of ion implantation is based on generating vacancies in the semiconductor. These vacancies capture the free carriers in the semiconductor layers necessary for the current transport, thus decreasing the conductivity and increasing the resistance of the layers.

For the fabrication of multimode diode-laser structures the insulation of only the highly doped semiconductor layers (the contact layer and the cladding-layer region next to the contact layer) is sufficient. This area can be isolated by a shallow ion implantation with low-energy ion doses. The average depth and the thickness of the created insulating layer is defined by the ion type and the implantation energy. The implantation dose defines the number of vacancies and consequently the degree of insulation. Figure 11 shows the calculated ion distribution of He^+ ions which were implanted into GaAs with an energy of 25 keV and a dose of $4 \times 10^{13}\,\text{cm}^{-2}$. Figure 12 shows the corresponding calculated distribution of the created Ga and As vacancies. The maximum of the vacancy distribution extends to about 250 nm below the surface. Hence, the contact layer usually has a thickness of 100–200 nm and a doping level of $2–4 \times 10^{19}\,\text{cm}^{-3}$; the adjacent cladding-layer area should be electrically isolated. It is worth noting that the ions do not penetrate deeper

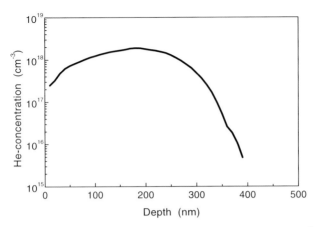

Fig. 11. Calculated concentration profile of implanted He$^+$ ions in GaAs, implanted with an energy of 25 keV and a dose of 4×10^{13} cm^{-2}, typical parameters, used to insulate defined areas of the highly doped GaAs contact layer. The projected range is 180 nm (depth with the highest He-ion concentration)

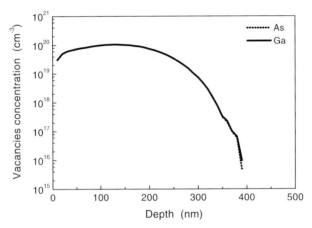

Fig. 12. Ga and As vacancies profile in a GaAs layer, produced by an He$^+$ implantation with an energy of 25 keV and a dose of 4×10^{13} cm^{-2}. This vacancy concentration is sufficient to make the 200 nm highly p-doped layer isolating

than 400 nm and thus are far away from the active zone. This fact is very important because the ion implantation creates crystal defects which cannot be annealed completely. Through annealing the insulating effect would likewise vanish.

Damage located in a region with high optical field intensity forms initial points for the growth of dislocations, leading to the degradation of the diode-laser characteristics and finally to the failure of the device. Therefore, an alignment of the implantation with a major crystallographic direction (with an angle lower than 7°) should be avoided. Due to this so-called channeling effect the implanted ions are guided between rows of atoms in the crystal lattice, and the penetration depth is up to 10 times higher than in the non-channeling case.

To isolate only selected areas of the semiconductor – outside of the contact windows – an appropriate mask should be used for the implantation. The implantation is a low-temperature process and therefore a large variety of masking materials can be used. Very suitable are masks of a commonly used photoresist or a metal. Because of the extremely low implantation energy applied for a depth of 250 nm, a positive resist (about 1.8 μm thick) or a thin metal layer (for example the layers of the metal contacts) can be used as a mask. Deeper implantation reduces current spreading but needs more stable masks, and the distance from the region of created defects to the active region becomes smaller.

The technology to fabricate a gain-guided diode laser with the help of ion implantation is illustrated in Fig. 13.

2.1.2   Insulating Layer

The technology to fabricate a gain-guided diode laser with a lateral isolated area formed by a dielectric layer is depicted in Fig. 14. The most common dielectric layers used as a current barrier in GaAs-based diode-laser structures consist of SiO_2, SiN_x or Al_2O_3. Depending on the dielectric layer used and the deposition technique applied (chemical vapor deposition, sputtering or electron-beam evaporation), the technology to structure this dielectric layer is different. In the presented technology, SiN_x deposited by radio-frequency sputtering is used. Because of the low thermal load imposed on this deposition process as compared to the CVD process, a lift-off technology for the openings of the contact windows can be used. Lift off is preferred to direct structuring as the electrical properties of the semiconductor contact layer are much less affected by a lift-off procedure.

Regarding the thickness of the dielectric layer, the ratio between the insulating and the contacting area is near to 1, and in view of the low voltage applied to the diode laser, a thickness of 100 nm for the dielectric layer is sufficient.

2.2   Processing of Mesa Structures

Another generic type for optical and electrical confinement is the index-guided structure. In this structure, part of the semiconductor layers outside the confined region is replaced by a material with a lower refractive index. In the simplest case, parts of the epitaxial layers (contact and cladding layers typically) outside the confined region are etched away. This mesa could be made very small (3–5 μm) so that a lateral waveguide having only a single lateral mode can easily be created. Therefore this structure is very often used

1. Resist mask for metallization (1st)

2. Metallization (1st)

3. Insulating implantation

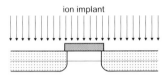

4. Metallization (2nd)

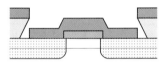

5. Finale structure

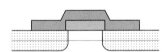

Fig. 13. Technology to produce a planar gain-guided laser diode. The gain-guiding is created by the current confinement, produced by the isolating implantation (step 3). The mask to structure the metal film is a lift-off mask made of an image-reversal resist. To prevent the contact area on the GaAs layer from damage, caused by the ion implantation, metallization stripes of a thickness of 70 nm are sufficient

for a fundamental-mode diode laser, the so-called ridge-waveguide laser. In more complicated structures the etched areas outside the confined region are filled with another semiconductor layer grown by a second epitaxy step. This so-called buried heterostructure is very well known from fundamental-mode diode lasers for the 1.3 μm and 1.55 μm wavelength regions but they are not used for high-power diode lasers based on GaAs substrates.

The laser diode with a mesa structure for high output power as shown e.g. in Fig. 1 is a special case of the ridge-waveguide laser. The lateral width of the mesa is 60 μm to 200 μm. The precision of the etching process to reach a stop at the exact depth is reduced as compared to the ridge-waveguide laser. The main advantage of the mesa structure is an excellent and defined current confinement and also a slightly better optical confinement which results in a somewhat lower threshold-current density, especially for stripe widths smaller than 100 μm.

1. Resist mask for insulator deposition

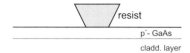

2. Insulator deposition (sputtering)

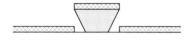

3. Metallization

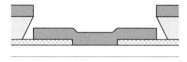

Fig. 14. Technology to fabricate a planar gain-guided laser diode. The current confinement is realized by a contact window, defined by an isolator on the highly doped GaAs layer. The best way to structure the isolator from the point of view of contact conditions is by using a lift-off mask. The insulator has to be deposited by a low-temperature process for the use of the lift-off process, for example by sputtering

4. Final structure

The technology to produce laser diodes with a mesa structure is illustrated in Fig. 15. The etching of the mesa is by wet or dry etching methods; these are well established processes with relatively large tolerances. The etching procedure should produce side walls which can be easily metallized. In the case of wet chemical etching only such etching solutions should be used which prevent underetching of a semiconductor layer, which would yield holes under, or breaks in, the metallization film. This happens if the etching solution has different etching rates for different semiconductor materials. Because of the anisotropic etching behavior of most wet chemical etchants, the window openings should be oriented only in one direction.

A lot of papers treat wet etching of GaAs and other semiconductors (e.g. [20,23,24,25]). An often-used etching solution for GaAs and AlGaAs is the $H_2SO_4/H_2O_2/H_2O$ system, having moderate etching rates and excellent etching profiles in the fundamental crystal-lattice directions. Another wet etchant is composed of HCl and water or H_3PO_4, often used for etching InGaP. Unfortunately, these systems have very high etching rates, making the etching process difficult to control. For InP, InGaAsP and related com-

1. Resist mask for mesa etching

2. Mesa etching (wet chemically)

3. Insolator deposition (sputtering)

4. Metallization

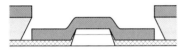

5. Final structure

Fig. 15. Mesa technology to fabricate a quasi-index-guided laser diode. The resist mask, used as the etching mask, consists of positive (especially for the following wet chemical etching) or negative resist (for a dry-etching process). Then the same mask can be used for structurization of the isolator

pounds an etching solution based on bromine, viz. Br_2 : $H_2O/HBr/H_2O$, is used.

In diode lasers with a mesa structure, electrical confinement should be supported by an additional insulating layer (SiN_x or Al_2O_3) which also protect layers with a high aluminum content from corrosion [26]. On the other hand, fabrication of low mesa structures allows the formation of an insulating layer on material with high aluminum content by wet chemical oxidation [27].

A scanning electron microscope picture of a diode laser with a low mesa structure is shown in Fig. 16.

2.3 Metallization

The purpose of metallizing the p- and n-sides of the diode laser is to provide an ohmic contact to allow electrical current to flow through the diode. The contact should have a linear I–V characteristic, and be stable in the time and temperature domain. The contribution to the series resistance of the

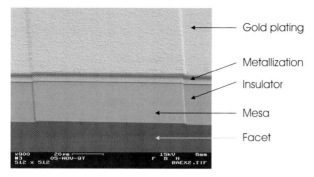

Gold plating

Metallization

Insulator

Mesa

Facet

Fig. 16. SEM picture of a high-power laser diode with a mesa structure for the optical and electrical confinement. The gold plating supports the heat transfer to the environment

diode should be as small as possible. On the other hand, the metallization is the basis for the mounting of the diode laser on a heat sink. Therefore the metallization should also allow soldering on a submount for heat transfer and wire bonding.

For high-power diode lasers based on GaAs substrates in the wavelength range between 650 nm and 1100 nm, the semiconductor contact layer consists typically of GaAs. If this contact layer is highly enough doped, almost any metal placed in intimate contact with the surface will result in an ohmic contact without having to be alloyed. The necessary doping level is at least $1 \times 10^{19}\,\mathrm{cm}^{-3}$. This high doping level can be realized during epitaxial growth of the contact layer. Usually a doping level of $2\text{–}4 \times 10^{19}\,\mathrm{cm}^{-3}$ can be realized with Zn or C as the dopant. More critical is the metallization of the n-doped GaAs substrate, where the second (n-) contact is formed. Typically used substrates only have doping levels around $10^{18}\,\mathrm{cm}^{-3}$. To fabricate an ohmic contact on these substrates an appropriate metallization is applied to the wafer and then alloyed with the GaAs. During alloying at an elevated temperature and the subsequent cooling period, a component of the metallization enters into the GaAs and creates a highly doped surface layer. Of all the metallizations used, gold-germanium-based systems (germanium as the dopant) were found to be most adequate for contacting n-type GaAs substrates over reasonable doping ranges. Gold–germanium is applied in proportions that represent a eutectic alloy (for example 88% Au, 12% Ge by weight) [19]. This alloy is deposited on the GaAs substrate by electron-beam evaporation. The AuGe alloy has a poor sheet resistance, is very difficult to wire-bond and is not solderable to a heat sink. This is why an additional gold layer is evaporated on the AuGe alloy. The thickness of this gold layer depends on the further contacting procedure. For wire-bonding on the n-contact (mounting p-side down), a relatively thick gold layer is required ($\approx 400\,\mathrm{nm}$). For soldering, a thin layer ($\leq 100\,\mathrm{nm}$) is preferred. To prevent gold diffusion into the GaAs during the alloying, an additional layer is necessary. Typically a platinum layer acts as a barrier with a titanium adhesion layer. The contact metallization system AuGe/Ti/Pt/Au is alloyed in a rapid thermal annealing chamber in a temperature region of 400° C to 430° C for 10 s to 20 s.

The p-contact is not as critical as the n-contact because of the high doping level of the GaAs contact layer. Evaporation of unalloyed metal directly on p-type GaAs can also produce acceptable ohmic contacts. The commonly used contact system for highly doped GaAs is the metallization system Ti/Pt/Au, which means that the same materials as for the n-contact can be applied. At a doping level of 2×10^{19} cm^{-3} of the GaAs contact layer a contact resistance of nearly 10^{-6} Ω cm^2 can be achieved. For the usual p-side down mounting the Au layer thickness is crucial. A value of 100 nm appears ideal. The thickness yields good wettability but is thin enough to avoid spurious metallurgical reactions with the Au/Sn or In solder material [28]. If the diode laser is mounted p-side up, an additional thick gold layer should be applied to support heat transfer to the surroundings and to improve the homogeneity of current injection. This 3 μm- to 5 μm-thick gold layer can be produced by electroplating.

As shown in Figs. 13–15, the various metallizations deposited by electron-beam evaporation can be patterned by applying a lift-off procedure.

2.4 Laser-Bar Preparation

One of the most common methods to generate more output power from a semiconductor-diode laser is to increase the width of the emitting area. However, if the width of a single emitter is increased, filamentation and lateral-mode instabilities become increasingly severe. This means that the optical power is no longer distributed homogeneously along the facet. So-called hot spots are likely to occur which lead to a fast degradation of the diode laser. As a matter of fact, the output power of single-stripe emitters cannot be enhanced significantly for stripe widths exceeding about 200 μm. Today, single-stripe emitter-diode lasers usually have a width of the emitting area which is below 300 μm. A practicable way to increase the output power is the monolithic integration of many (20–70) emitters into one diode-laser bar. These emitters are isolated optically and electrically from each other. Each emitter can be driven at a moderate output power, which does not lead to a accelerated degradation of the output facet. By summing the power of all single emitters in the laser bar a total output power of 100 W can be achieved [29]. In spite of the high power loss (about 50% of the electrical power) in such integrated array structures the thermal management – the soldering procedure and the heat sink architecture – play an important role to achieve a stable high-power operation with long lifetimes on the order of 10 000 h. In recent years, a standard length of the laser bars of 10 mm has been established. The resonator length is typically 600 μm and 1000 μm and the distance between the single emitters is about 50 μm and 200 μm. The filling factor (ratio of the

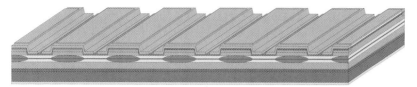

Fig. 17. Scheme of a part of a laser bar, consisting of electrically and optically separated single emitters. Each emitter has a mesa structure. The length of the bar is normally 10 mm, the resonator length between 600 μm and 1000 μm. The fill factor, the ratio between the optically active area to the whole area of the bar, is between 30 and 50%

active to the whole area of the laser bar) is between 30% and 80%. Figure 17 shows a section of such a laser bar.

The technology to fabricate such diode-laser bars is quite similar to the one described for broad-area single-stripe-diode lasers. In general, high-power laser bars based on gain-guided or quasi-index-guided diode-laser structures and for CW operation are always mounted p-side down on a heat sink. Therefore the p-metallization sequence should be terminated with a 100 nm-thick gold layer, separated from the final metallic contact layer by a barrier layer of platinum. Figure 18 shows a mounted diode-laser bar on the right-hand side, a diode-laser bar and an enlarged section with a few single emitters of the bar.

Since the technology to process single-stripe emitters and diode-laser bars is almost identical, the fabrication of diode-laser bars means only a change of the mask layout. However, there are special demands on the process technology of laser bars due to the high number of single emitters integrated in

260 μm

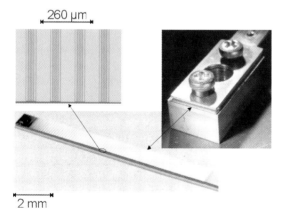

2 mm

Fig. 18. Optical microscope picture of a complete laser bar with 70 emitters, with some details, showing the separation channels between the emitters. The separation channels contain a high-absorption layer of germanium, to suppress some spurious modes. Also shown is a completely mounted bar in a micro-channel cooler

a laser bar. Each process step must be carried out on a very high reliability level. A defect of one emitter can be the reason for the failure of the whole bar by current or thermal effects.

Another problem, which arises from the integration of so many broad-area emitters having the same waveguide layers, is the appearance of so-called spurious modes. These modes have a propagation direction 'perpendicular' to the normal resonator modes. Such modes, which decrease the overall efficiency of the bar, will appear more in bars with a high fill factor, and have to be suppressed by drastically increasing the losses for such modes.

There are two approaches to suppress these modes. The first is to etch deep grooves parallel to the resonators between the single emitters. To increase the absorption at the side walls of the etched channels, an additional absorber can be deposited in these grooves which can be, e.g., a germanium layer having a sufficiently high absorption coefficient. To prevent diffusion of the p-side soldering material into the substrate or the n-doped region which would invariably lead to a short circuit, an additional insulating layer (Si_3N_4) covering the side walls and the bottom of the channels is necessary.

The second approach to suppress the spurious modes is to use the mesa structure. If the channels on both sides of the mesa are etched sufficiently deep (in general at least up to the waveguide layer) the optical field of a mode propagating in this region has an asymmetrical field distribution. Consequently, this mode has a high leakage towards to the substrate and cannot reach laser threshold. Using such kinds of mesa structures, no additional technological steps for the suppression of spurious modes are necessary.

2.5 Facet Coating

One of the most crucial aspects of high-power semiconductor-laser technology is that of generating long-term stable laser mirrors at the facets of the cleaved wafer. The energy heating the facet region is concentrated in a small volume and the power density easily reaches values on the order of $10\,MW/cm^2$ at the output mirror, which is near the damage threshold of a laser facet. The mirrors are realized by a three-step process:

- cleaving the processed wafers into laser bars,
- passivation of the cleaved surfaces,
- coating for the desired reflectivity.

The process has to fulfill mainly three requirements. First, the desired reflectivity, typically $> 90\%$ at the rear facet and values between 3% and 20% at the front facet, should be realized. Second, the process has to yield a high stability of the facet to ensure a device lifetime of several thousand hours. Third, it must be reproducible and cheap technology suitable for mass production.

2.5.1 Passivation

Figure 19 shows the feedback loops at the cleaved diode-laser facets leading to the damage of the facet and consequently to the failure of the device. All the processes leading to a degradation of a diode laser are described in detail in [30].

The reason for the initial absorption at facets is that there are deep centers (interface states) at the semiconductor–insulator interface which are multiplied by oxidation of the semiconductor material. The absorption of stimulated emission generates electron–hole pairs which recombine nonradiatively in the facet region. This nonradiative surface recombination heats the facet and as a result the band gap is reduced. This band-gap reduction increases the light absorption at the facets, and a positive-feedback loop develops. This effect will be enhanced by current crowding at the facet due to the lower band gap [31]. If the absorbed energy is high enough, a self-supporting process occurs which leads to thermal damage of the facet, the so-called Catastrophic Optical Mirror Damage (COMD).

To reduce the probability that this process might occur, various approaches are being pursued and reported. Whatever method is applied is a well-protected secret of every diode-laser manufacturer. A few approaches are given here.

- Decreasing the initial light absorption at the facet: to avoid the absorption, the band gap near the facet has to be widened. As an example, strain relaxation near the facets of tensile-strained quantum wells increases the

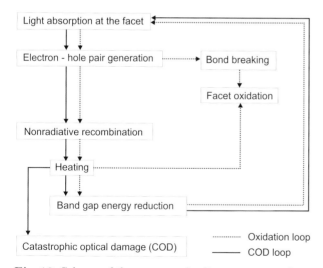

Fig. 19. Scheme of the processes leading to catastrophic optical damage of the laser facets (Reprinted with permission from: M. Fukuda: *Reliability and Degradation of Semiconductor Lasers and LED's* (Artech House, London 1991) [26])

gap value in the facet region [32]. But the effect is not very pronounced and tensile-strained quantum wells cannot always be implemented. Another possibility to achieve a larger band gap in the facet region is diffusion- or implantation-enhanced interdiffusion between the quantum wells and the surrounding material [33]. The absorption edge of the changed material is shifted several nanometers to shorter wavelengths and hence the absorption of stimulated emission in the facet region is reduced. The interdiffusion process requires a relatively high temperature of more than $800°$ C thus being a process step not easily integrated into diode-laser manufacturing. A third but more exotic technology is the re-growth of a window section at the facet with a semiconductor material having a higher band gap, which also requires high temperature and processing steps connected with the single diode-laser bars instead of a full wafer processing [34].

- Reducing surface-recombination velocity: this involves a rather complicated technology for semiconductor lasers which usually is not in line with an industrial low-cost production. One possibility is to cleave the wafer into bars in ultrahigh vacuum or in a protective atmosphere and to evaporate an appropriate passivation layer on the surface [35]. A second variant is surface treatment with sulfuric reagents to replace the nonstable oxide by a more stable compound. The influence of sulfur-stabilized facets on the electroluminescence power of a ridge-waveguide laser is shown in Fig. 20. The sulfur treatment leads to a decrease of the nonradiative recombination velocity and an increase in the COMD level [36,37]. However, the introduction of such a process in off-the-shelf diode-laser production has not yet been reported. A further approach is to deposit very thin layers of Al on the facet to getter the oxygen at the surface [38]. Using Al-free active regions, at least the area of the quantum well and waveguides inherently results in a reduced surface-recombination velocity. This is one of the advantages of an Al-free material system.

- Reducing current crowding at the surface: not only the absorption of light initiates a temperature rise at the surface, but also part of the forward current heats the surface due to nonradiative recombination. The latter will be increased as the band gap is decreased near the heated surface. One of the simplest measures to decrease the nonradiative recombination of electron–hole pairs at the facet is to prevent the supply of carriers to the area close to the facet [39]. The contacts should be set apart from the facet, at least over a distance larger than several times the diffusion length of the carriers in the diode.

The COMD level per stripe width is of course also determined by the light density at the facet. A reduction of the light density through an appropriate vertical waveguide structure led to the broadened-waveguide concept, resulting in so-called large optical cavity (LOC) structures which support a higher output power per stripe width. The highest COMD levels for multimode broad-area diode lasers are now between 15–$20\,\mathrm{MW/cm^2}$ [40]. This results in

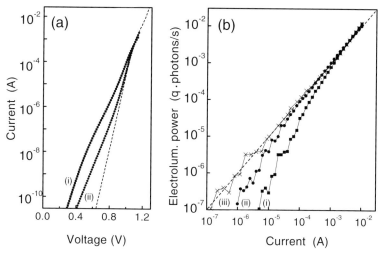

Fig. 20. Nonradiative current–voltage characteristics **(a)** and electroluminescence power **(b)** of a ridge-waveguide fundamental-mode laser without **(i)** and with **(ii)** sulfur treatment of the facets. It is clearly seen that the sulfur treatment of the facets reduces the nonradiative current by two decades. This reduction of the nonradiative current is connected with an increase of the electroluminescence power (Reprinted with permission from: G. Buster: Simple Method for Examining Sulphur Passivation of Facets in InGaAs-AlGaAs (0.98 µm) Laser Diodes, Appl. Phys. Lett. **68**, 2467–2468, (1996) American Institute of Physics)

maximum output powers of about 100 mW per µm stripe width. Similar and slightly higher values are achieved with fundamental-mode lasers [41]. Nevertheless, COMD and long-term facet degradation are often still the limiting factors for higher output powers of edge-emitting high-power diode lasers.

2.5.2 Reflectivity

The reflectivity of pure cleaved facets is nearly 30%. This value is easily modified by coating of the facets with an appropriate layer or a stack of layers. Especially the coating of the front facet should have a negligible absorption for the laser light. Moreover, the coating material should exhibit long-time stability to prevent the diffusion of components from the environmental atmosphere into the active layer under high-power operation. From a technological point of view, the coating material should be depositable on the facets without causing any damage of the underlying semiconductor material and without removing the passivation material and hence passivation effects. On the other hand the adhesion has to be very good.

To minimize the reflectivity with a single layer, the coating material should have a refractive index close to the geometric mean between the effective index of the waveguide and the index of air. But for the typically used values of reflectivity between 3% and 20% there are a lot of materials possible for the facet coating. Typically a single layer of Al_2O_3 is used or combinations with ZnSe are also reported.

For increasing the reflectivity, pairs of quarter-wavelength layers with a second material having a high refractive index should be used. Al_2O_3, Si_3N_4 and sometimes SiO_2 are utilized as a coating material with low refractive index. Commonly used coating materials with higher refractive indices are Si, ZnSe, TiO_2 or Ta_2O_5. All these materials can be deposited by secondary-ion-beam sputtering or evaporation, which can be ion-beam assisted. However, ion-beam assisted techniques require a trade-off between very good adhesion of the layer and low damage at the facet.

Figure 21 shows how the facet reflectivity is modified by a single layer of Al_2O_3 of various thicknesses as a function of the lasing wavelength around 808 nm. A layer of optical thickness (d): $dn_{die} = \lambda/2$, (n_{die} is the refractive index of the dielectric material used), does not change the reflectivity of the facets, but is very often used to protect the pure semiconductor surface at the facet against degradation in the case of low-power operation. The desired reflectivity in the range between 0.5% and 30% can easily be adjusted by choosing the right thickness of such a single layer . For a lower reflectivity Si_3N_4 is preferred because the refractive index is closer to the square root of the refractive index of the semiconductor material. For stable broadband AR coatings a stack of Al_2O_3 and TiO_2, or ZnSe, with appropriate thicknesses can be applied.

For the rear facet, typically a multilayer stack consisting of quarter-wavelength layers with low- and high-index materials is used. Figure 22 shows

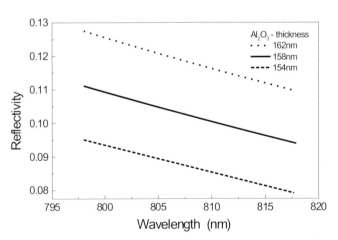

Fig. 21. Calculated reflectivity (at $\lambda = 808$ nm) of an Al_2O_3-coated facet with an AR coating of 10%, as used for a real device. The *two other curves* show the change in the reflectivity, if the layer thickness varies by $\pm 2.5\%$ (the tolerance of normal coating equipment)

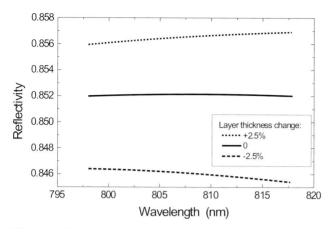

Fig. 22. Calculated reflectivity (at $\lambda = 808\,\mathrm{nm}$) of a multilayer system to realize a facet reflectivity of 85%. The layer system consists of two pairs $\mathrm{Al_2O_3}$ ($117\,\mathrm{nm}$)/ α-Si ($39\,\mathrm{nm}$) and a protection layer of $20\,\mathrm{nm}$ $\mathrm{Al_2O_3}$

the calculated reflectivity of a multilayer stack consisting of $\mathrm{Al_2O_3}$/Si for a wavelength around 808 nm. Silicon (amorphous) has the advantage of a very high refractive index at this wavelength. Only two pairs are necessary to achieve a reflectivity of about 90% at 808 nm. However, the absorption is quite high and, although the light intensity at the rear facet is much lower, there is some evidence for possible facet damage in high-power devices [42]. Instead of Si, other high-index materials such as ZnSe, $\mathrm{TiO_2}$ or $\mathrm{Ta_2O_5}$ are used to reduce the probability of damage. In this case at least three pairs of layers are necessary to achieve a reflectivity of nearly 90%.

3 Behavior of High-Power CW Diode Lasers

3.1 Influence of Heat on Laser Performance

The conversion efficiency of high-power diode lasers is high in comparison with other types of lasers; values approaching 70% seem possible. In practice, the conversion efficiency is around or slightly above 50%. Nevertheless, a considerable amount of heat is generated in high-power diode lasers. The small size of diode lasers results in a relatively high thermal resistance R_{th}, which is usually defined as the temperature rise of the active region divided by the difference of input power minus optical output power

$$R_{\mathrm{th}} = \frac{\Delta T}{P_{\mathrm{in}} - P_{\mathrm{out}}} \, . \tag{1}$$

Depending on the thermal resistance, this heat generation results in a higher temperature of the active region. Increasing the temperature of the active region of diode lasers reduces the carrier confinement and increases the nonradiative recombination processes. Both effects lead to a higher threshold current and a lower differential efficiency. The maximum output power is thus limited by the temperature rise of the active region, the so-called thermal rollover. In practice it might be eventually advantageous to have such reversible rollover effects before the COMD ends the lifetime of the diode laser. On the other hand, the degradation of diode lasers depends on temperature, and therefore the temperature of the active region should be as low as possible. The calculation of the thermal resistance is not straightforward because the effective sources of heat depend on the operation conditions. At low current below and near threshold the generated light is distributed across the crystal and will be absorbed in the higher-doped regions, contact layer and substrate. Also possible is a considerable amount of nonradiative recombination at the active region. Far above threshold, conversion to the usable radiation is high, and the heating stems from the series resistance R_s and from absorption of the laser light in the resonator especially in the vicinity of the mirrors. In either case localization of the heat sources is not straightforward. In addition, the temperature is not constant along the active region. This subject will be discussed in detail in the following sections.

Here, we will briefly discuss the impact of heating on the maximum output power. In a first approach the heat source is located in the active region and the thermal resistance can be calculated from the thermal properties of the materials and the geometry of the device. A first comprehensive study of the underlying principles and facts was published in 1975 [43]. Calculations using more sophisticated numerical methods were published in the 1980s and 1990s [44]. Published experimental data for the thermal resistance of broad area single-stripe emitters are between 5–15 K/W depending on stripe width, resonator length and mounting scheme [45].

The impact of temperature rise on threshold and differential efficiency depends on the given structure and resonator quality of the diode laser considered. It is common to describe this behavior by an exponential relation in the practically used temperature range $10°$ C to $50°$ C, according to [46]

$$I_{\text{tha2}} = I_{\text{tha1}} \exp\left(\frac{T_{a2} - T_{a1}}{T_0}\right) \tag{2}$$

and

$$\eta_{\text{da2}} = \eta_{\text{da1}} \exp\left(\frac{T_{a1} - T_{a2}}{T_1}\right), \tag{3}$$

where I_{tha1} and I_{tha2} are the threshold currents at ambient temperatures T_{a1} and T_{a2}; η_{da1} and η_{da2} are the differential efficiencies at ambient temperatures T_{a1} and T_{a2}; and T_0 and T_1 as the characteristic temperatures for the threshold current and the differential efficiency respectively.

The equations describe the behavior in the pulsed regime with short pulses (below 1μs) and at a low duty cycle (1:1000), thus neglecting the heating of the diode laser while it is being measured.

The higher the characteristic temperatures T_0 and T_1, the more stable is the operation in the CW regime at elevated temperatures. The characteristic temperature T_0 for the threshold current roughly depends on the barrier height and the necessary threshold-current density, which depends for a given active region (QW) on the resonator losses and the optical confinement factor. The values for T_0 are around 200 K for wavelengths at 980 nm and they decrease to around 140 K at 800 nm and to typically below 100 K for shorter-wavelength (630 nm – 760 nm) diode lasers based on GaAs. The values for T_1 are usually a factor three to five higher. The wavelength dependence reflects the change of the typical barrier height.

Using the temperature dependence of threshold and differential efficiency the power–current characteristic can be described approximately by the following expression:

$$P = \eta_{\mathrm{d}} \exp\left(-\frac{R_{\mathrm{th}}\left[I(V_{\mathrm{d}} + IR_{\mathrm{s}}) - P\right]}{T_1}\right)$$
$$\times \left[I - I_{\mathrm{th}} \exp\left(\frac{R_{\mathrm{th}}\left[I(V_{\mathrm{d}} + IR_{\mathrm{s}}) - P\right]}{T_0}\right)\right], \tag{4}$$

with P the optical power, I the current, R_{s} the series resistance, and V_{d} the voltage across the p-n junction.

The temperature rise of the active region is calculated as the product of the thermal resistance and the waste input energy (4). In Fig. 23 the influence of R_{th} on the power–current and efficiency–current characteristics of a single-stripe diode laser at 810 nm is shown.

The parameters (T_0, T_1, I_{th}, η_{d}) are typical for high-power diode lasers in this wavelength range at room temperature. A higher thermal resistance does not only limit the maximum output power but also leads to a strong increase of the temperature of the active region. As can be seen from Fig. 23 at 1 W output power a conversion efficiency of 50% and a temperature rise of 15 K can achieved for a thermal resistance of 15 K/W. In the case of a higher thermal resistance the temperature rise is strong also increased by the drop of conversion efficiency at higher temperatures. Therefore, in the case of 55 K/W the temperature of the active region rises about 110 K for 1 W output power. This demonstrates the necessity of a low thermal resistance. In Fig. 24 results of a similar calculation for a laser bar are shown. Due to the large area of semiconductor material the thermal resistance of the laser bar is usually sufficiently low. However, for the thermal performance of a diode-laser bar, heat spreading induced by the mounting scheme is of the utmost importance.

3.2   Simulation of Temperature Distribution

The temperature behavior of high-power diode lasers is investigated by modeling based on a rate-equation approach as well as the Finite-Element Method

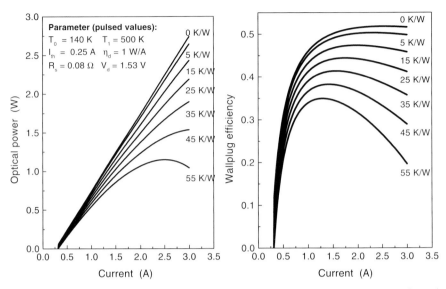

Fig. 23. Calculated optical output power (*left*) and conversion efficiency (*right*) versus current of a typical 1 W diode laser at 810 nm; the parameter is the thermal resistance of the diode laser

(FEM). The theoretical results obtained are compared with experimental data from laser-emission measurements (which provide information on the average temperature of the optically active layer) as well as micro-Raman facet temperature measurements.

3.2.1 Balance-Equation Approach

Several models have been developed for determining laser parameters such as the temperature dependence of threshold current, external quantum efficiency, lasing wavelength as well as the facet-heating processes in a self-consistent way [47,48,49,50,51]. As an example we discuss a two-dimensional model which self-consistently calculates the time dependence of the axial distribution of the photon density, carrier density and temperature profiles in laser diodes. The model includes a microscopic description of gain for a given QW structure as well as radiative and nonradiative recombination processes in both bulk and facet regions. It also takes into account the effects caused by temperature-dependent leakage currents and Joule heating (heating caused by the electrical resistance) in all regions of the laser structure.

The model starts with the equations for the photon densities of laser waves propagating in the forward (S^+) z direction:

$$\frac{\partial S^+}{\partial t} + \nu_\mathrm{g}\frac{\partial S^+}{\partial z} = \nu_\mathrm{g}[\Gamma g(\hbar\omega, n, T) - \alpha_\mathrm{tot}(n)]S^+ + 10^{-4}B_\mathrm{sp}(T)n^2 \,, \qquad (5)$$

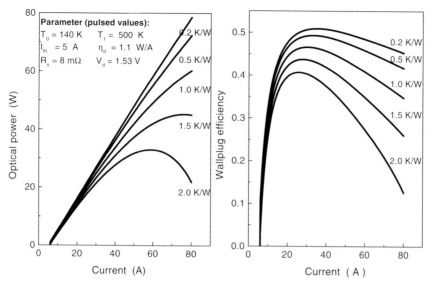

Fig. 24. Calculated optical output power (*left*) and conversion efficiency (*right*) versus current of a diode-laser bar at 810 nm; the parameter is the thermal resistance of the diode-laser bar

and backward (S^-) z-direction over the length of the resonator:

$$\frac{\partial S^-}{\partial t} + \nu_{\mathrm{g}}\frac{\partial S^-}{\partial z} = \nu_{\mathrm{g}}[\Gamma g(\hbar\omega, n, T) - \alpha_{\mathrm{tot}}(n)]S^- + 10^{-4}B_{\mathrm{sp}}(T)n^2 , \qquad (6)$$

where ν_{g} is the group index, n is the carrier concentration, Γ is the optical confinement factor, $g(\hbar\omega, n, T)$ is the gain function, α_{tot} is the total optical loss within the resonator, and $B_{\mathrm{sp}}(T)$ describes the temperature dependence of the spontaneous emission. The terms in (5) and (6) take into account the light-intensity distributions caused by stimulated emission, optical absorption, and spontaneous emission that couples into the coherent wave. The gain functions $g(\hbar\omega, n, T)$, α_{tot} and $B_{\mathrm{sp}}(T)$ are given in [50]. The boundary conditions for the waves are defined by the reflection coefficients at the front (R_1) and the rear facet (R_2).

The following relations hold:

$$S^+(0, t) = R_1 S^-(0, t)$$
$$S^-(L, t) = R_2 S^+(L, t) . \qquad (7)$$

For $t = 0$ there are no photons within the resonator. The diffusion equation is written as

$$\frac{\partial n}{\partial t} = D\frac{\partial^2 n}{\partial z^2} + \frac{J(z, t)}{ed} - \nu_{\mathrm{g}}[g(\hbar\omega, n, T) - \sigma_{\mathrm{fc}}n][S^+ + S^-] - \frac{n}{\tau_{\mathrm{e}}(n, T)} , \qquad (8)$$

where D is the diffusion coefficient, J is the injection current density, σ_{fc} is the cross section of free-carrier absorption, e is the electron charge, d is the thickness of the active region, and τ_e is the total carrier lifetime which is determined by processes such as Auger recombination, interface recombination and spontaneous emission. The carrier density is influenced by diffusion in the axial direction, carrier injection from the contact sheets, stimulated emission, free-carrier absorption, and spontaneous recombination. Additionally, surface recombination at the facets acts as a sink for the carriers.

The rate of carrier recombination at the facets is assumed to be proportional to the surface-recombination velocity, ν_{sur}, leading to the boundary conditions at the front facet ($z = 0$) and at the rear facet ($z = L$):

$$- D\frac{\partial n}{\partial z(0,t)} = -\nu_{sur}n(0,t)$$

$$-D\frac{\partial n}{\partial z(L,t)} = -\nu_{sur}n(L,t) \ , \tag{9}$$

where L is the length of the laser diode.

The temperature rise of the active region is calculated by a modified heat-conduction equation

$$\rho_m c_p \frac{\partial T}{\partial t} = \tilde{k}_z \frac{\partial^2 T}{\partial z^2} + q(n,t) - \gamma(T - T_{hs}) \ , \tag{10}$$

where ρ_m is the mass density, c_p is the specific heat capacity, k_z is the thermal conductivity within the resonator along the z-axis and T_{hs} is the temperature of the heat sink where the diode is mounted. The function q describes the power density of heat production by Auger recombination which is proportional to the cube of the carrier density n^3. The last term in (10) phenomenologically describes the thermal coupling between the semiconductor chip and the heat sink. The numerical value of the parameter γ is assumed to be that value for which the calculated steady-state temperature of the resonator reproduces the experimental values obtained from the emission-wavelength measurements [52]. The boundary conditions for (10) are determined by the assumption that for each electron–hole recombination act a specified amount of energy is dissipated which equals the band gap energy E_q at the facets. Thus the heat flux density is the product of E_q and the recombination rate

$$- k_z\frac{\partial T}{\partial z(0,t)} = +E_q\nu_{sur}n(0,t)$$

$$-k_z\frac{\partial T}{\partial z(L,t)} = -E_q\nu_{sur}n(L,t) \ . \tag{11}$$

The surface-recombination velocity ν_{sur} is a key parameter in the model, quantifying the influence of the quality of the facets on the temperature change of the facet region. Auger recombination acts as a bulk heat source. Optical coupling between individual emitters of the array, appearing at high injection currents, is neglected. The system of coupled nonlinear partial differential equations is solved by a modified relaxation technique described in [53,54].

Figure 25 shows the photon density S, the carrier density n, and the temperature T of the active region along the z axis for a typical high power laser design with a cavity length of 600 μm. The profiles are calculated for injection currents of 0.7 A, 1.2 A and 1.7 A and for a low surface recombination velocity of 5×10^4 cm/s. A range around 15 μm at each facet is excluded from the figure. Otherwise, the steep profiles close to the facets would hide the relatively low variations inside the cavity. The distribution of photon density, is asymmetric due to the different reflection coefficients of the facets. Stimulated recombination of carriers is strong within regions of high photon density resulting in a lower carrier density there [55,56]. Consequently, the profiles of carrier density (central part of Fig. 25) exhibit a spatial profile opposite to the photon density (top). In contrast to the facets, the internal regions of the resonator are mainly heated by Auger recombination. The temperature distribution follows the profile of the carrier density (bottom). As a consequence, the temperature within the resonator near the rear facet is higher than in the respective region near the front facet.

Temperature profiles in the region close to the facets are shown in Fig. 26. The curves were calculated for an injection current of 0.7 A $(I_{\mathrm{th}} = 0.5\,\mathrm{A})$ and for two different values of the surface recombination velocity. At low ν_{sur} values the higher temperature is surprisingly found at the rear facet (cf. dotted line in Fig. 26). In this case the optical gain remains positive within the facet region; therefore no additional reabsorption occurs. However, this does not take place when ν_{sur} approaches 5×10^5 cm/s or becomes even higher and a formation of an absorbing region close to the mirrors becomes

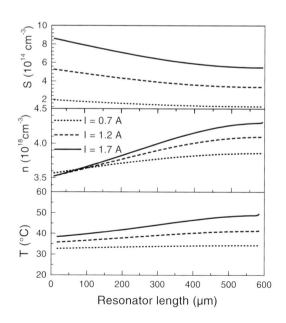

Fig. 25. Axial distribution of photon density, carrier density and temperature of a DQW-AlGaAs/GaAs laser diode for three injection currents. Facet regions are excluded

relevant. Reabsorption causes an additional heating source which is stronger at the front facet because of the higher photon density there.

The influence of ν_{sur} on the heating of both facets is shown in Fig. 27, where the temperature rise is plotted versus ν_{sur}. The temperature of the rear facet is higher than that of the front facet for lower values of the surface recombination velocity (lower than 2×10^5 cm/s) and no reabsorption in the facet region occurs. However, at $\nu_{sur} \geq 2 \times 10^5$ cm/s the temperature of the front facet exceeds the rear facet temperature. This is due to the formation of the absorbing region when the gain becomes negative. Now, the higher photon density at the front facet causes the stronger heating of this facet. According to the calculations, COMD is expected at the rear or front facet depending on operation conditions and the value of surface recombination velocity.

Now the model calculations are compared with experimental values of facet temperatures obtained by micro-Raman spectroscopy. The solid line in Fig. 28 gives the result of the calculation assuming a surface recombination velocity value of $\nu_{sur} = 5 \times 10^4$ cm/s. The squares are facet temperatures determined at a single emitter in the center of the emitting area of a array structure. Facet-temperature measurements on different emitters of multi-stripe arrays demonstrate a variation of the facet temperatures which correlates with the near-field intensity distribution [57]. Good agreement between calculation and data (Fig. 28) was only found on 'fresh' diode lasers. After a certain operation time at high output power the devices show higher facet temperatures, i.e. the front-facet temperature cannot be described by assuming exclusively intrinsic heating mechanisms such as reabsorption due to thermal band-gap shrinkage. Thus, additionally, extrinsic processes such as formation of absorbing defects must be relevant.

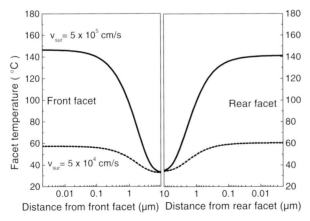

Fig. 26. Profiles of temperatures in a 50 μm region of both facets for high- and low-surface recombination velocity. Front facet AR coated, rear facet HR coated

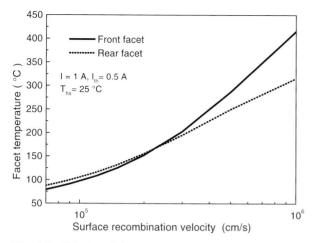

Fig. 27. Calculated facet temperature as a function of surface recombination velocity of a quantum well AlGaAs/GaAs laser-diode array

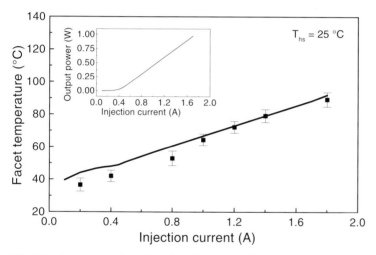

Fig. 28. Calculated (*solid line*) and measured (*symbols*) temperature of the front facet of a GaAlAs/GaAs laser diode versus injection current with a surface recombination velocity of 5×10^4 cm/s

3.2.2 Finite-Element Modeling

Numerical simulations of temperature distributions in diode lasers by Finite-Element Modeling (FEM) are an efficient tool for the optimization of the thermal properties [58,59,60,61]. A variation of parameters of the semiconductor-chip architecture as well as different thermal architectures of the heat sink allow us to get an insight into the processes determining the thermal behavior (see also Sect. 3.1).

Steady-state two- and three-dimensional as well as time-dependent FEM computations of temperature distributions in diode lasers were carried out for AlGaAs/GaAs 808 nm and GaInAs/GaAs 940 nm high-power diode lasers. The applicability of the FEM method was tested by comparing its results with analytical solutions of the heat-conduction equation [60]. Thermal device modeling usually considers strongly localized heat sources such as the p-n junction region of the diode where the main pump-power consumption occurs. Based on 'distributed heat sources' [62] a refined approach is introduced. In addition to ohmic and p-n junction losses this treatment includes ultrafast heat-transfer processes by photons, e.g. by reabsorption within the structure and subsequent radiationless recombination. Such a model requires knowledge of the contributions of all radiative (stimulated and spontaneous emission) and nonradiative mechanisms (Auger, interface recombination) as calculated from the balance-equation approach (see Sect. 3.2). The recombination mechanisms are temperature-dependent and discussed in detail in [62]. These numerical results of the rate-equation model are used as input parameters for the FEM in order to consider the heating caused by reabsorption of spontaneous emission. The thermal analysis was based on the FEM programs PATRAN 3/MacNeal-Schwendler installed on a workstation. The key issue of all FEM work is the appropriate choice of the FEM grid. For modeling thermal transients of array structures, this problem is of particular significance: the smallest dimension (the width of the quantum wells is 6–10 nm) is 6 orders of magnitude smaller than the lateral width of the laser bar. This requires a high aspect ratio in the finite-element calculations. But this ratio is limited for reasons of numerical stability. An equidistant three-dimensional grid with the lowest grid size of few nanometers width would exceed the potential of current computer systems. Therefore, most computations were done as two-dimensional simulations with an element size of 10 nm × 1 µm neglecting effects in the z-direction along the resonator axis. This grid size was increased step by step in order to find to which extent element sizes can be enlarged without significant changes of the simulation results. Methodical investigations show the finite-element dimension without significant changes in the calculated values must to be chosen lower than 90 nm [61,62]. This ensures that the temperature error as compared to calculations with 10 nm element size does not exceed 5%. The relative temperature error considering temporal heat variation is even smaller. For the solder and the different heat sinks and heat spreaders, element sizes have been varied from 1 µm × 0.2 µm to 200 µm × 200 µm. The complete grid of a '20 stripe array' (without heat spreader) on a copper heat sink contains at least 13 210 nodes. This number is increased according to the geometry when 'cm-bars' or different heat spreaders are considered. In this way significant model simplifications of the investigated diode-laser structures were possible.

Figure 29 illustrates the effect caused by the introduction of the concept of distributed heat sources into the model. Temperature profiles of an array

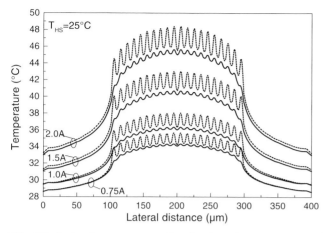

Fig. 29. Lateral temperature distribution along a 20-stripe laser array for different injection currents. *Dashed line*: p-n junction losses concentrated within the active region only, *full line*: p-n junction losses with distributed heat sources

emitting 1 W at 808 nm wavelength are calculated for different operating currents. Dashed lines mark the conventional approach, whereas the full line depicts profiles taking into account the distributed heat sources. Note the smoothing effect which was experimentally verified [57].

Transients of bulk and facet temperatures are also modeled by FEM. Here again the temperature-dependent loss processes in the active region, which are calculated with the balance-equation approach, are included [50]. For a given laser-array architecture the influence of different heat-spreader materials such as copper, silicon and diamond on the transient active-layer temperature of the diode laser is calculated [63]. These numerical results are presented in Fig. 30. They are compared with averaged temperatures of the active layer obtained by spectral emission measurements [62] which are given as symbols. In the time window between 10 ns and several μs the temperature transient is exclusively determined by the design and the thermal properties of the semiconductor-laser array. Different kinds of packaging do not significantly influence the thermal behavior. In the next time window between several μs and some ten to hundred ms the heat flows away from the p-n junction and the temperature rises strongly. At some 100 ms a nearly steady-state thermal distribution is reached, which is characterized by the CW temperature. For larger array structures such as 'cm-bars' these characteristic time windows are shifted, with the equilibrium state being reached after seconds [63].

The COMD is one of the mechanisms limiting the lifetime of high-power diode lasers. It is a result of reversible (e.g. thermal energy-gap shrinkage) and irreversible (defect creation) microscopic processes at the facets. All COMD scenarios include a continuous increase of the front-facet temperature due to

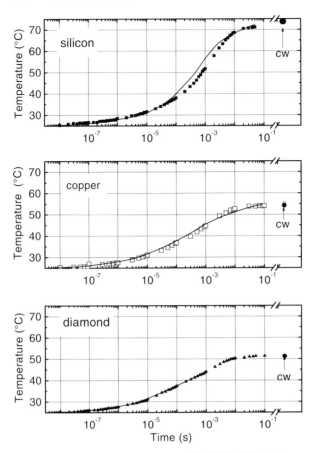

Fig. 30. Temperature transients of DQW-AlGaAs laser array mounted on different heat spreaders. *Symbols*: experimental values, *full line*: FEM calculations

an interplay of surface recombination, heating and increased interband re-absorption due to energy-gap shrinkage. Finally, this positive-feedback cycle (thermal runaway model of COMD) leads to damage or even facet perforation (see also Sect. 2.5). Therefore facet-temperature measurements under high-power operation can provide more insight into failure mechanisms [55,64]. Results of FEM calculations (lines) and micro-Raman temperature measurements (full symbols) are given in Fig. 31. Averaged active-layer temperatures obtained from emission data are added as open circles. Note that the full circles represent facet temperatures of the 'fresh' device, whereas the triangles are data obtained from the same device after 50 h of operation. The dotted line gives the FEM result assuming uniform heat-source distribution along the z axis. The excellent agreement between the open circles and the FEM results indicates that only a few percent (less than 10%) of the total power is consumed at the facets. The dashed lines were obtained by FEM assuming

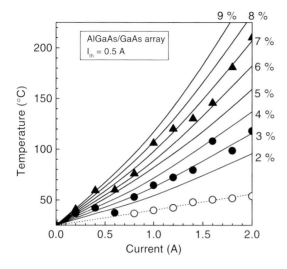

Fig. 31. Facet temperature (front facet) as a function of injection current (FEM and micro-Raman data). *Full circles*: 'fresh' diode laser; *triangles*: diode laser after 50 h of operation at a current of 2 A; *dotted line*: heat production uniformly distributed along the whole resonator; *dashed lines*: FEM computations with the given percentage of heat production concentrated in the facet region

that a certain percentage (given in Fig. 31) of the total heat power is concentrated as a two-dimensional heat source within the facet. Note that the 3−4% difference between the Raman facet temperatures before and after 50 h of operation (triangles and full circles) can only be understood assuming that the total laser-power reduction of 3.2% solely contributes to the facet-heating effect. Thus, the key role of the facets is also demonstrated for high-power diode-laser arrays. Furthermore, it is shown that numerical simulation with FEM is a good approach to optimize thermal device management.

3.2.3 Reliability

State-of-the-art high-power diode lasers have a lifetime of some thousand hours. Nevertheless, reliability issues are still a matter of concern limiting their practical use as promising radiation tool.

Standard device-test methods, such as statistical reliability tests for constant-power or constant-current operation, are applied for quantifying the aging behavior of high-power diode lasers [65,66,67]. Improvement of device performance, however, requires insight on a microscopic scale into the processes occurring within the device during aging especially in the case of operation under high thermal load. Destructive preparative methods, such as photoluminescence mapping of aged high-power diode-laser arrays after removing the substrate, provide valuable insight into aging scenarios [68].

In order to reduce the costs of testing, additional *nondestructive* test methods are required, e.g. optical methods that allow device diagnosis before and after certain aging steps. Among others, photocurrent (PC) spectroscopy meets specific demands of aging analysis in relatively large devices such as cm-bars. It turned out that photocurrent spectra are strongly affected by gradual aging processes [69]. The basic principle of this method is

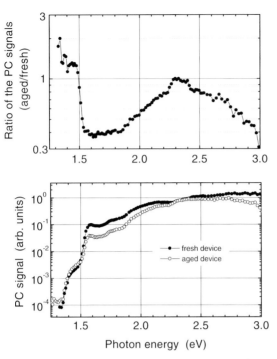

Fig. 32. Photocurrent spectra of an AlGaAs/GaAs high-power diode laser (*top*) and the aging pattern (*bottom*)

quite simple. The high-power diode laser itself serves as detector. Thus the photocurrent spectrum reflects the spectral characteristics of this device. Figure 32 shows photocurrent spectra from a high-power diode laser before and after 100 h of regular operation which reduced the output power by about 3% (top). The ratio of photocurrent signals measured before and after aging processes is called the 'aging pattern' (bottom). These patterns are very typical for a given laser design and allow us to choose special wavelengths where the spatial distribution of aging effects such as defect accumulation or increased surface recombination can be monitored by the Laser-Beam-Induced-Current (LBIC) technique [70].

Figure 33 shows deep-level LBIC line scans of an Al-free 20-stripe array as a cm-bar emitting at 808 nm wavelength measured with an excitation wavelength of 940 nm, i.e., 215 meV below the effective energy gap of the structure. The grid lines separate the positions of the emitters. Note the photocurrent minima for the 'fresh' devices in the center of the emitters, where the contact stripes are located. This is due to surface damage at the emitter center during burn-in. After 770 h of high-power operation defect accumulation and hence increased defect absorption at the positions of the emitter centers relatively overcompensate the current lows into current highs.

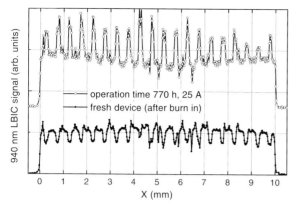

Fig. 33. LBIC scans along the front facet of an 808 nm emitting Al-free cm-bar array. The excitation wavelength was 940 nm. Thus only defect centers were excited and contribute to the signal

Mapping techniques are of special interest for high-power device analysis as they help to *locate* the starting points of device-degradation processes.

The Near-field Optical-Beam-Induced-Current (NOBIC) technique is the extension of the LBIC technique to a spatial resolution on the order of 100 nm, i.e. beyond the diffraction limit [71]. The application of the NOBIC technique to high-power diode-laser arrays allows aging analysis of the epitaxial layer along the growth direction of the epitaxial process. Thus, the spatial distribution of defects can be determined on a nanoscopic scale by exciting their absorption bands [72].

PC spectroscopy in the spectral region below the effective energy gap of the structures is feasible using a Fourier-Transform (FT) spectrometer as excitation source and for data processing. Thus a new kind of optical deep-level spectroscopy in devices became available [73]. Defect-generation dynamics of various deep levels in devices have been monitored versus operation time. It was demonstrated that different aging conditions, such as regular operation and accelerated aging, are responsible for different dynamics of defect generation in high-power diode lasers [73,74].

3.3 Recent Results

Research on high-power diode lasers has considerably enhanced the optical output power per stripe width in the last years. However, not only do 'record' values, verified in most cases at temperatures slightly below room temperature, climb to new highs but there is also much work done towards higher reliability. Steady improvements in various respects as described in the preceding sections are responsible for the progress high-power diode lasers have experienced a number of additional applications. These improvements include the utilization of strained quantum wells for lower threshold-current density and higher internal efficiency as well as the use of waveguide structures with broadened near fields and of Al-free waveguides to reduce the facet load. In addition to these more fundamental aspects published in the literature, of course there are lots of measures addressing improvements in crystal growth, in handling and processing, in facet passivation, etc. which are intellectual property not reported in the current literature but likewise crucial for laser performance, especially for reliability.

Table 4. Output power P of high-power diode lasers of various wavelengths λ and various apertures W and resonator lengths L

λ (nm)	P (W)	W (μm)	T_a ($^\circ$C)	L (μm)	Waveguide material	Year
670	1.3	40	RT	1200	AlInGaP	1999 [75]
730	7.0	100	10	4000	AlGaAs	1999 [76]
810	5.3	100	5	1230	InGaAsP	1991 [77]
810	8.8	100	10	1250	InGaP	1998 [78]
810	8.5	100	20	1800	AlGaAs	1998 [79]
840	7.3	55	10	1500	AlGaAs	1998 [80]
870	11.3	100	10	1500	AlGaAs	1998 [81]
915	10.9	100	10	2000	AlGaAs	1998 [82]
970	9.3	100	10	2000	AlGaAs	1997 [83]
970	10.6	100	10	2000	InGaAsP	1998 [84]
980	6.0	100	RT	500	InGaAsP	1998 [85]

In this section we focus on single emitters. Table 4 summarizes recent achievements with respect to optical-output power which we consider impressive if one takes into account the small size of a single-stripe high-power diode laser as compared to solid-state or gas lasers. We will not evaluate the presented device performance as it differs in various features such as vertical divergence, lateral far-field width, temperature stability, conversion efficiency, and lifetime expectation for the relevant applications. It is obvious that the 'record' values have been achieved only recently, demonstrating the efforts undertaken to exploit the potential of high-power diode lasers in the laser marketplace.

Record values achieved with very long resonators are primarily of academic interest, nevertheless demonstrating future potential. Short diode lasers with good reliability yielding output powers of 1 W to 2 W at 100 µm stripe width are preferred due to higher wafer yields and lower production costs. From a practical point of view, the 6 W output power achieved with a relatively short resonator length of only 500 µm (last row of Table 4) is remarkable. In Fig. 34 the power–current characteristics of this diode laser are shown. The high conversion efficiency of 60% at an output power of 1 W is impressive.

Highest output powers of more than 10 W for 100 µm stripe width could be achieved in the wavelength range between 870 and 980 nm. These struc-

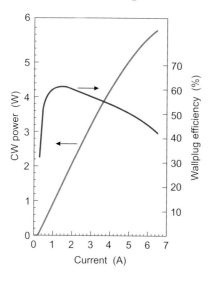

Fig. 34. Power–current characteristics of a high power Al-free diode laser with a resonator length $L = 500\,\mu$m; emission wavelength $\lambda = 960\,$nm. (Reprinted with permission from: D. Botez: 6 W CW front facet power from short-cavity (0.5 mm), 100 μm-stripe Al-free 0.98 μm-emitting diode laser, Electron. Lett. **33**, 2037–2039 (1997) [85])

tures benefit from the quality of strained GaInAs quantum wells and the higher barriers as compared to shorter-wavelength diode lasers. In the peak power there is practically no difference between Al-free and AlGaAs waveguide structures. The AlGaAs structures have a longer tradition and are easier to grow but processing, especially mirror passivation, seems more difficult than for Al-free structures although very long lifetimes have also been demonstrated for diodes with Al-containing material.

In the wavelength range around 800 nm the maximum output power that has been reached so far is nearly 9 W for a stripe width of 100 μm. This output power is somewhat lower than that for longer-wavelength devices due to a lower COMD level. Similar to the longer-wavelength diode lasers there is no difference in peak power between AlGaAs and Al-free structures. Reliability tests and commercial devices at output powers over 1 W per front facet for 100 μm stripe width are available only for AlGaAs structures until now [84]. Since higher barriers are achievable, the temperature stability is higher for AlGaAs waveguides compared to GaInP or GaInAsP waveguides. In Fig. 35 the power–current characteristics of a diode laser with an AlGaAs structure and a tensile strained GaAsP quantum well is shown for different temperatures [85]. At a temperature of 85° C an output power of more than 3 W from 60 μm stripe width was demonstrated. Considering the conversion efficiency of about 45% and a thermal resistance of about 12 K/ W of the diode laser the temperature of the active region rises to more than 120° C.

With an AlGaAs structure at a wavelength of 840 nm, promising lifetime data were demonstrated for an output power of 22 mW/ μm (1.2 W per 55 μm stripe width) at a heat-sink temperature of 60° C. Hence, driving high-power diode lasers at elevated temperatures has some potential [86].

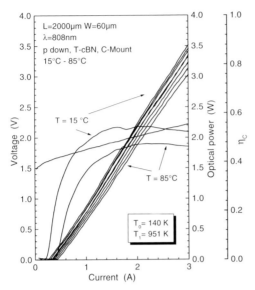

Fig. 35. Temperature dependence of power–current characteristics of a high-power LOC-AlGaAs diode laser; emission wavelength $\lambda = 810\,\mathrm{nm}$

At the shorter wavelengths below 700 nm the maximum output power drops to 3 W per 100 μm stripe width. The optical output power is three times lower than in the longer-wavelength region. This ratio hold also for single-mode diode lasers with a small stripe width of 3–5 μm. At 800 nm and 980 nm reliability has been demonstrated for output powers of 200 mW to 300 mW [42] whereas for the red-emitting diode lasers good results have been obtained only for an output power between 50 mW and 100 mW [87].

As mentioned above the 'record' values which have been achieved recently also demonstrate the very high potential for further improvement, and finally reliable single-stripe emitters at power levels around 5 W per 100 μm stripe width and over 100 W for laser bars should be achieved.

List of Symbols

α_{tot}	Total optical losses within the resonator
α_{i}	Absorption losses
B_{sp}	Spontaneous emission coefficient
c_{p}	Specific heat capacity
d	Thickness of active layer
$d\text{-QW}$	Thickness of quantum well
$\Delta\lambda$	Wavelength tolerance
ΔT	Temperature difference
e	Elementary charge

E_q	Band-gap width
η_d	Slope efficiency
η_{da}	Internal efficiency at ambient temperature (T_a)
η_i	Internal efficiency
G_0	Material gain coefficient
g, g_{max}	Optical gain, maximum of optical gain
Γ	Confinement factor
ΓG_0	Modal gain coefficient
γ	Parameter for thermal coupling of semiconductor to heat sink
$\hbar$	Planck's constant ($\hbar = h/2\pi$)
$\hbar\omega$	Energy of emitted laser photons
I	Current
I	Injection current density
I_{th}	Threshold current
I_{tha}	Threshold current at ambient temperature (T_a)
j_T	Transparency current density
k_z	Thermal conductivity
λ	Wavelength
L	Cavity length
n	Carrier density in active region
N	Number of quantum wells
P	Optical output power
P_{in}	Input power (electrical)
q	Power density of heat production by Auger recombination
R_1, R_2	Reflection coefficients of front and rear facets
R_s	Series resistance
R_{th}	Thermal resistance
ρ_m	Mass density
S^+, S^-	Photon density of forward and backward propagating laser waves
T_0	Characteristic temperature for the threshold current
T_1	Characteristic temperature for the differential efficiency
T	Temperature
T_a	Ambient temperature
T_{hs}	Heat-sink temperature
t	Time
τ_e	Total carrier lifetime
$\Theta_\perp$	Vertical far-field angle
ν_g	Group velocity index
ν_{sur}	Surface recombination velocity
V	Volume of active zone
V_d	Voltage across the p-n junction
W	Ridge or mesa width
z	Axial coordinate

References

1. W. Koechner: *Solid-State Laser Engineering*, 4th edn., Springer Ser. Opt. Sci. **1** (Springer, Berlin, Heidelberg 1996) p. 302
2. D. Botez, L. J. Mawst, A. Bhattacharya, J. Lopez, J. Li, T. F. Kuech, V. P. Iakovlev, G. I. Suruceanu, A. Caliman, A. V. Syrbu: 6% CW wallplug efficiency from Al-free 0.98 μm emitting diode lasers, Electron. Lett. **32**, 2012–2013 (1996)
3. J. Haden, J. Endriz, M. Salzmann, D. Dawson, G. Browder, K. Anderson, D. Mundinger, P. Worland, E. Wolak, D. Scifres: Advances in high-average power long life laser-diode pump array architectures, SPIE Proc. **2382**, 2–21 (1995)
4. D. Botez: Design considerations and analytical approximations for high continuous-wave power, broad-waveguide diode lasers, Appl. Phys. Lett. **74**, 3102 (1999)
5. D. Z. Garbusov, J. Abeles, N. A. Morris, P. D. Gardner, A. R. Triano, M. G. Harvey, D. B. Gilbert, J. C. Conolly: High power separate confinement heterostructure AlGaAs/GaAs laser diodes with broadened waveguide, SPIE Proc. **2682**, 20 (1996)
6. G. Erbert, F. Bugge, A. Oster, J. Sebastian, R. Staske, K. Vogel, H. Wenzel, M. Weyers, G. Traenkle: High CW power diode lasers with unstrained and compressively strained InGaAsP QWs in AlGaAs waveguides emitting at 800 nm, IEEE Proc. LEOS 97, 199–200 (1997)
7. G. Erbert, F. Bugge, A. Knauer, J. Maege, A. Oster, J. Sebastian, R. Staske, A. Thies, H. Wenzel, M. Weyers, G. Traenkle: Diode lasers with Al-free quantum wells embedded in LOC AlGaAs waveguides between 715 nm and 840 nm, SPIE Proc. **3628**, 19–28 (1999) and unpublished work
8. P. W. A. Mc Illroy, A. Kurobe, Y. Uematsu: Analysis and application of theoretical gain curves to the design of multi-quantum well lasers, IEEE J. QE **21**, 1958 (1985)
9. G. Erbert, F. Bugge, A. Knauer, J. Sebastian, A. Thies, H. Wenzel, M. Weyers, G. Traenkle: High-power tensile-strained GaAsP-AlGaAs quantum well lasers emitting beween 715 nm and 790 nm, Proc. ISLC'98, IEEE J. Sel. Topics Optoelectron. (1999) in press
10. J. K. Wade, L. J. Mawst, D. Botez, M. Jansen, F. Fang, R. F. Nabiev: High continuous power, 0.8 μm band, Al-free active-region diode lasers, Appl. Phys. Lett. **70**, 149–151 (1997)
11. L. J. Mawst, A. Bhattacharya, J. Lopez, D. Botez, D. Z. Garbuzov, L. De Marco, J. C. Conolly, M. Jansen, F. Fang, R. F. Nabiev: 8 W continuos wave front facet power from broad-waveguide Al-free 980 nm diode lasers, Appl. Phys. Lett. **69**, 1532–1534 (1996)
12. A. Knauer: unpublished results; Ferdinand-Braun-Institut Berlin, Germany (1998)
13. L. J. Mawst, A. Bhattacharya, M. Nesnidal, J. Lopez, D. Botez, A. V. Syrbu, V. P. Yakovlev, G. I. Suruceanu, A. Z. Mereutza, M. Jansen, R. F. Nabiev: MOVPE-grown high CW power InGaAs/InGaAsP/InGaP diode lasers, J. Cryst. Growth **170**, 383–389 (1997)
14. W. Koechner: *Solid-State Laser Engineering*, 5th edn., Springer Ser. Opt. Sci. **1** (Springer, Berlin, Heidelberg 1999) pp. 78–80

15. H. Tada, A. Shima, T. Utakoji, T. Motoda, M. Tsugami, K. Nagahama, M. Aiga: Uniform fabrication of highly reliable, 50-60 mW class, 685 nm, window-mirror lasers for optical data storage, Jpn. J. Appl. Phys. **36**, 2666–2670 (1997)
16. J. Kuhn, C. Geng, F. Scholz, H. Schweizer: Low-threshold GaInP/AlGaInP ridge waveguide lasers, Electron. Lett. **33**, 1707–1708 (1997)
17. M. Mannoh, T. Fukuhira, O. Imafuji, M. Yuri, T. Takayama: High-power red lasers for DVD-RAM drives, SPIE Proc. **3628**, 186–195 (1999)
18. J. Köngas, P. Savolainen, M. Toivonen, S. Orsila, P. Corvini, M. Jansen, R. F. Nabiev, M. Pessa: High-efficiency GaInP-AlGaInP ridge waveguide single-mode lasers operating at 650 nm, IEEE Photon. Technol. Lett. **10**, 1533–1535 (1998)
19. R. Williams: *Modern GaAs Processing Techniques*, 2nd edn. (Artech House, London 1990)
20. S. Sze: *Semiconductor Devices, Physics and Technology* (Wiley, New York 1985)
21. C. Lin: *Optoelectronic Technology and Lightwave Communications Systems* (Van Nostrand Reinhold, New York 1989)
22. O. Wada: *Optoelectronic Integration: Physics, Technology and Applications* (Kluwer Academic, Dordrecht 1994)
23. M. Köhler: *Ätzverfahren für die Mikrotechnik* (Wiley-VCH, Weinheim 1998)
24. H. Löwe, P. Keppel, D. Zach: *Halbleiterätzverfahren* (Akademie-Verlag, Berlin 1990)
25. P. H. L. Notton: *Etching of III-V Semiconductors: An Electrochemical Approach* (Elsevier Advanced Technology, Oxford 1991)
26. M. Fukuda: *Reliability and Degradation of Semiconductor Lasers and LED's* (Artech House, London 1991)
27. H. Burkhard, V. Pitaev, W. Schlapp: High-power high-To native oxide stripe-geometry 980-nm laser diodes, SPIE Proc. **2682**, 11–19 (1996)
28. K. Mitsuichi: Some aspects of bonding-solder deterioration observed in long-lived semiconductor lasers: solder migration and whisker growth, J. Appl. Phys. **55**, 289–295 (1984)
29. F. Daiminger, S. Heinemann, J. Näppi, M. Toivonen, H. Asonen: 100 W CW Al-free 808 nm linear bar arrays, CLEO 97 Tech. Dig. Ser. **11**, 482–483 (1997)
30. R. G. Waters: Diode laser degradation mechanisms: a review, Prog. Quant. Electron. **15**, 153–174 (1991)
31. W. C. Tsang, H. J. Rosen, P. Vettiger, D. J. Webb: Evidence for current-density-induced heating of AlGaAs single-quantum well laser facets, Appl. Phys. Lett. **59**, 1005–1007 (1991)
32. A. Valster, A. T. Meney, J. R. Downes, D. A. Faux, A. R. Adams, A. A. Brouwer, A. J. Corbijn: Strain-overcompensated GaInP-AlGaInP quantum well laser structures for improved reliability at high-output powers, IEEE J. Sel. Topics Quantum Electron. **3**, 180–187 (1997)
33. P. Collot, J. Arias, V. Mira, E. Vassilakis, F. Julien: Non-absorbing mirrors for AlGaAs quantum well lasers by impurity-free interdiffusion, SPIE Proc. **3628**, 260–266 (1999)
34. M. Watanabe, K. Tani, K. Takahashi, K. Sasaki, H. Nakatsu, M. Hosoda, S. Matsui, O. Yamamoto, S. Yamamoto: Fundamental-transverse-mode high-power AlGaInP laser diode with windows grown on facets, IEEE J. Sel. Topics Quantum Electron. **1**, 728–733 (1995)

35. L. W. Tu, E. F. Schubert, M. Hong, G. J. Zydik Meyer: In-vacuum cleaving and coating of semiconductor laser facets using thin silicon and a dielectric, J. Appl. Phys. **80**, 6448 (1996)

36. V. N. Bessolov, M. V. Lebedev, B. V. Tsarenko, Yu. M. Shernyakov: Increase in the degree of catastrophic optical degradation of InGaAs/AlGaAs (977 nm) laser diodes after sulfidization in solutions based on isopropyl alcohol, Tech. Phys. Lett. **21**, 561–562 (1995)

37. A. J. Howard, C. I. H. Ashby, J. A. Lott, R. P. Schneider, R. F. Corless: Electrochemical sulfur passivation of visible (670 nm) AlGaInP lasers, J. Vac. Sci. Technol. A **12**, 1063–1067 (1994)

38. P. Tihany, D. R. Scifres, R. S. Bauer: Reactive outdiffusion of contaminants from (AlGa)As laser facets, Appl. Phys. Lett. **42**, 313–315 (1983)

39. M. Ohkubo, T. Ijichi, A. Iketani, T. Kikuta: Aluminium-free In-GaAs/GaAs/InGaAsP/InGaP GRINSCH SL-SQW lasers at 0.98 μm, Electron. Lett. **28**, 1149–1150 (1992)

40. D. F. Welch, B. Chan, W. Streifer, D. R. Scifres: High power 8 W CW single quantum well laser-diode array, Electron. Lett. **24**, 113–115 (1988)

41. T. Fujimoto, Yu. Yamada, Yo. Yamada, A. Okubo, Y. Oeda, K. Muro: High-power InGaAs/AlGaAs laser diodes with decoupled confinement heterostructures, SPIE Proc. **3628**, 38–45 (1999)

42. A. Oosenbrug, C. S. Harder, A. Jakubowicz, P. Roentgen: Power integrity of 980 nm pump lasers at 200 mW and above, LEOS 96 IEEE Proc. **1**, 348–349 (1996)

43. W. B. Joyce, R. W. Dixon: Thermal resistance of heterostructure lasers, J. Appl. Phys. **46**, 855–862 (1975)

44. E. Marin, I. Camps, M. Sanchez, P. Diaz: Thermal resistance of double heterostructure separate confinement GaAs/AlGaAs semiconductor lasers in stripe geometry configuration, Rev. Mex. Fis. **42**, 414–424 (1996)

45. S. O'Brien, H. Zhao, R. J. Lang: High power wide aperture AlGaAs-based lasers at 870 nm, Electron. Lett. **34**, 184–186 (1998)

46. G. H. B. Thompson: *Physics of Semiconductor Laser Devices* (Wiley, Chichester 1980)

47. S. R. Chin, P. D. Zory, A. R. Reisinger: A model for GRINSCH-SQW diode lasers, IEEE J. QE **24**, 2191–2214 (1988)

48. W. Nakwaski: Three dimensional time dependent thermal model of catastrophic mirror damage in stripe geometry double-hetero GaAs-(AlGa)As diode lasers, IEEE J. QE **21**, 331–334 (1988)

49. G. Chen, C. L. Tien: Facet heating of quantum well lasers, J. Appl. Phys. **74**, 2167–2174 (1993)

50. U. Menzel, A. Bärwolff, P. Enders, D. Ackermann, R. Puchert, M. Voß: Modelling the temperature dependence of threshold current, external differential efficiency and lasing wavelength in QW laser diodes, Semicond. Sci. Technol. **10**, 1382–1392 (1995)

51. C. H. Henry, P. M. Petroff, R. A. Logan, F. R. Merritt: Catastrophic damage of $Al_x Ga_{1-x}As$ double-heterostructure laser material, J. Appl. Phys. **50**, 3721–3732 (1979)

52. M. Voss, C. Lier, U. Menzel, A. Bärwolff, T. Elsaesser: Time-resolved emission studies of GaAs/AlGaAs laser-diode arrays on different heat sinks, J. Appl. Phys. **79**, 1170–1172 (1996)

53. U. Menzel: Self consistent calculation of facet heating in asymmetrical coated edge-emitting diode lasers, Semicond. Sci. Technol. **13**, 265–275 (1998)
54. H. W. Press, B. P. Flannery, S. A. Teukolski, W. T. Vetterling: *Numerical Recipes in Pascal* (Cambridge Univ. Press, Cambridge 1989)
55. U. Menzel, R. Puchert, A. Bärwolff, A. Lau: Facet heating in high power GaAlAs/GaAs edge emitting laser diodes, Proc. 26th European Solid State Device Research Conference Bologna 1996 (Editions Frontiers, Gif-sur-Yvette 1996) pp. 749–752
56. U. Menzel, R. Puchert, A. Bärwolff, A. Lau: Heating of the front and rear facets of GaAlAs/GaAs edge emitting laser diodes, SPIE Proc. **2994**, 591–599 (1997)
57. R. Puchert, A. Bärwolff, U. Menzel, A. Lau, M. Voß, T. Elsaesser: Facet and bulk heating of GaAs/AlGaAs high-power laser arrays studied in spatially resolved emission and micro-Raman experiments, J. Appl. Phys. **80**, 5559–5563 (1996)
58. R. P. Sarzala, W. Nakwaski: An appreciation of usability of finite element method for the thermal analysis of stripe geometry diode lasers, J. Therm. Anal. **36**, 1171–1189 (1990)
59. R. P. Sarzala, W. Nakwaski: Finite-element thermal analysis of buried-optical-guide lasers, J. Therm. Anal. **39**, 1297–1309 (1993)
60. A. Bärwolff, R. Puchert, P. Enders, U. Menzel, D. Ackermann: Analysis of thermal behavior of high power semiconductor laser arrays by means of the finite-element method (FEM), J. Therm. Anal. **45**, 417–436 (1995)
61. R. Puchert, M. Voß, Ch. Lier, A. Bärwolff: Transient temperature behavior of GaAlAs/GaAs high power laser arrays on different heat sink, SPIE Proc. **2997**, 26–34 (1997)
62. R. Puchert, U. Menzel, A. Bärwolff, M. Voß, C. Lier: Influence of heat source distribution in GaAs/GaAlAs quantum well high power laser arrays on temperature profile and thermal resistance, J. Therm. Anal. **48**, 1273–1282 (1997)
63. R. Puchert, A. Bärwolff, M. Voß, U. Menzel, J. W. Tomm, J. Luft: Transient thermal behavior of high-power diode-laser arrays, IEEE Transaction on Components, Packaging and Manufacturing Technology, Part **A**,23, 95–100 (2000)
64. R. Puchert, J. W. Tomm, A. Jaeger, A. Bärwolff: Irreversible emitter failure and thermal facet load in high-power laser-diode arrays, Appl. Phys. A **66**, 1–4 (1998)
65. A. Ovtchinnikov, J. Näppi, J. Aarik, S. Mohrdiek, H. Assonen: Highly efficient 808 nm range Al-free lasers by gas-source MBE, SPIE Proc. **3004**, 34–42 (1997)
66. F. X. Daiminger, F. Dorsch, S. Heinemann: Aging properties of AlGaAs/GaAs high-power diode lasers, SPIE Proc. **3244**, 587–593 (1998)
67. D. Botez, D. R. Scifres: *Diode-Laser Arrays* (Cambridge Univ. Press, Cambridge 1994)
68. M. Baeumler, J. L. Weyher, S. Müller, W. Jantz, R. Stibal, G. Herrmann, J. Luft, K. Sporrer, W. Späth: Investigation of degraded laser diodes by chemical preparation and luminescence microscopy, IOP-Publ. Inst. Phys. Conf. Ser. **160**, 467–470 (1997)
69. J. W. Tomm, A. Bärwolff, U. Menzel, M. Voß, R. Puchert, T. Elsaesser, F. X. Daiminger, S. Heinemann, J. Luft: Monitoring of aging properties of AlGaAs high-power laser arrays, J. Appl. Phys. **81**, 2059–2063 (1997)

70. J. W. Tomm, A. Bärwolff, Ch. Lier, T. Elsaesser, F. X. Daiminger, S. Heinemann: Laser based facet inspection system, SPIE Proc. **3000**, 90–98 (1997)

71. A. Richter, J. W. Tomm, Ch. Lienau, J. Luft: Optical near-field photocurrent spectroscopy: a new technique for analysing microscopic aging processes in optoelectronic devices, Appl. Phys. Lett. **69**, 3981–3983 (1996)

72. Ch. Lienau, A. Richter, J. W. Tomm: Near-field photocurrent spectroscopy: a novel technique for studying defects and aging in high-power semiconductor lasers, Appl. Phys. A **64**, 341–351 (1997)

73. J. W. Tomm, A. Jaeger, A. Bärwolff, T. Elsaesser, A. Gerhardt, J. Donecker: Aging properties of high-power laser-diode arrays analysed by Fourier-transform photocurrent measurements, Appl. Phys. Lett. **71**, 2233–2235 (1997)

74. J. W. Tomm, A. Bärwolff, A. Jaeger, T. Elsaesser, J. Bollmann, W. T. Masselink, A. Gerhard, J. Donecker: Deep level spectroscopy of high-power laser-diode arrays, J. Appl. Phys. **84,** 1325–1332 (1998)

75. S. Orsila, M. Toivonen, P. Savolainen, V. Vilokkinen, P. Melanen, M. Pessa, M. Saarinen, P. Uusimaa, P. Corvini, F. Fang, M. Jansen, R. S. Nabiev: High power 600 nm range lasers grown by solid source molecular beam epitaxy, SPIE Proc. **3628**, 203–208 (1999)

76. A. Knauer, G. Erbert, H. Wenzel, A. Bhattacharya, F. Bugge, J. Maege, W. Pittroff, J. Sebastian: 7 W CW power from tensile-strained GaAsP/AlGaAs ($\lambda = 735$ nm) QW diode lasers, Electron. Lett. **35**, 638–639 (1999)

77. D. Z. Garbuzov, N. Y. Antonishkis, A. D. Bondarev, A. B. Gulakov, S. Z. Shigulin, N. I. Katsavets, A. V. Kochergin, E. V. Rafailov: High-power 0.8 μm InGaAsP-GaAs SCH SQW lasers, IEEE J. QE **27**, 1531–1536 (1991)

78. J. K. Wade, L. J. Wawst, D. Botez, J. A. Morris: 8.8 W CW power from broad-waveguide Al-free active region ($\lambda = 805$ nm) diode lasers, Electron. Lett. **34**, 1100–1101 (1998)

79. Y. Oeda, T. Fujimoto, Y. Yamada, Y. Yamada, H. Shibuya, K. Muro: High-power 0.8 μm band broad-area laser diodes with a decoupled confinement heterostructure, CLEO 98 Tech. Dig. Paper CMD 7 (1998) p. 10

80. S. O. Brien, H. Ransom, M. Hagberg, E. Zucker, H. Zhao: High-power narrow-aperture multimode AlGaAs-based lasers at 840 nm, CLEO 98 Tech. Dig. paper CMD 7 (1998) p. 15

81. S. O'Brien, H. Zhao, R. J. Lang: High-power wide-aperture AlGaAs-based lasers at 870 nm, Electron. Lett. **34**, 184–186 (1998)

82. X. He, S. Srinivasan, S. Wilson, C. Mitchell, R. Patel: 10.9 W continuous-wave optical power from 100 μm aperture InGaAs/AlGaAs (915 nm) laser diodes, Electron. Lett. **34**, 2126–2127 (1998)

83. S. O'Brien, H. Zhao, A. Schoenfelder, R. J. Lang: 9.3 W CW (In)AlGaAs 100 μm wide lasers at 970 nm, Electron. Lett. **33**, 1869–1870 (1997)

84. A. Al-Muhanna, L. J. Mawst, D. Botez, D. Z. Garbuzov, R. U. Martinelli, J. C. Conolly: High-power (> 10 W) continuous-wave operation from 100 μm aperture 0.97 μm emitting Al-free diode lasers, Appl. Phys. Lett. **73**, 1182–1184 (1998)

85. D. Botez, L. J. Mawst, A. Bhattacharya, J. Lopez, J. Li, T. F. Kuech, V. P. Iakovlev, G. I. Suruceanu, A. Caliman, A. V. Syrbu, J. Morris: 6 W CW front-facet power from short-cavity (0.5 mm), 100 μm stripe Al-free 0.98 μm emitting diode lasers, Electron. Lett. **33**, 2037–2039 (1997)

86. W. Pittroff, F. Bugge, G. Erbert, A. Knauer, J. Maege, J. Sebastian, R. Staske, A. Thies, H. Wenzel, G. Traenkle: Highly reliable tensily strained 810 nm QW laser diode operating at high temperatures, IEEE Proc. LEOS'98 (1998) paper WQ3

87. M. Watanabe, H. Shiozawa, O. Horiuchi, Y. Itoh, M. Okada, A. Tanaka, K. Gen-ei, N. Shimada, H. Okuda, K. Fukuoka: High-temperature Operation (70° C, 50 mW) of 660-nm-Band InGaAlP Zn-Diffused Window Lasers Fabricated Using Highly Zn-Doped GaAs layers, IEEE Journal of Selected Topics in Quantum Electronics, **5**, 750–755 (1999)

Properties and Frequency Conversion
of High-Brightness Diode-Laser Systems

Klaus-Jochen Boller[1], Bernard Beier[2], and Richard Wallenstein[3]

[1] Universiteit Twente, Faculteit der Technische Natuurkunde
Postbus 217, 7500 AE Enschede, The Netherlands
[2] Heidelberger Druckmaschinen AG,
Kurfürsten-Anlage 52-60, D-69115 Heidelberg, Germany
[3] Universität Kaiserslautern, Fachbereich Physik,
Erwin-Schrödinger-Straße, D-67663 Kaiserslautern, Germany

Abstract. An overview of recent developments in the field of high-power, high-brightness diode-lasers, and the optically nonlinear conversion of their output into other wavelength ranges, is given. We describe the generation of continuous-wave (CW) laser beams at power levels of several hundreds of milliwatts to several watts with near-perfect spatial and spectral properties using Master-Oscillator Power-Amplifier (MOPA) systems. With single- or double-stage systems, using amplifiers of tapered or rectangular geometry, up to 2.85 W high-brightness radiation is generated at wavelengths around 810 nm with AlGaAs diodes. Even higher powers, up to 5.2 W of single-frequency and high spatial quality beams at 925 nm, are obtained with InGaAs diodes. We describe the basic properties of the oscillators and amplifiers used. A strict proof-of-quality for the diode radiation is provided by direct and efficient nonlinear optical conversion of the diode MOPA output into other wavelength ranges. We review recent experiments with the highest power levels obtained so far by direct frequency doubling of diode radiation. In these experiments, 100 mW single-frequency ultraviolet light at 403 nm was generated, as well as 1 W of single-frequency blue radiation at 465 nm. Nonlinear conversion of diode radiation into widely tunable infrared radiation has recently yielded record values. We review the efficient generation of widely tunable single-frequency radiation in the infrared with diode-pumped Optical Parametric Oscillators (OPOs). With this system, single-frequency output radiation with powers of more than 0.5 W was generated, widely tunable around wavelengths of 2.1 μm and 1.65 μm and with excellent spectral and spatial quality. These developments are clear indicators of recent advances in the field of high-brightness diode-MOPA systems, and may emphasize their future central importance for applications within a vast range of optical wavelengths.

What has attracted scientists and engineers most about lasers is, certainly, that strongly directed and monochromatic light beams can be generated with high power. Such radiation with high spatial beam quality and narrow spectral bandwidth is commonly termed high-brightness radiation. The general properties of diode lasers, i.e. small size, robustness, potentially low costs, and their high wall-plug efficiency have so far been the central issues in designing diode lasers for applications. Diode lasers with low brightness, though with

R. Diehl (Ed.): High-Power Diode Lasers, Topics Appl. Phys. **78**, 225–263 (2000)
© Springer-Verlag Berlin Heidelberg 2000

high power, have become important sources of radiation. The low brightness, however, limits their use to relatively coarse applications which do not exploit the main property of a laser – its coherence. Typical examples are high-power diode-lasers designed for pumping solid-state lasers, or diode-lasers for materials processing.

On the other hand, it is essentially the brightness of radiation which makes the key difference between a lamp-like source and a laser source. It is well known that powerful diode-laser radiation with high brightness would open new possibilities of applications. Examples are free-space communications, micro-machining, and high-speed, high-resolution printing. In addition the output from powerful high-brightness diode-lasers is suitable for a direct frequency conversion into new wavelength ranges by optically nonlinear crystals. Efficient generation of ultraviolet and visible beams is of high interest for holographic data storage, deposition process control, or laser display. Near- and mid-infrared output in wavelength ranges not directly accessible with diode-lasers is of high relevance for sensitive trace gas detection or gas monitoring in chemical or power plants. These possibilities have long since been realized and have thus initiated intense international research and development with the goal to provide advanced high-power diode-lasers with high brightness. This review is therefore devoted to such lasers, to characterize, understand and optimize their properties and to investigate a direct nonlinear application.

Most of the laser oscillators that have long formed the basis of laser technology, such as Ar-ion lasers or single-frequency solid-state lasers, naturally emit Continuous-Wave (CW) light of high brightness. Ironically this is the case because the named oscillators are low-gain devices, i.e. the light-amplification factor per resonator roundtrip is on the order of one percent. In such lasers the low gain is compensated by an optical resonator of correspondingly low loss, so-called high-Q resonators. To provide a lower loss per roundtrip than gain, the resonator mirrors have to be coated for a high reflectivity. This results on the order of about 100 roundtrips of the light before output coupling. Additionally, the mirrors have to possess a proper curvature, fitting their distance, to make sure that no light spills over the edges of the mirrors even after a large number of roundtrips. Such resonators are called optically stable [1]. Due to the high number of roundtrips it is the resonator which defines the spatial distribution of light inside the laser and thus also the shape of the output beam. The beam inside such a low-gain laser with a stable resonator is of Gaussian–Hermite or Gaussian–Laguerre shape, and such lasers emit a so-called Gaussian beam [2,3]. Given a high number of roundtrips, weak spatial filtering inside the resonator is very effective to restrict laser oscillation to the lowest-order Gaussian mode, which corresponds to the maximum spatial quality a light beam can attain. Similarly, even weak intracavity spectral filtering, such as by Lyot filters or etalons, is very effective to restrict the optical-emission spectrum to a single frequency. In summary, a laser beam with high spatial and spectral quality is the result of resonator properties, i.e. its spatial and spectral filtering effect during a high number of roundtrips. The gain medium merely acts to sustain the oscillation of light in a preferred mode which experiences the lowest loss.

The situation becomes different if high output power is to be generated while maintaining a high spectral and spatial quality. In general, the spatial and spectral properties deteriorate as the output coupling is increased for more efficiency and the oscillator is simply pumped harder for more output. The reason for this behavior is that the radiation now performs only a small number of roundtrips before output coupling. As a result the maximum output tends to be emitted in undesired directions and at undesired frequencies. These are the directions and frequencies experiencing the maximum gain–length product in a single or a few passes. In short, high gain in a laser reduces the influence of the resonator, its contribution to spatial and spectral filtering, and thus the brightness of the output beam.

Diode lasers are usually high-gain oscillators. Accordingly, it is difficult to control their spatial and spectral properties. Diode lasers offer roundtrip gains in the range 10^3 to 10^6 and are thus operated with correspondingly high roundtrip losses of more than 99%. A second complication, making the generation of high-brightness radiation particularly difficult with high-power diode-lasers, is the strong optically nonlinear response of the amplifying semiconductor material. An investigation of these effects in high-power diode-lasers and amplifiers in its full complexity is presented in this book by *Gehrig* and *Hess*. A more phenomenological picture of such effects and its consequences is used by *Mikulla*. Both approaches aim on gaining valuable information on how to optimize the brightness of such lasers by an improved design.

Nonlinear effects in laser diodes are described, e.g., in the books of *Petermann* [4] or *Agrawal* [5]. Here we give just a brief and simplified review. The refractive index in the diode laser increases with decreasing charge-carrier density through the Henry factor, α_H. Since the carrier density is reduced by stimulated emission, this corresponds to an intensity-dependent refractive index. In low-power single-frequency diode oscillators this is of less importance for the spatial quality of the output because here the light is confined in a narrow (single spatial mode) optical waveguide. One merely observes a slight broadening of the spectral bandwidth of the laser [6]. In high-power diodes (diode arrays, broad-area diodes, or diode amplifiers), which generally do not have such a confinement, the intensity-dependent index can lead to self-focusing, seen as an increasing distortion of the output beam with increasing power. At sufficiently high power densities the internal light field starts to form filaments which quickly lead to a destruction of the laser. This is true for diode-lasers which provide output powers well above 100 mW such as diode arrays and broad-area diodes, as well as for diode bars and stacks with even higher powers. Radiation with high spectral and spatial quality is available only from single-stripe lasers, which are low-power oscillators delivering a maximum output of 100 mW. In conclusion, achieving high-brightness radiation from diode-lasers requires specific methods which combine the high quality of radiation available from single-stripe lasers with the large quantity of radiation from high-power diodes.

This review concentrates on the recently developed diode Master-Oscillator Power-Amplifier (MOPA) systems which generate CW output beams with high brightness. Advanced prototype diode amplifiers have already shown several watts of CW output with an excellent spatial quality. AlGaAs and InGaAs diode amplifiers with lower power of typically 500 mW and 1 W at wavelengths around 810 nm and 980 nm, respectively, have been commercially available for several years [7]. Here, we report on 810 nm AlGaAs systems with higher output powers of up to 2.85 W. Improved MOPA systems based on InGaAs will be presented, which deliver more than 5 W at wavelengths around 925 nm.

In Sect. 1 we review current techniques to measure the spatial beam quality of a diode-laser beam. Section 2 gives an overview of the properties of low-power diode oscillators as they are used for the different MOPA systems described here. Section 3 presents the properties of standard high-power diode-lasers with low brightness and summarizes the main techniques that have been investigated so far for a better spatial and spectral control. The following Sect. 4 gives a general description of diode-MOPA systems. Examples of multi-watt systems based on GaAlAs (emitting at 810 nm) and based on InGaAs (925 nm) will be presented in Sects. 5 and 6. The final Sect. 7 reviews recent experiments on direct and efficient nonlinear conversion of diode-MOPA radiation, made possible by the high brightness of the diode output. We describe examples of CW ultraviolet and visible second-harmonic generation with output powers reaching the 1 W level, and hundreds-of-milliwatts widely tunable CW single-frequency IR generation using diode-pumped Optical Parametric Oscillators (OPOs). Such power levels have so far been the exclusive domain of powerful solid-state lasers.

1   Beam Quality

Of highest importance for the application of a diode-laser are the spatial properties of the output beam. An overview of various ways to quantify the spatial quality of laser beams, such as via the M^2 diffraction parameter, Power-In-the-Bucket [PIB], Strehl's ratio, Times-Diffraction-Limit number [TDL], and beam-quality number, as well as a comparison, can be found in [8]. An ideal laser with a stable, high-Q resonator would generate the well-known Gauss–Laguerre or Gauss–Hermite modes [9]. A unique property of such modes is that in their lowest order they describe a laser beam with a Gaussian intensity distribution in both transverse dimensions at any point along the beam [2]. It can be shown that such lowest-order modes represent light with the highest possible spatial beam quality in a particular sense: focusing of the beam generates the smallest beam cross section in the focus, and with a given focal cross section a Gaussian beam possesses the lowest divergence upon propagation. The term 'beam quality' is widely used by experimentalists, however, in many cases lacking an appropriate definition. Moreover, even with a well-defined expression of beam quality, this may be valid or useful only for a certain purpose. This can be seen, e.g., by considering frequency doubling of light in a single pass through an optically nonlinear crystal [10]. If one focuses the fundamental beam through the crystal with a Bessel-function-shaped transverse-intensity distribution, one may expect a higher conversion efficiency than with a Gaussian beam [11], although contrary results have also been presented [12]. A Gaussian beam can be perfectly mode-matched into a stable, high-Q power enhancement resonator because it is the eigenfunction of such resonators. Therefore, the conversion efficiency of second-harmonic generation with a nonlinear crystal placed inside such a cavity for power enhancement is highest with a Gaussian fundamental beam.

Having the choice between different definitions of beam quality, one should also consider that these definitions characterize a beam in different details. A complete characterization of a laser beam is achieved only by a measurement of the two-dimensional intensity and phase distribution across some plane through which the beam propagates. This reference plane can be chosen for convenience because the intensity and phase distribution of the beam in any other plane behind (or in front of) the reference plane can be calculated using Huygen's diffraction integrals [1,2]. Although this characterization of a beam provides the maximum information, it is clearly not very practical. A large amount of data is to be recorded and processed, before even a simple result is obtained, such as the beam diameter or a divergence angle at some other plane, say downstream behind a lens. This explains why, in practice, it is generally more useful to apply simpler methods of characterization, even if they provide only limited information on the beam.

A particularly practical specification of beam quality was common in the initial experiments to improve the spatial quality of the radiation from diode-laser arrays by various methods, such as by external cavities or injection locking [13]. In such experiments, a measure of beam quality was obtained as follows. The Full-Width-at-Half-Maximum (FWHM) diameter d_D of the beam intensity is measured at a sufficiently large distance D from the emitting facet, called the far field. The corresponding (FWHM) divergence angle calculated from $\theta_D = d_D/D$ can then be compared to the theoretical minimum value of far-field divergence, θ_{th}. The latter value is estimated from the transverse width of the active area in the facet, as specified by the manufacturer, assuming a specific intensity distribution in the facet and a monochromatic wave with specific phasefronts in the emitting area. As an example, a rectangular, $200\,\mu m = d_0$-wide facet emitting a plane wave with $810\,nm$ wavelength would diffract into a sinc^2-shaped angular distribution in the far field, with a divergence angle (FWHM) of $\theta_{th} = 2\sqrt{2}/\pi \times \lambda/d_0 = 0.20°$. As a figure of beam quality, the ratio of θ_D and θ_{th} gives the TDL number, i.e. how many times diffraction-limited the observed beam is. The advantage of this method is its simplicity and that it quickly gives a realistic impression of the beam quality. Due to the underlying uncertainties, however, such as assuming a specific intensity and phase distribution in the facet plane, the TDL number remains merely a coarse estimate of beam quality.

1.1   The Diffraction Parameter M^2

A highly general method to quantify beam quality is based on the second moments σ_x^2 and σ_y^2 of the transverse intensity distribution [14,15,16,17]. This moment is determined by integrating the transverse intensity distribution at some point z along the beam, weighted with the square of the distance from the beam axis, and normalizing to the total power in the beam. Defining $W = 2\sqrt{\sigma^2}$ gives an effective $1/e^2$-radius which can be determined for any beam, i.e. also for real, arbitrary, nongaussian beams. The radius can be evaluated along z in both of the two transverse dimensions, x or y, yielding $W_x(z)$ or $W_y(z)$. These radii are of particular importance because they can be used to describe an important invariance of any light beam under propagation. Consider the transverse dimension x. It can be shown that W_x varies hyperbolically with z, i.e.

$$W_x^2(z) = W_{0x}^2 \left[1 + \left(\frac{z - z_{0x}}{z_{Rx}} \right) \right] , \qquad (1)$$

where W_{0x} is measured in the waist (focus plane) of the beam at location z_{0x}. z_{Rx} is the Rayleigh length describing the length of the focus, i.e. at what distance from the waist the radius increases by a factor of $\sqrt{2}$. A corresponding equation is valid in the y dimension.

A replacement of W_{0x} and z_{Rx} in (1) with the $1/e^2$-power beam radius w_0 of the Gaussian beam with wavelength λ and its Rayleigh length $z_R = \pi/(w_0^2\lambda)$, respectively, shows that the radius of an arbitrary, real beam, defined via the second moment, propagates much like the beam radius of a Gaussian beam with only a slight difference: if W_{0x} is given, the real beam possesses a far-field divergence which is bigger than for a Gaussian beam of same radius in the waist by a characteristic factor, M_x^2. This factor is called the diffraction parameter (or sometimes the beam-quality factor) for the x direction. Alternatively, if the far-field divergence is the same, focusing of the real beam leads to a waist radius which is larger by M_x^2 compared to the waist radius of a focused Gaussian beam. In summary of these properties some prefer to think of a real beam as a Gaussian beam with a longer wavelength of λM^2 (and thus diffraction-increased by M^2). With the M^2 definition of beam quality, a poor beam correspond to a large M^2 while the best possible beam quality is represented by $M_x^2 = M_y^2 = 1$, which are the diffraction parameters of a Gaussian beam.

This scaling of real beams with respect to a Gaussian beam explains the high usefulness of the concept of second moments: although (1) describes the simplest case of propagation in free space, it can easily be extended to propagation through complex optical systems (e.g. a set of lenses). Just as for Gaussian beams, optical systems are described by a simple matrix formalism [2,3]. This is a highly useful property to design lens systems for real beams without actually having to know further details of the beam profile.

A correct measurement of the M^2-parameter can be achieved as follows: the beam under inspection is sent through a focusing lens to form a beam waist. With a spatially resolving detector, such as a two-dimensional Charge-Coupled Device (CCD) camera, with a moving pinhole, or crossed moving slits in front of a photodiode, the two-dimensional transverse intensity distribution is measured at various positions (planes) z along the beam, e.g., in front of the focus, in the focus and behind. The second moment is determined numerically for each position, and plotted versus the position. As a result, the data follow the shape given by (1), such that a hyperbola can be fitted to the data. The parameter M^2 is then given by the product of far-field divergence angle (half width at $1/e^2$) and waist radius (W_0) derived from the fit, divided by λ/π (the radius–divergence product obtained for a Gaussian beam). The position of the waist, z_0, and the waist radius, W_0, are also obtained from the fit. Performing this procedure in both dimensions, x and y, the beam quality is fully characterized to second order. In analogy to Gaussian beams, different values for z_{0x} and z_{0y} indicate an astigmatism, whereas different values for W_{x0} and W_{y0} (with $z_{x0} = z_{y0}$) indicate an elliptic beam cross section.

The spatial beam quality of the high-power diode-laser systems described in Sects. 5 and 6 is specified mostly in terms of the M^2 parameter. Transverse intensity profiles were recorded with $5\,\mu m$ spatial resolution using a two-dimensional beam profiler with two moving, vertically arranged slits. An example of such a measurement is shown in Fig. 1, where we recorded the beam diameter ($2\,W$) of an $810\,nm$ AlGaAs master oscillator tapered-amplifier system as a function of the propagation distance z behind a focusing lens. The beam diameter was recorded in the direction which is parallel to the amplifier's active area. The data show that the beam waist is located at $101\,mm$ having a beam diameter of approximately $45\,\mu m$. The solid curves shows a least-squares fit of the hyperbola function of (1), which agrees well with the experimental data. The fit yields an M^2 parameter of 1.1 ± 0.1, where the given error is the statistical uncertainty of the least-squares fit. The slight and partially systematic deviations are caused by the following effect: the profiler data processor operates in a simplified mode which performs a least-squares fit of a Gaussian-shaped transverse intensity profile to the measured beam profiles and provides the $1/e^2$ power diameter of the Gaussian as the read out. This profiler evaluation of the second moment is sufficiently good only for low M^2 values, such as present here. We note, however, that significant errors would occur with high M^2 values [14,18,19,20,21]. In this

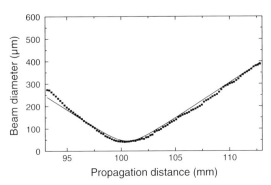

Fig. 1. $1/e^2$-power diameter of the beam from an AlGaAs MOPA system measured as function of the propagation distance behind a focusing lens with a beam profiler. The CW output power of the investigated diode-MOPA system is 600 mW at a wavelength of 810 nm. The *solid curve* is a least-squares fit of (1) to the data, which gives $M^2 = 1.1$

case, an evaluation via the second moment is strongly recommended instead of the method of a $1/e^2$ Gaussian fit diameter. Obviously, the method of $1/e^2$ Gaussian diameters fails also if the hyperbolic fit gives M^2 values below unity.

1.2   Measurement of the Wavefront

The hyperbolic-fit method, applied either to the second moment, or to the $1/e^2$-Gaussian beam diameter at low M^2 values, is certainly a powerful method to describe beam quality. A characterization to more detail has to include, however, higher orders than the second moment [22]. Such characterization is possible with a measurement of the shape of the wavefront using interferometry [23]. The basic arrangement for a method which we applied to characterize the output from our tapered-amplifier systems, dynamic shearing interferometry, is depicted in Fig. 2. Further details on this method and alternative interferometric techniques are described in [24].

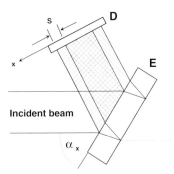

Fig. 2. Arrangement for dynamic shearing interferometry. On the detector (D) two reflections from a glass plate (E) generate an interference pattern which is evaluated to obtain the wavefront of the incident beam

The beam under inspection is sent through a plane, parallel, glass plate E. The purpose of the plate is to reflect a first part of the beam from the front surface and to superimpose an identical but sheared copy of the beam from the back surface to form an interference pattern. This pattern can be recorded with a spatially resolving detector, such as a two-dimensional CCD camera, and contains the desired information on the wavefront of the incident beam. The principal ideas and steps involved in a wavefront measurement will be reviewed.

In the detector plane the phasefront of the original beam, $P(x, y)$, can be expressed as its deviation from a plane wave, where x and y are the transverse coordinates. For an illustration, consider a diverging Gaussian beam with vertical incidence on the detector D. In this case the wavefront is spherical and can be well approximated by a second-order parabolic function, $P(x, y) = (x^2 + y^2)/R$, where the coefficient R expresses the wavefront curvature or defocusing [3]. Similarly, the front of an unknown incident beam, reflected from the front of E, can be expressed by a two-dimensional Taylor series,

$$P(x, y) = \sum_{n=0}^{k} \sum_{m=0}^{n} B_{nm} x^m y^{n-m} \tag{2}$$

(which had been restricted to $N = (k+1)(k+2)/2$ terms). The lowest-order coefficients in this equation are known as the tilt of the front with respect to the detector plane (B_{01} and B_{10}), the curvature or defocusing (B_{02}, B_{20}), primary coma (B_{12} and B_{21}), and primary spherical aberration (B_{04}, B_{22}, and B_{40}) [25,26].

The beam reflected from the back side of the plate is an identical copy of the incident beam except that it is sheared by a distance S into the x direction. The sheared beam can thus be written using the same coefficients B_{nm} as for the beam

$$P(x + S, y) = \sum_{n=0}^{k} \sum_{m=0}^{n} B_{nm} (x + S)^m y^{n-m} , \tag{3}$$

from the front of the plate, if S is included in the argument.

The optical path difference between the incident and sheared wavefronts in the detector plane, $\Delta P(x, y)$, is obtained by subtracting (2) from (3). The result can again be expanded in a power series,

$$\Delta P_S(x, y) = P(x + S, y) - P(x, y) = \sum_{n=0}^{k} \sum_{m=0}^{n} C_{nm} (x)^m y^{n-m} , \tag{4}$$

where the C coefficients are related to the wavefront coefficients B by a set of linear equations,

$$C_{nm} = \sum_{j=1}^{k-n} \binom{j+m}{j} S^j B_{j+n,j+m} . \tag{5}$$

Equations (4) and (5) state that finding the wavefront of the incident beam can be traced back to measuring the path difference between the beam and its sheared copy, i.e. to measuring $\Delta P_S(x, y)$.

For finding the path difference let us refer to Fig. 2 again. Both the shearing distance in the x direction, S, and the propagational phase lag, δ, of the sheared beam in a particular point in the detector plane, depend on the optical thickness of the plate and on the angle of incidence, α_x. It is important to note that δ sweeps over many radians with only a tiny variation of α_x while S remains approximately unchanged. This can be used for an easy measurement of $\Delta P_S(x, y)$ as follows: a slight variation of α_x over a small angular interval of typically a few mrad is used to generate a time-varying interference pattern on the detector. Four interference patterns are recorded, $I_1(x, y)$ to $I_4(x, y)$, where the phase lag in an arbitrary but fixed point, (x_0, y_0), on the detector is $\delta_1(x_0, y_0) = 0$, $\delta_2(x_0, y_0) = \pi/2$, $\delta_3(x_0, y_0) = \pi$, and $\delta_4(x_0, y_0) = 3\pi/2$, respectively. The corresponding instances, t_1 to t_4, at which the electronics reads out the interference patterns $I_i(x, y)$ from the CCD camera, are selected by an electronic trigger circuit. Having recorded the four patterns, the path difference between the original and sheared wavefronts can be calculated,

$$\Delta P_S(x, y) = \frac{\lambda}{2\pi} \tan^{-1} \left\{ [I_4(x, y) - I_2(x, y)] / [I_1(x, y) - I_3(x, y)] \right\} . \quad (6)$$

In summary, the wavefront of the incident beam is determined with four steps: (i) recording of four suitably phased interferograms $I_i(x, y)$, (ii) calculation of $\Delta P_S(x, y)$ with (6), (iii) calculation of the C coefficients with a fit of (4) to $\Delta P_S(x, y)$, and (iv) solving (5) for the B coefficients. The described procedure gives the full information on the incident wavefront if the beam possesses rotational symmetry. Since in particular the beam from a diode laser may lack such symmetry these four steps are repeated with additional shearing of the beam vertically in the y direction by a distance T. The wavefront of any laser beam can thus be recorded with a total of eight steps and the result is coefficients describing the strength of the different types of aberrations present. These coefficients are also known as Kingslake coefficients from classical optics describing the wavefront distortions caused by a lens system [24].

We used dynamical shearing interferometry to characterize the spatial beam properties of high-power diode systems. The experiments were performed with an interferometer which displays the shape of the wavefront, and calculates the expansion coefficients as described above. An example of wavefront measurements with a near-diffraction-limited beam from an In-GaAs diode-MOPA system is displayed in Fig. 8, where the lowest-order coefficients for tilt and curvature have been set to zero. Figure 8 thus corresponds to the wavefront measured in a beam waist. In comparison to a beam characterization via the M^2 parameter, the advantage of shearing inteferometry is that the full amount of information can be obtained. Different types of aberrations contributing to a nonunity M^2 value are displayed separately with their corresponding strengths. This additional information is helpful to identify (and thus remove) the underlying physical mechanisms causing these aberrations. In diode-laser systems such mechanisms are aberrations in the collimator and other optics placed behind the diode amplifier, but also the spatially inhomogeneous refractive-index distribution within the active area of the amplifier.

A wavefront measurement can also be used to assign a single quality factor to the inspected beam, as does an M^2 measurement. For this purpose one determines the Root-Mean-Square (RMS) deviation of the measured phase from that of an ideal beam (which in Fig. 8 is a plane located at $z = 0$). For a calculation of the RMS value, the phase deviations are spatially integrated over the beam cross section, by weighting each deviation with its corresponding intensity. The integral is normalized to the total power in the beam. The weighting of phase deviations accounts for the fact that even strong phase distortions do not alter much the further propagation of the beam, if such distortions occur in the far wing of the beam profile, where the intensity is low. There is also a practical meaning associated with the RMS value. Multiplying the RMS value by $\lambda/2\pi$ one obtains the usual figure which is used to specify the surface quality of optical components. As an example, if $\lambda/2\pi$ times the RMS value yields $\lambda/20$, the quality of the laser beam can be viewed as that of a perfect beam reflected from a mirror with a surface roughness of $\lambda/10$ across the illuminated area. There should further be a close correspondence of the RMS value to the M^2 parameter, at least for high-quality beams. To our knowledge, such correspondence has not been investigated so far.

2 Single-Stripe Diode Lasers

Single-stripe diode-lasers are a basic component of diode-MOPA systems where they are used to form a low-power oscillator emitting with high spectral and spatial quality. The working principle of single-stripe diodes and their basic properties are treated in most textbooks on lasers. More details and a description of the various possible designs can be found, e.g., in the books of *Petermann* [4] or *Agrawal* [5].

Diode lasers are based on stimulated emission in a p-n-junction of semiconductor material. If the junction is supplied with a forward current of sufficient density, the electrons in the conductance band and holes in the valence band form an inversion which amplifies light.[1] The thickness of this laser-active zone is about $1\,\mu m$ as given by the typical diffusion length of the charge carriers. In the other dimension the size of the active zone can be defined by the shape of the metallic electrode on top of the chip. The first diode lasers were realized in 1962 [27]. The injection current was applied via

---

[1] The recently demonstrated quantum-cascade lasers are based on transitions within only one of the bands

a planar electrode and the junction was made of a single type of semiconductor material (GaAs) with p- and n-doping (homojunction). Optical feedback then was provided by the Fresnel reflectivity of the cleaved end facets of the chip to form a Fabry–Perot (FP) cavity with plane mirrors. With this simple a structure, the laser showed a number of severe disadvantages, particularly a high current density at threshold, the need for cryogenic temperatures, and the poor spectral and spatial quality of its output limited the interest in such devices.

Now, after decades of intense research, advanced manufacturing techniques are available which allow one to define a variety of complicated structures within the diode-laser, designed to avoid all of the above-mentioned disadvantages. Examples are heterojunctions built from several types of semiconductor material. The resulting modifications of the band structure provide a spatially Graded Refractive INdex and Separate Confinement of carriers and the optical wave (GRINSCH) for a stronger interaction and lower threshold currents. The energy and density of electronic states can be engineered with quantum structures, which confine the position of charge carriers to small regions comparable in size to the DeBroglie wavelength of the carriers. Single Quantum Well (QW) and Multiple Quantum Well (MQW) structures are now routinely used, but recently also quantum dot lasers have been realized [28,29]. Other diodes provide a reliable single-frequency output by spectral filtering of the optical feedback within the diode-laser using internal Distributed BRagg grating (DBR) and Distributed FeedBack grating (DFB) structures. Vertical Surface-Emitting Lasers (VCSELs) with cavity lengths short compared to the laser wavelength have been built to restrict lasing to a single mode.

A large number of different diode-laser types are available now, designed for specific applications or properties. An overview can be found in [30,31]. Among different materials, AlGaAs diodes have been developed to a comparatively high level. The reason is a traditionally high interest in the wavelength range emitted (750 nm to 830 nm) which covers the pump-absorption bands of neodymium-doped solid-state laser crystals. Further developments with respect to the growth of various other semiconductor materials have made diode-lasers available within a vast spectral range, from lead sulfide lasers in the mid-IR [32] to the recent realization and the beginning of commercial availability of GaN lasers in the blue–ultraviolet range [33,34].

The most basic technique to improve the spatial properties of the diode-laser output beam is to form a so-called single-stripe laser. In this case the pump current is applied via a thin stripe electrode, which is only a few μm wide and extends over the entire length of the laser chip. The purpose of the stripe is to confine the transverse distribution of charge carriers, and thus the width of the light amplifying (active) zone to typically 3 μm. This suppresses transverse-mode laser oscillation and favors oscillation in longitudinal modes. The absence of transverse-mode oscillation gives single-stripe lasers their high spatial beam quality. Our own measurements revealed typical diffraction parameters $M^2 = 1.05$ in both dimensions. A further proof of spatial quality is that a high percentage (typically up to 90%) of the diode output can be spatially mode-matched into the fundamental Gaussian mode of a high-finesse power build-up cavity.

Due to the simple longitudinal mode structure of single-stripe lasers, also their spectral bandwidth is easy to control. Even lasers with a plain Fabry–Perot (FP) resonator based on the reflectivity of the chip's end facets often show emission of a single, main frequency with a spectral bandwidth of typically 10 MHz, while other FP side modes are strongly suppressed by tens of dB through gain competition [35]. If required, the spectral bandwidth of a single-stripe laser can be further narrowed into the kilohertz range with optical feedback from an external mirror [6]. A continuous tuning of the laser frequency over most of its gain spectrum is possible if one of the laser facets is AR coated and feedback is provided with an external grating in Littman configuration [36,37]. Such lasers have been used as well in our experiments described below. The collimated output from the laser chip is sent at grazing incidence onto a diffraction grating. The first diffraction order is retro-reflected from a plane mirror to the grating and from there, via a second diffraction, back into the chip. Output coupling from the Littman laser is obtained as the zero-order diffracted beam. The laser wavelength is widely tuned by tilting the plane mirror. A fine tuning of the laser frequency can be achieved by mounting the plane mirror on a piezo actuator which controls the optical cavity length. If required, a proper synchronization of the mirror's tilt angle and displacement, either mechanically or electronically, can be used to tune continuously the output frequency (without mode hops) over thousands of GHz. A long-term stability is obtained by electronically locking the laser frequency to an absolute reference, such as an atomic absorption line. So far, the lowest spectral bandwidth of a diode-laser of 40 Hz has been achieved with a combination of optical feedback and electronic locking to a reference line [38].

With these spatial and spectral properties single-stripe lasers may be considered as perfect for most applications. Unfortunately the output of such lasers is limited at present to powers of approximately 100 mW. The reason for this is that such modest power already corresponds to a high intensity of several MW/cm^2 since the emitting area of a single-stripe laser ($1\,\mu m \times 3\,\mu m$) is microscopically small. Above the given intensity, residual absorption of light in surface states of the facets leads to Catastrophic Optical Damage (COD) of the laser. Given this limit and the fixed $1\,\mu m$ thickness of the junction, the output power of a diode-laser can be increased only by increasing the lateral dimensions of the emitting area significantly above $3\,\mu m$, e.g., by increasing the lateral dimensions of the active zone. This approach is used to build high-power diode-lasers but it re-introduces transverse-mode oscillation and thus a poor spectral and spatial beam quality.

3   High-Power Diode Lasers

Numerous types of high-power lasers are available. For a review and comparison of different approaches in this field the reader may refer to the books by *Botez* and *Scifres* [39] or *Carlson* [40]. Most high-power diode-lasers are designed as diode arrays or broad-area diodes. Typically the emitting area of the facet is increased to about 200 μm in the plane of the junction to avoid COD. As a result the maximum output power of such diodes is on the order of 1 W. Much higher powers can be generated by further extension of the emitting area, by manufacturing typically 20 high-power diodes close to each other on the same chip to form a diode bar. Vertical packaging of such bars, each providing tens of watts output is used to form a so-called diode stack. Such stacks, with an emitting area of about 1 cm × 1 cm, deliver power at the kilowatt level.

A severe disadvantage of the increased transverse size of the active zone in a high-power laser is that it enables transverse-mode oscillation. In most cases transverse-mode oscillation clearly dominates over longitudinal modes because the latter have the smallest gain–length product per roundtrip. Further, all modes are mutually coupled via gain competition and eventually phased via scattering and nonlinear effects. Correspondingly, for a better description it is more appropriate to consider the oscillation of so-called supermodes, which are phased superpositions of high-order transverse modes [41,42,43], or to treat such lasers numerically [44]. The most obvious consequence of supermode oscillation is a low quality of the emitted laser beam. For example, the far field of a diode-laser array or broad-area diode with a 200 μm-wide emitting area shows a large divergence angle of about 10° (FWHM) in the plane of the junction, which is about 50 times larger than that of a diffraction-limited beam emitted from an emitter of the same width. Further, the presence of a large number of supermodes, each having a different emission frequency, gives high-power lasers a considerable spectral bandwidth. For GaAlAs high-power diode-lasers emitting around 810 nm, this bandwidth is typically one to several nanometers, i.e. it is on the order of 10^3 GHz.

It is instructive also to quantitatively compare the overall quality of the output from such a high-power diode-laser to that of a single-stripe laser in terms of brightness. To express the brightness of a light source one can refer the output power (in mW) to the emitting area (in μm^2), to the solid angle of the far-field emission (in sr), and to the spectral bandwidth (in MHz). In these units the brightness of a standard single-stripe laser with a spectral bandwidth of 10 MHz and an output power of 100 mW is approximately 30 mW/ μm^2/sr/ MHz. The brightness of a high-power diode-laser is lower by many orders of magnitude. For the above example of a diode array one obtains a brightness of 5×10^{-6} mW/ μm^2/sr/ MHz, due to the 60-times wider emitting area and the 10^5-times larger spectral bandwidth. The brightness of diode bars and diode stacks with output powers in the range 10 W to some kW can be calculated accordingly. It turns out that for these lasers the brightness is even lower than that of a diode array or broad-area diode.

The spectral and spatial beam quality of high-power diode-lasers can be substantially improved with the objective of obtaining a brightness much higher than that of single-stripe lasers. Methods under investigation have been optical feedback by external cavities [45,46,47], the formation of Anti-Resonant Ridge Optical Waveguides in laser arrays (ARROWs) [48,49,50], or the recently developed α-DBR broad-area lasers [51]. A method of practical importance to many experimentalists is injection-locking, because it can be performed with a commercially available, standard high-power diode array or broad-area diode [52,53,54]. For locking to occur, the output from a single-frequency laser (which might be a single-stripe diode-laser) is injected into the high-power diode and tuned into the locking range, i.e. close to the free-running frequencies of the supermodes of the high-power laser. In this case the oscillation frequency of most supermodes becomes identical with that of the injected laser. With a variation of the injection angle, the relative temporal phases of the locked supermodes can be adjusted to maximize constructive interference in a particular angle of the far field. As a result a major fraction of the total power is emitted as a single, near-diffraction-limited output beam with a narrow spectral bandwidth (roughly that of the single-stripe diode-laser). The resulting improvement of brightness has been shown to reach factors of 10^8 and higher although, in general, not all of the supermodes can be properly locked. So far, the highest output power of 640 mW was achieved by injection-locking the 800 mW free-running output of a diode-laser array with 45 mW of power from a grating-stabilized single-frequency, single-stripe diode-laser [13].

4   Diode Amplifiers

The approaches to generate high-brightness radiation with high-power diode-lasers described so far suffer from several fundamental limits. These are either an undesired multi-lobed output (such as with ARROWs), a broad spectral bandwidth (α-DBR lasers), or a low power stability of the output (with external cavities or injection-locking) due to low tolerances for mechanical perturbations and thermally induced drifts. In the mid-1980s researchers started to concentrate on a most promising approach to overcome the named disadvantages. This approach is the so-called MOPA system.[2] A single-stripe diode master-oscillator generates radiation with high spatial and spectral quality, though with low power on the order of tens of milliwatts. A high output power without significant reduction of the beam quality is achieved by amplification of the oscillator radiation in a high-power broad-area diode amplifier. Most important for a proper operation of the broad-area diode as an amplifier is

---

[2] This approach has proven successful already with a variety of other sources, e.g., for pulsed dye lasers, excimer lasers, solid-state lasers, and ns-pulsed OPOs

that at least one, preferably both, of its facets are AR-coated. This suppresses optical feedback and increases the threshold current for laser oscillation to above the maximum current rating. Using different rectangular geometries for the metallic electrodes and thus the active zone (widths of 400 to 600 μm, lengths of 500 to 2300 μm), a maximum CW output power of 3.3 W had been demonstrated [55]. A rectangular broad-area amplifier (600 μm wide and 1100 μm long) had been used to amplify the 400 mW output of TEM$_{00}$-mode radiation from a Ti:sapphire laser to a near-diffraction-limited CW output beam of up to 3.7 W power [56]. Meanwhile, a number of different diode MOPA designs have been realized, where the master-oscillator and one or a chain of several amplifiers are manufactured on the same chip [39,40].

To achieve higher optical output powers at lower optical input powers, rectangular broad-area amplifiers have been used in a double-pass. This also has the advantage that a high-quality AR coating is required for only one of the facets while the other facet is coated for High Reflectivity (HR, moderate quality is sufficient here). In a prior experiment we generated up to 780 mW of 810 nm single-frequency radiation in a near-diffraction-limited output beam (diffraction parameter $M^2 = 1.1$), by injecting 90 mW from a free-running, single-frequency single-stripe diode laser into the 100 μm × 750 μm active area of a GaAsAs broad-area amplifier in double-pass [57].

The power limit of broad-area amplifiers is given by self-focusing effects in the active zone. A double-pass amplifier, however, imposes an even more stringent limit. The standing waves in such devices create a much stronger inhomogeneity of carrier depletion and thus of the refractive index compared to a single-pass device. Consequently, double-pass amplifiers are very susceptible to self-focusing and defocusing, i.e. to spatial beam distortions and a destruction of the amplifier facets by filamentation [56]. This mechanism is further enhanced by the residual reflectivity of the input-output-facet, even if high-quality AR coatings of less than 10^{-3} reflectivity have been deposited.

An amplifier design which significantly alleviates such problems with beam distortions, and also requires less optical input power from the oscillator, is the tapered amplifier [58,59,60]. The design and its general purpose is briefly presented here as this device is part of the MOPA systems described below.

The tapered amplifier is a single-pass device of a typical total length of 1 mm to 3 mm. The transverse width of the electrode is tapered, i.e. it increases linearly along the propagation direction to a typical width of 150 μm at the output facet. In this tapered section the internal light wave can freely diffract in the plane of the junction. Although the total power grows linearly with saturated amplification and propagation, the power density may remain approximately constant because diffraction widens the cross section of the beam. With a proper design the local intensity remains limited such that severe beam distortion by self-focusing or filamentation is suppressed, and a near-diffraction-limited output of high power is generated from a diffraction-limited input beam. For a proper operation of the amplifier, both facets have to be AR coated (or tilted in addition) to suppress laser oscillation at high pump currents. Considerable CW output powers with high beam quality have been generated with such amplifiers. In a hybrid system with a diffraction-limited TEM_{00} beam from a (spectrally multi-mode) CW Ti:sapphire master-oscillator, emitting 200 mW at 860 nm, and an AlGaAs tapered amplifier with a 250 µm-wide electrode at the entrance facet tapered to 500 µm at the exit facet, the diode amplifier emitted a maximum output power of 5.25 W in a near-diffraction-limited beam [61].

The setup of a diode-MOPA system and that for injection-locking a high-power diode-laser appears quite similar because the difference is essentially the AR coatings on the facets of the amplifier. Nevertheless, these systems posses very different properties. Injection-locking requires only a low input power from the master-oscillator, typically much less than a few mW. Also, the residual amount of spectrally broadband Amplified Spontaneous Emission (ASE) of a locked diode is low, because the locked high-power diode forms a laser oscillator operating well above threshold and thus with a strongly saturated gain. Diode-MOPA systems, on the other hand, are able to amplify all wavelengths within the approximately 30 nm-wide gain bandwidth and there is no restriction due to a limited locking range. Such systems are thus more appropriate for the amplification of tunable or pulsed sources [62]. With the above geometry, however, a tapered amplifier requires a relatively high input power from the master-oscillator to obtain an efficient depletion of the amplifier gain. The minimum input power can be as high as several hundred mW for rectangular single-pass broad-area amplifiers. The design of a tapered amplifier thus often includes a single-stripe amplifier section beginning at the entrance facet and extending over several hundreds of µm in length, just as in a single-stripe laser. This section serves to pre-amplify a weak master-oscillator beam with only a few mW of power to more than 100 mW before the radiation enters the second, tapered section. Residual reflections from the output facet, further amplified upon backward propagation, are blocked by etched grooves surrounding the transition between single-stripe and tapered sections.

5   AlGaAs Diode-MOPA Systems

In this section we present the design and the properties of a double-stage MOPA system, based on AlGaAs diodes, as shown in Fig. 3. The oscillator is a 100 mW single-stripe, single-frequency AlGaAs index-guided multiple-quantum-well laser emitting at a wavelength of 810 nm. For the first amplifier stage we use a 2 mm-long AlGaAs tapered single-pass amplifier device, with AR coatings on both facets. At the entrance side the amplifier chip contains a 1 mm-long single-stripe gain section, which serves for pre-amplification of weak input powers from the oscillator. The remaining length of 1 mm is ta-

Master-oscillator

Broad-area amplifier

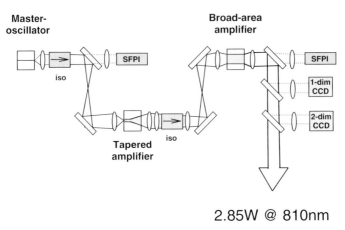

2.85W @ 810nm

Fig. 3. Setup of a double-stage diode-MOPA system with an AlGaAs single-stripe oscillator, followed by a tapered amplifier and a broad-area amplifier (iso: 60 dB Faraday isolators, SFPI: scanning Fabry–Perot interferometer, CCD: one- or two-dimensional charge-coupled-device camera)

pered, extending from a width of a few µm to a width of 130 µm at the exit facet. The second amplifier is a 1.1 mm-long broad-area single-pass amplifier with a 600 µm-wide active area and AR coatings on both facets.

The amplifiers are mounted on Peltier coolers such that the temperature of the amplifiers can be varied independently, to match the wavelength of maximum gain to the wavelength emitted by the oscillator. Separate housings equipped with silica-gel reservoirs reduce the humidity and prevent contamination of the facets. The housings are sealed using AR-coated windows, and lenses of short focal length which focus the input and collimate the output beams. To prevent feedback from the amplifiers into the oscillator, the diodes are separated by 60 dB Faraday isolators. A cylindrical telescope (lenses with focal lengths of $f = 50$ and 300 mm) is used to match the beam from the tapered amplifier into the input facet of the rectangular amplifier. The angle of incidence on this facet is approximately $5°$ to prevent the build-up of standing waves which would cause filamentation effects. Additional devices used are CCD cameras to monitor the near and far field of the rectangular amplifier for an adjustment. Scanning confocal Fabry–Perot Interferometers (SFPIs) with 2 GHz Free Spectral Range (FSR) and a finesse of 100 monitor the input spectrum of the master-oscillator and the output spectrum after amplification. The spatial quality of the output beam is measured as described in Sect. 1, either via the M^2 parameter, or via a recording of the wavefront.

With their input beams blocked, the output from both the tapered and the rectangular amplifiers consists entirely of broadband ASE. The ASE spectrum was recorded with a double-grating optical spectrum analyser (resolution: 0.05 nm). The ASE power from the tapered amplifier was 50 mW with a pump current of 1 A and a temperature of 5° C. The ASE center wavelength of 807 nm can be tuned with a coefficient of 0.3 nm/ K. The FWHM of the spectrum is 9 nm, indicating the wide amplification bandwidth of the device. The ASE power from the rectangular amplifier was 100 mW with a 15 nm bandwidth measured at the maximum pump current of 8 A and a temperature of 15° C. We notice, however, that in both amplifiers the amount of ASE is reduced to a small fraction of about 1% of the total output, if the amplifiers are saturated by injection of radiation from the master-oscillator, as described below.

With the oscillator beam injected, we measured the output power of high-brightness radiation from the tapered amplifier as a function of the optical input power as shown in Fig. 4. The different data sets correspond to different pump currents for the amplifier. As can be seen, the output power increases both with the pump current and with the input power from the oscillator. For a fixed current and small oscillator powers, the output power increases proportionally with the oscillator power. In this small-signal regime the gain attains values between 23 dB (with 0.5 A pumping) and 35 dB (with 1.5 A). A further increase of the oscillator power leads into a transition regime at around 1.5 mW, followed by a regime of saturated amplification with input powers above 3 mW.

Similar measurements were performed with the rectangular amplifier (Fig. 5). It can be seen that the output increases with both the pump current and the input from the tapered amplifier. The two lowest traces, recorded with pump currents of 1 A and 2 A, show that the output power can even be below the input power. However, a maximum small-signal gain of 12 dB is found at 8.0 A. At this current the maximum output power generated was 3 W with an input power of 500 mW.

The internal light amplification, ASE generation, gain competition and saturation within the amplifiers were calculated with a simple model which

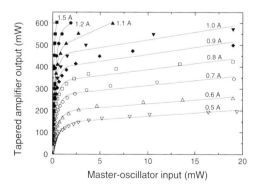

Fig. 4. Measured output power from the tapered amplifier as a function of the input power from the master-oscillator for various amplifier pump currents

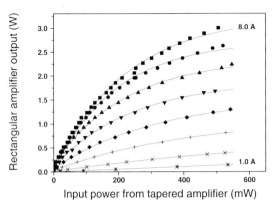

Fig. 5. Output power from the rectangular broad-area AlGaAs amplifier as a function of the optical input power from the tapered-amplifier stage, recorded at various amplifier pump currents

combines geometrical beam propagation in the plane of the junction with two-level-laser rate equations. This model had so far shown excellent agreement with the experimental data in the cases of an AlGaAs rectangular double-pass broad-area amplifier yielding 780 mW at 810 nm [57] and an AlGaInP tapered amplifier yielding 500 mW at 670 nm [63]. The theoretical prediction from this model is shown in Figs. 4 and 5 as solid lines and shows a good agreement with the experimental data. This agreement indicates that the general performance of the investigated amplifiers can reasonably well be described even with a simple model. We note, however, that for a detailed understanding, and thus also for a consequent improvement and optimization of diode amplifiers, better models are required.

The spatial beam quality of the tapered-amplifier and the rectangular-amplifier outputs was determined via a measurement of the M^2 value as described in Sect. 1. At an output power of 2.85 W the measurements yielded $M_x^2 = 1.1$ and $M_y^2 = 1.1$, and $M_x^2 = 1.6$ and $M_y^2 = 1.4$, respectively.

The emission spectrum of the double-stage diode-MOPA system of Fig. 3, measured with a SFPI, is shown in Fig. 6. It can be seen that the output consists of single-frequency radiation with a spectral bandwidth of less than 14 MHz, which is given by the resolution limit of the SFPI. Within this limit, the measured bandwidth is equal to that of the master-oscillator such that amplification in both stages does not lead to a noticeable spectral broadening.

These values show that the two-stage MOPA system provides a high-brightness output, i.e. a powerful near-diffraction-limited beam of narrow spectral bandwidth. In terms of the units given in Sect. 3, the brightness is approximately 1000 mW/sr/ μm^2/ MHz, which is about a factor of 30 above that provided by single-stripe lasers (and many orders of magnitude above that of standard high-power diode-lasers). With this brightness the output from the diode MOPA should be particularly useful for an efficient nonlinear

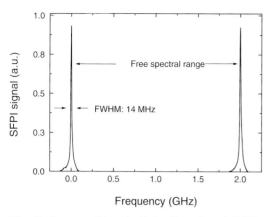

Fig. 6. A spectral bandwidth of less than 14 MHz of the AlGaAs double-stage diode-MOPA system is measured with a scanning confocal Fabry–Perot interferometer

conversion into other wavelength ranges. An example, where UV radiation is generated, is described in Sect. 7.1.

6   Diode-MOPA Systems Based on InGaAs

In further experiments we investigated the properties of diode-MOPA systems based on a different semiconductor material, InGaAs. This material is of interest because it contains no Al and gives such diodes a better reliability and higher power capability. The setup of the diode MOPA was similar to that shown in Fig. 3, however, with the rectangular broad-area amplifier omitted.

The master-oscillator is based on a 750 μm-long single-stripe laser with a 4 μm-wide active zone operating at wavelengths around 925 nm. The back facet is HR coated and the direction of the active stripe is tilted by a few degrees with respect to the surface normal of the AR-coated front facet, which suppresses feedback and laser oscillation in the chip. The output is collected with a short focal length collimator lens. In fact, the emission spectrum of the free-running single-stripe diode (without optical feedback) consists of a broad peak of 20 nm FWHM, with no further spectral structures visible. The center wavelength of gain can be varied with the temperature by 0.15 nm/K. For tunable single-frequency operation we used a Littman-type resonator [36,37] with feedback from a grating with 1800 lines per mm, and with 20% output coupling provided by the zero-order diffracted beam. At the maximum pump current of 130 mA the oscillator output power was 17 mW with a spectral bandwidth of less than 2 MHz (resolution-limited).

Two prototype versions of InGaAs tapered amplifiers were available [7]. The first had a 1 mm-long single-stripe section, followed by a 1 mm-long tapered section of 200 μm width at the exit facet. The second amplifier had a

1 mm-long single-stripe section followed by a 2 mm-long tapered section with a 400 μm width at the exit facet. To avoid back reflections and suppress oscillation, the facets are AR coated at a wavelength of 920 nm to a specified reflectivity of less than 10^{-4}. Additionally, the longer amplifier has its input facet wedged so that optimum input coupling is observed at approximately 17° angle of incidence. The oscillator beam is sent with a 1:1 spherical telescope ($f = 300$ mm) through two 30 dB Faraday isolators, and then focused onto the entrance facet of the amplifier with a collimator lens. The amplifier output is collimated with the same type of lens.

With this setup we measured the output power from the tapered amplifier as a function of the oscillator input power and the amplifier pump current. An example for a 3 mm-long amplifier is shown in Fig. 7 which was recorded at a pump current of 8 A. With the oscillator beam blocked, the amplifier output consists entirely of ASE, with a broad spectral bandwidth of approximately 10 nm, centered at 925 nm. We measured an ASE power of 1.8 W, which corresponds to the square symbols in Fig. 7 at zero oscillator input power. With the oscillator beam injected and at an input power of 14.5 mW the total output power from the tapered amplifier reaches a maximum of 4.2 W (see triangular symbols). In order to determine what fraction of the output actually consists of amplified oscillator radiation of high brightness and what fraction is residual ASE, we used a small part of the output for a simultaneous spectral analysis with the optical spectrum analyser. In the spectrum, ASE is apparent as a broad spectral background while the amplified oscillator radiation corresponds to a sharp peak of narrow spectral bandwidth (0.05 nm, as given by the resolution of the analyser). From such spectra the power ratio of the two types of radiation is determined by integration over the respective

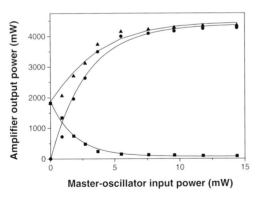

Fig. 7. Total output power (*triangular symbols*) from a 925 nm-emitting InGaAs master-oscillator tapered-amplifier system as a function of the input power from the oscillator. The amplifier pump current is 8 A. Spectrally resolved measurements reveal the increasing power of amplified, spectrally narrowband oscillator radiation (*dots*) while ASE (*square symbols*) becomes suppressed with increasing oscillator input. The *solid lines* represent the prediction of the model described in Sect. 5

line shapes. A scaling with this ratio yields the power of amplified oscillator radiation in Fig. 7 (circular symbols) and the power of ASE (squares). Fig. 7 shows that up to an input power of 2 mW the amount of ASE is higher than that of the amplified oscillator beam. When the input is increased to about 15 mW, however, ASE becomes strongly suppressed to below a relative power of 2%. In this case the maximum power of amplified oscillator radiation with narrow spectral bandwidth is 4.1 W. A measurement of the spectral bandwidth with a SFPI yields a value of 2 MHz, which is the resolution limit of the SFPI.

The model as described in Sect. 5 was used to calculate the power of amplified radiation and ASE as a function of the input power. The result, shown as the solid curves in Fig. 7, shows reasonable agreement with the experimental data. The model is thus suitable also to describe diode-MOPA systems of InGaAs, which may be useful for a further power scaling and optimization of the investigated amplifiers, supporting more elaborate models.

The spatial beam quality of the InGaAs-MOPA system was measured in a first step by measuring the M^2 parameter as in Fig. 1. A beam waist was produced approximately 140 mm behind a spherical focusing lens ($f = 150$ mm) with the pump current set to 5 A. The M^2 measurements were carried out in both transverse dimensions and also as a function of the amplifier pump current. The beam quality was found to be excellent and approximately independent of the pump current, as expressed by a diffraction parameter close to unity, $M_x^2 = 1.1 \pm 0.2$ (vertical to the plane of the active zone) and $M_y^2 = 1.1 \pm 0.3$ (parallel). As seen in Fig. 1, the measurements also yield the location of the beam waist and the beam radius. In the x direction, both the waist position and beam radius were found to be independent of the pump current of the amplifier. They varied, however, in the y direction by approximately 3 mm/ A and by 2.3 µm/ A, respectively. This effect is equivalent to a power-dependent astigmatism of the output beam. The reason for such a dependence is probably that the transverse refractive-index profile in the tapered amplifier depends on the pump current and the internal power density of light via a transversely inhomogeneous distribution of the inversion and temperature in the plane of the active zone.

A practical conclusion can be drawn from the presence of this power-dependent astigmatism. In many situations the application of a diode-MOPA system is sensitive to the beam-focusing conditions while one needs to vary the optical power. In this case one should operate the MOPA system at constant, maximum pump current and instead use a variable attenuator in the output beam to adjust the power. A typical application of this kind is optically nonlinear frequency conversion by second-harmonic generation or optical parametric oscillation. Both processes are based on parametric amplification and therefore depend critically on a spatial matching of the involved wavefronts. A power-dependent spatial mode matching would result in a dramatically reduced nonlinear conversion efficiency in certain ranges of input power.

An analysis of the spatial beam quality properties in higher detail than with an M^2 measurement should be possible with a wavefront measurement. We performed such measurements using dynamical shearing interferometry as described in Sect. 1.2, to inspect the wavefront of a diode-MOPA system. The InGaAs MOPA was based on a 3 mm-long tapered amplifier, emitting near a wavelength of $\lambda_0 = 925$ nm. In a first step the amplifier pump current was set to 8 A and the system emitted a total output of 3 W with an oscillator input power of 15 mW. The B coefficients obtained for spherical aberration and coma are $-0.3\lambda_0/\,\mathrm{mm}^4$ and $0.1\lambda_0/\,\mathrm{mm}^3$ respectively, and were approximately independent of the amplifier pump current. The coefficient for astigmatism varies approximately linearly with the pump current from a value of $0.2\lambda_0/\,\mathrm{mm}^2$ at 2 A, to zero at 3.5 A, and then increases again to $0.5\lambda_0/\,\mathrm{mm}^2$ at 8 A.

Figure 8 shows the shape of the measured wavefront. Note that this phasefront is displayed by omitting the B coefficients for tilts and spherical curvatures in both transverse dimensions. The displayed front thus corresponds to a measurement in a beam waist as if the beam were passed through aberration-free spherical and cylindrical optics to compensate spherical curvatures and astigmatism.

It can be seen that the remaining higher-order phasefront distortions in Fig. 8 are different in the x and the y directions. The negative fourth-order parabolic shape in the y-direction (vertical to the plane of the junction) did not vary with the pump current. It may thus be caused by aberrations in the collimator lens which occur due to the high divergence of the radiation

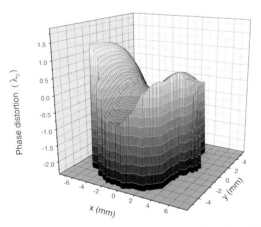

Fig. 8. Measured wavefront emitted by the 925 nm InGaAs-MOPA system at an output power of 3 W. The x direction is parallel to the plane of the amplifier junction, the y direction is vertical. The displayed front is corrected for tilt and spherical curvatures, and also for astigmatism

in that plane. The positive parabolic shape in the x direction varies with the current. It may thus be caused in part by a variation of the refractive-index distribution within the plane of the junction. The RMS phase deviation from a beam with a purely spherical wavefront (which would be represented by a plane at $z = 0$ in Fig. 8) was numerically calculated for amplifier currents in the interval from 3.5 A to 8 A. This RMS deviation showed a constant value of $\lambda_0/10$ independent of the current. This measurement proves a high beam quality at all power levels, as was also found from the M^2 measurements.

In summary, the results described above demonstrate that advanced diode-MOPA systems are capable of generating CW output radiation at power levels of several watts with near-perfect spatial beam quality, spectral bandwidths on the order of a few MHz, and less than 2% ASE background. Our recent experiments have so far increased the power of such high-brightness radiation from InGaAs-MOPA systems with a single tapered amplifier to above 5.4 W, by injecting 15 mW of single-frequency radiation from the master-oscillator into a 3 mm-long tapered amplifier pumped with a current of 10 A. This is so far the highest output from an all-diode-MOPA system or for a single-frequency system. Nevertheless, it can be expected that several tens of watts will become available in the near future, due to intense research and development in this field, and stimulated by applications, such as described throughout this book.

7 Nonlinear Frequency Conversion with High-Brightness Diode-MOPA Systems

High-power diode-laser radiation with high spatial beam quality or high brightness is of advantage for numerous direct applications as briefly described in the Introduction. Perhaps the most challenging application and strict proof of quality for the generated radiation is its direct nonlinear conversion. So far up to 156 mW of blue (486 nm) and 2.1 mW of ultraviolet (UV, 243 nm) radiation had been generated by frequency doubling the output from a monolithically integrated diode MOPA [64]. In the following we review recent experiments, where the diode-MOPA systems described in the previous sections have been applied to generate even more powerful CW light in the UV and blue spectral ranges, or to generate widely tunable light in the near- and mid-infrared.

7.1 Second-Harmonic Generation

As a result of the excellent brightness of the MOPA systems employed, the nonlinearly generated output is the highest power obtained so far by direct frequency conversion of diode radiation. Efficient generation of UV and visible light from infrared diode-lasers is usually based on Second-Harmonic Generation (SHG) in nonlinear crystals placed in an external power-enhancement resonator. The setup used by *Woll* and coworkers to generate UV [65] and blue radiation [66] is shown in Fig. 9. The pump source for UV generation was the AlGaAs-MOPA system tuned to a wavelength of 806 nm. At that time the second amplifier stage with the rectangular broad-area diode was not available. Nevertheless, the maximum single-frequency output power from the first stage with the tapered amplifier generated up to 600 mW of fundamental radiation with a beam of high spatial quality ($M^2 = 1.1$ in both planes).

To generate a maximum intensity in the crystal and thereby increase the conversion efficiency, the fundamental radiation was mode-matched with spherical and cylindrical telescopes into the power-enhancement ring cavity containing a suitable nonlinear crystal. To avoid back reflections into the amplifier a 40 dB Faraday isolator was used such that the power available in front of the ring was 400 mW. The power-enhancement ring cavity comprised two plane and two curved mirrors (M_1 to M_4 in Fig. 9) with HR coatings centered at 806 nm for a high finesse. A 0.7% transmission of the input mirror M_3 was found to provide optimum impedance-matching, i.e. it gave a maximum input coupling efficiency of the fundamental beam. At the resonator internal beam waist between the mirrors M_1 and M_2 a 16 mm-long Lithium-tri-Borate (LBO) crystal was placed which was cut for type-I critical phase matching of SHG, to obtain radiation with 403 nm wavelength. With the

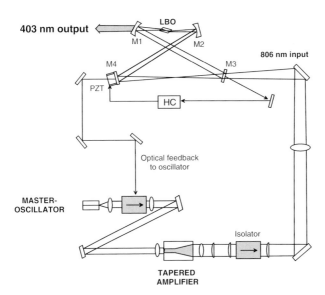

Fig. 9. Experimental setup for frequency doubling the output from an 806 nm AlGaAs diode-MOPA system in a power-enhancement cavity containing a nonlinear LBO crystal, for the generation of CW single-frequency UV radiation at 403 nm ([65]). The components labeled HC and PZT are a Hänsch–Couillaud-type electronic servo stabilizer and a piezo-electric transducer

crystal present, power-enhancement factors of around 70 were observed with a maximum input coupling efficiency of 70% of the fundamental beam into a single TEM_{00} mode of the ring cavity. This corresponds to a CW intracavity power of approximately 30 W and a power density of more than $1 MW/cm^2$ in the crystal, given the small mode radius of approximately 30 μm. We note that such numbers can only be achieved due to the excellent spatial and spectral quality of the diode radiation, since the power density in the crystal is based on good spatial and spectral mode matching of the diode-MOPA radiation into a single TEM_{00} mode of the cavity.

For a stabilization of the fundamental power in the cavity versus acoustic or thermally induced perturbation of the cavity length, a combination of two techniques was used, which keep the fundamental light frequency in resonance with the frequency of a particular TEM_{00} mode. The first is a well-known electronic technique developed by *Hänsch* and *Couillaud* (HC) [67]. A detuning of the cavity-mode frequency with respect to the frequency of the fundamental laser mode causes a change in the polarization of light reflected from the cavity. This is monitored with a polarization-sensitive detector (HC in Fig. 9) and converted into an electronic error signal. After suitable amplification the error signal is applied to a piezo-electric actuator (PZT) which counteracts perturbations of the cavity length. The response time of such servo loops is usually limited by the PZT to approximately 1 ms. To further shorten the response time, an optical stabilization technique was applied in addition. A small fraction of resonant fundamental radiation, leaking out of the end mirror M_4, is fed back through the side port of the optical isolator in the MOPA system and is injected into the master-oscillator diode. This feedback locks the oscillator frequency to a cavity resonance with a fast response time of about 10 ns as given by the length of the optical feedback path. Further, it reduces the spectral bandwidth of the master-oscillator (and thus the MOPA output) to well below the approximately 10 MHz bandwidth of the cavity resonance for improved spectral mode matching.

The UV power generated is shown in the upper part of Fig. 10. The maximum output power of single-frequency 403 nm light is 100 mW for a diode pump power of 400 mW. The lower part of Fig. 10 shows the UV output as generated with a 12 mm-long Brewster-cut β-Barium BOrate (BBO) crystal for critical type-I phase matching [65]. It can be seen that a similar output power is obtained although the walk-off angle in BBO of 3.9° is about four times larger than in LBO. This result is explained by the approximately two times higher nonlinear coefficient of BBO ($d_{eff} = 2pm/V$). A comparison of both sets of experimental data with theory (solid curves) shows good agreement, i.e. the UV output grows approximately quadratically at low pump powers and approaches a linear growth at higher powers, as expected for such systems. These results represent so far the highest UV output generated by diode-laser frequency doubling. Without saturation, such as via thermal effects in the crystal, the theoretical curves predict an output of 0.5 W, if

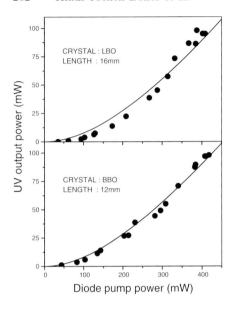

Fig. 10. UV output power generated by frequency doubling the 806 nm radiation from an AlGaAs MOPA system ([65]

the rectangular broad-area power amplifier is added to the pump source as shown in Fig. 3.

Frequency doubling of the InGaAs-MOPA system described above looks even more promising, simply because that source can deliver high-brightness radiation of much higher power. *Woll* et al. [66] additionally stabilized the master-oscillator of that MOPA system with grating feedback in a Littman configuration, and used the 3 mm-long tapered amplifier. The MOPA output wavelength was grating-tuned to a wavelength of 930 nm and provided a maximum power of 4.1 W with less than 2 MHz spectral bandwidth and high spatial quality ($M^2 < 1.2$ in both planes). The MOPA output was sent through two optical isolators (30 dB each) and spatially mode-matched to the power-enhancement ring cavity as in Fig. 9. A 4 mm-long Brewster-cut LBO crystal was placed in the cavity beam waist having a 15 µm beam radius. With the maximum power of 3.2 W in front of the cavity the input coupling efficiency was 55% and the power-enhancement factor was 48. These values include the passive roundtrip losses of the cavity and the losses due to nonlinear conversion in the crystal. The intracavity power calculated from these numbers is 155 W and corresponds to a power density of 30 MW/cm^2 in the crystal. As for the UV generation experiments described above, the fundamental intracavity power was stabilized with a HC servo loop and optical feedback. The output power of blue light at 465 nm is shown in Fig. 11 as a function of the fundamental power coupled into the enhancement cavity.

At a pump power of 1.7 W the generated blue output was 1 W, which corresponds to a conversion efficiency of 58%. The experimental data are in good agreement with a theoretical model (see solid line). The blue radiation

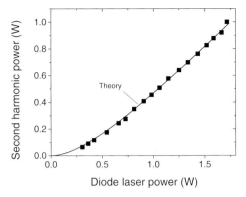

Fig. 11. 1 W of single-frequency blue radiation at 465 nm is generated by resonator internal frequency doubling the output from an InGaAs diode-MOPA system in a LBO crystal ([65]

is emitted as a high-quality beam, according to $M^2 = 1.2$ in the walk-off plane of the crystal and $M^2 = 1.1$ vertically. Long-time stable operation was demonstrated over 30min with less than 10% instability of the 465 nm output power. So far, no saturation of the output power due to thermal effects in the crystal had been observed, such that a further increase can be expected with future improvements of InGaAs diode-MOPA systems.

7.2   Diode-Pumped Optical Parametric Oscillators

CW Optical Parametric Oscillators (OPOs) [68,69] are well suited for the generation of widely tunable signal and idler radiation in the near- and mid-infrared spectral regions with narrow spectral bandwidth. Among possible applications are molecular spectroscopy [70,71], trace-gas detection [72] or phase-coherent division of optical frequencies for precision measurements and future optical time standards [73,74,75]. Basically, an OPO consists of an optical resonator containing an optically nonlinear crystal. Above a certain threshold power, an injected pump-laser beam is efficiently converted into two other waves of longer wavelengths, called the signal and idler waves. Depending on how many of the three waves are power-enhanced by the OPO resonator such OPOs are termed Triply, Doubly, or Singly Resonant (TRO, DRO, and SRO, respectively). TROs offer the advantage of a low threshold pump power, typically on the order of tens of mW, but a continuous wavelength tuning of the output (without spectral mode hops) is not possible due to spectral clustering [76]. SROs are of much higher interest, because a continuous tuning of the signal and idler output frequencies can be achieved. A drawback is, however, the high pump power at threshold which is typically in the range of several to ten watts. For this reason SROs have been operated so far only with powerful solid-state lasers as pump source.

Of particular interest are diode-pumped OPOs. A main advantage of direct diode pumping is the wide and easy tunability of the diode pump laser, which can be used for an easy tuning of the OPO output wavelengths. A further option of direct diode-pumped OPOs is a compact design with a high overall efficiency. Diode-pumped OPOs were first demonstrated by *Scheidt* et al. using single-stripe diodes [77], however the OPO was triply-resonant which makes frequency tuning difficult. Subsequent work thus aimed on increasing the available diode power and to reduce the number of resonant waves, but relied on conventional nonlinear crystals based on birefringent phase matching. With Potassium-Tytanyl-Phosphate (KTP) and Rubidium-Titanyl-Asenate (RTA) crystals diode-pumped DROs were realized [77,78,79,80] and showed an improved stability, tunability and output power. Besides the improvement of the output and beam quality of diode-lasers, new crystals with high nonlinearity have been developed. These are Periodically Poled (PP) crystals with Quasi-Phase Matching (QPM) [81], such as PP-Lithium-Niobate (PPLN) [82], PP-Lithium-Tantalate (PPLT) [83], PP-Potassium-Titanyl-Phosphate (PPKTP) [84], and PP-Potassium-Niobate (PPKN) [85]. A main advantage of QPM in such PP crystals is the free choice of the phase-matching wavelength via the poling period. Further, due to a free choice of the polarization of the interacting waves one can access the biggest element of the susceptibility tensor (with parallel polarizations) and also avoid walk-off. PP crystals thus offer a high nonlinearity which has been used, e.g., to lower the pump power at threshold for SROs into the range 1–2 W [86]. The combination of PP crystals with advanced diode-MOPA systems, such as presented in Sect. 6, should therefore also allow the operation of SROs with direct diode pumping [87].

Recently, the first diode-pumped SRO was demonstrated by *Klein* et al. [88]. The setup is similar to the one shown in Fig. 9, however with the electronic and optical feedback omitted. The pump source was an InGaAs-MOPA system, as described in Sect. 6, with a tapered amplifier operating at wavelengths near 925 nm. The oscillator was a single-stripe diode with an external grating–mirror feedback in Littman configuration. The system provided up to 2.5 W single-frequency radiation in a near diffraction-limited beam, wavelength-tunable via the grating. The OPO consisted of a 38 mm-long PPLN crystal in a four-mirror ring resonator. Singly resonant oscillation on the signal wave had been achieved by using mirrors with high reflectivity for the signal wave ($R > 99.8\%$ at $1560 - 1770$ nm) but with high transmission for the pump and idler waves ($T = 99\%$ at 925 nm, $T > 98\%$ at $2.07 - 2.19$ μm, all mirrors). The power roundtrip losses thus exceed 99.9999% for both the pump and the idler wave.

Figure 12 shows the output power of the idler wave at 2.1 μm measured in dependence on the pump power. The data indicate a threshold pump power of about 1.6 W. The idler wave is emitted as a single-frequency beam. As the signature of singly resonant oscillation, *Klein* et al. observed a contin-

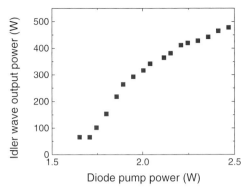

Fig. 12. Idler-wave output power at 2.1 μm wavelength of a singly resonant CW OPO (CW-SRO) pumped with a high-power InGaAs diode-MOPA system operating at 925 nm. The idler wave (and 120 mW of residually transmitted signal wave emitted in addition) had spectral bandwidths of a few MHz ([88])

uous tunability of the OPO output frequencies over nearly an entire FSR (1.3 GHz) of the OPO cavity, achieved by piezo-electrically scanning the cavity length. The maximum idler output power in a single beam was 480 mW at the maximum pump power of 2.5 W. The power stability of the output was better than 0.3% (RMS) over 1min and better than 0.9% over 17min, although no electronic or optical stabilization had been used, neither for the diode MOPA nor for the OPO.

The signal and idler wavelengths of the diode-pumped SRO were tunable in the ranges 1.55 μm to 1.70 μm and 2.03 μm to 2.29 μm, respectively, via a variation of the crystal temperature over the range around 12°. A particular advantage of a diode-laser as the OPO pump source (in comparison, e.g., to a diode-pumped Nd solid-state laser), is the wide and easy tunability of the pump wavelength. Figure 13 shows the wavelength of the signal wave in dependence on the wavelength of the pump radiation. As seen from this figure, tuning of the MOPA output from 924 nm to 925.4 nm changes the signal wavelength (and correspondingly the idler wavelength) over the same intervals as those covered by temperature-tuning (1.55 μm to 1.70 μm and 2.03 μm to 2.29 μm, respectively). One should mention that tuning of the OPO via the diode-laser wavelength bears a significant potential for high-speed tuning, e.g. to obtain a quick and random wavelength access for multi-component gas monitoring. Resent results show a 56 GHz continuous tuning of the diode pumped SRO and its application to molecular spectroscopy [89]. Also the first synchronously-pumped OPO driven directly with a diode laser has been realized, by replacing the CW master-oscillator in Sec. 6 with an actively mode-locked (picosecond emitting) oscillator diode [90].

In conclusion, the diode-pumped OPOs have now entered a field which has previously been dominated by powerful solid-state laser systems. Again, since the threshold power of OPOs critically depends on the spatial and spectral

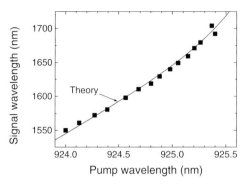

Fig. 13. Wavelength of the SRO signal wave in dependence on the wavelength of the diode-pump MOPA system measured for a fixed temperature of 175.8° C ([88]). The experimental data (*squares*) are in good agreement with the values calculated by the Sellmeier equations of Lithium Niobate

mode quality of the pump source, all of these achievements have resulted from the high brightness of the diode pump systems used.

8   Summary

We gave an overview on recent developments in the field of high-power high-brightness diode-laser systems. We described the generation of CW-laser beams at power levels of several hundreds of mW to several watts, with excellent spatial and spectral quality. With single-stage or double-stage MOPA systems, employing diode amplifiers in tapered or rectangular geometry, up to 2.85 W of high-brightness radiation can be generated at wavelengths around 810 nm with AlGaAs diodes or at powers exceeding 5 W with InGaAs diodes. We described the setup of such diode systems, reported on their basic properties, and reviewed methods to characterize the spatial quality of the output beam. As a strict proof-of-quality for the excellent spatial and spectral quality of the diode output we reviewed recent experiments with direct nonlinear optical conversion of near-infrared diode-MOPA radiation into other wavelength ranges with high power and efficiency. Examples are the generation of single-frequency CW ultraviolet radiation with powers of up to 100 mW, and the generation of single-frequency CW blue light with powers of up to 1 W, both achieved by intracavity frequency doubling of diode-laser radiation. Another example is the generation of widely tunable single-frequency infrared radiation with powers above 0.5 W with CW OPOs which are directly pumped by high-brightness diode-laser systems. The reviewed experiments demonstrate that current diode-MOPA systems have reached a considerable degree of maturity. Already, it can be expected that certain applications would benefit by replacing solid-state lasers in their traditional applications. Most promise, however, lies in the ongoing research and development of high-brightness diode systems assisted by efficient nonlinear conversion, giving room for new applications in a vast range of the electromagnetic spectrum.

Acknowledgements

We thank A. Robertson and G. Anstett for critically reading the manuscript. This work was supported by the German Federal Ministry of Education and Research (BMBF) under Contract No. 13N7021 and Contract No. FKZ 16 SV 5286.

List of Symbols and Abbreviations

α_H	Henry linewidth enhancement factor
M^2, M_x^2, M_y^2	Diffraction parameter (beam-quality factor)
Q, high-Q	Angular frequency times ratio of energy and energy loss rate in an optical resonator
D	Distance from the diode-laser front facet
d_D	FWHM power beam diameter measured at a distance D from the diode-laser facet
w_D	$1/e^2$-power power radius of a Gaussian beam measured at a distance D from the waist
d_0	Beam diameter in the plane of the diode facet
θ_D	Experimentally determined FWHM divergence angle of a laser beam
λ, λ_0	Laser wavelength
W, W_x, W_y	$1/e^2$ power radius of a real laser beam
z	Longitudinal coordinate of beam propagation
x, y	Transverse coordinates, vertical to beam propagation
z_0	Position of beam waist
Z_R, Z_{Rx}, Z_{Ry}	Rayleigh length of a focus in a real beam
w_0	$1/e^2$-power radius in the waist of a Gaussian laser beam
$P(x, y)$	Wavefront of a laser beam
R	Radius of curvature of a wavefront
B_{nm}	Kingslake (Taylor-expansion) coefficients of a wavefront
C_{nm}	Taylor-expansion coefficients of optical path difference between two wavefronts
S, T	Transverse displacements of a wavefront
$\Delta P(x, y)$	Optical path difference between two wavefronts
δ	Propagational phase lag between two wavefronts
α_x	Angle of incidence
$I(x, y)$	Transverse intensity distribution (interference pattern)
σ^2, σ_x^2, σ_y^2	Second moment of the transverse intensity distribution in a laser beam
θ_{th}	Calculated FWHM divergence angle of a laser beam
z_R	Rayleigh length of the focus in a Gaussian beam
f	Focal length of a lens

α-DBR laser	Diode-laser with tilted internal distributed
AR	Anti-Reflection
ARROW	Anti-Resonant Ridge Optical Waveguide (diode-laser) feedback structure
ASE	Amplified Spontaneous Emission
BBO	β-Barium-BOrate
CCD	Charged-Coupled Device (camera)
COD	Catastrophic Optical Damage
CSFPI	Confocal Scanning Fabry–Perot Interferometer
CW	Continuous Wave
DBR	Distributed Bragg grating
DFB	Distributed FeedBack
DRO	Doubly Resonant OPO
FP	Fabry-Perot
FSR	Free Spectral Range of an optical resonator
FWHM	Full Width at Half Maximum
GRINSCH	Graded Refractive Index and Separate Confinement Heterostructure
HC	Hänsch–Couillaud-type servo loop
HR	High-Reflection
HWHM	Half-Width Half-Mean
KTP	Potassium-Titanyl-Phosphate
LBO	Lithium-tri-BOrate
MOPA	Master-Oscillator Power-Amplifier
OPO	Optical Parametric Oscillator
PIB	Power-In-the-Bucket number to describe spatial beam quality
PPLN	Periodically Poled Lithium Niobate
PZT	PieZoelectric actuator (Transducer)
QPM	Quasi-Phase-Matching
QW	Quantum Well
RMS	Root-Mean-Square value to describe spatial beam quality
RTA	Rubidium-Titanyl-Arsenate
SFPI	Scanning Fabry–Perot Interferometer
SHG	Second-Harmonic Generation
SRO	Singly Resonant OPO
TDL	Times-Diffraction-Limit number to describe spatial beam quality
TEM_{00}	Lowest-order Transverse ElectroMagnetic (mode)
TRO	Triply Resonant OPO
VCSEL	Vertical Cavity Surface-Emitting Laser

References

1. O. Svelto: *Principles of Lasers* (Plenum, New York 1998)
2. A. E. Siegman: *Lasers* (Univ. Science Books, Mill Valley, CA 1986)
3. P. W. Milonni, J. H. Eberly: *Lasers* (Wiley, New York 1988)
4. K. Petermann: *Laser Diode Modulation and Noise* (Kluwer Academic, Dordrecht, KTK Scientific Publishers, Tokyo 1988)
5. G. P. Agrawal (ed.): *Semiconductor Lasers: Past, Present, and Future* (American Institute of Physics, Woodbury 1995)
6. C. H. Henry: Line Broadening of Semiconductor Lasers, In: Y. Yamamoto (Ed.) *Coherence, Amplification, and Quantum Effects in Semiconductor Lasers* (Wiley, New York 1991)
7. Spectra Diode Laboratories (SDL) Inc., San Jose, CA, USA
8. R. M. Hofstra: *On the Optical Performance of the Long Pulse XeCl* Excimer Laser* (Koninklijke Bibliotheek, Den Haag 1999)
9. H. Kogelnik, T. Li: Laser Beams and Resonators, Appl. Opt. **5**, 1550 (1966)
10. G. D. Boyd, D. A. Kleinman: Parametric Interaction of Focused Gaussian Light Beams, J. Appl. Phys. **39**, 3597 (1998)
11. K. Shinozaki, C.-Q. Xu, H. Sasaki, T. Kamijoh: A Comparison of Optical Second-Harmonic Generation Efficiency Using Bessel and Gaussian Beams in Bulk Crystals, Opt. Commun. **133**, 300–304 (1997)
12. J. Arlt, K. Dholakia, M. J. Padgett: Second-Harmonic Generation Efficiency of Bessel Beams, Quantum Electron. and Laser Science Conference, QELS, Baltimore, MA (May 1999) paper QTuB4
13. B. Beier, J.-P. Meyn, R. Knappe, K.-J. Boller, G. Huber, R. Wallenstein: A 180 mW Nd:LaSc$_3$(BO$_3$)$_4$ Single-Frequency TEM$_{00}$ Microchip Laser Pumped by an Injection-Locked Diode Laser Array, Appl. Phys. B **58**, 381–388 (1994)
14. M. W. Sasnett: Properties of Multimode Laser Beams - The M^2 Factor, In D. R. Hall, P. E. Jackson (eds.): *The Physics and Technology of Laser Resonators* (Institute of Physics-Publishing, London 1992)
15. G. Nemes, A. E. Siegman: Measurement of All Ten Second-Order Moments of an Astigmatic Beam by the Use of Rotating Simple Astigmatic (Anamorphic) Optics, J. Opt. Soc. Am. A **11**, 2257–2264 (1994)
16. Y. Champagne: Second-Moment Approach to the Time-Averaged Spatial Characterization of Multiple-Transverse-Mode Laser Beams, J. Opt. Soc. Am. A **12**, 1707–1714 (1995)
17. C. Paré, P. A. Bélanger: Propagation Law and Quasi-Invariance Properties of the Truncated Second-Order Moment of a Diffracted Laser Beam, Opt. Commun. **123**, 679–693 (1996)
18. J. M. Fleischer: Calibration Standard for Laser Beam Profilers: Method for Absolute Accuracy, Appl. Opt. **35**, 1719 (1996)
19. J. M. Fleischer: Gaussian Beam Profiling: How and Why, in C. B. Hitz: *Laser & Optronics* **6**, 61 (1987)
20. J. M. Fleischer: Standardizing the Measurement of Spatial Characteristics of Optical Beams, SPIE Proc. **888**, 60 (1988)
21. D. L. Wright, P. Greve, J. Fleischer, L. Austin: Laser Beam width, Divergence and Beam Propagation Factor – An International Standardization Approach, Opt. Quantum Electron. **24**, S993 (1992)

22. A. E. Siegman: Analysis of Laser Beam Quality Degradation Caused by Quartic Phase Aberrations, Appl. Opt. **32**, 5893–5901 (1993)

23. J. M. Geary: *Introductions to Wavefront Sensors* (SPIE, Washington 1995)

24. D. Malacara: *Optical Shop Testing*, 2nd edn. (Wiley, New York 1992)

25. R. Kingslake: Lens Design Short Course, SPIE Proc. **380**, 485 (1983)

26. R. Kingslake: Some Interesting and Unusual Lenses, SPIE Proc. **237**, 448 (1980)

27. R. N. Hall, G. E. Fenner, J. D. Kingsley, T. J. Soltys, R. O. Calson: Coherent Light Emission from Gaas Junctions, Phys. Rev. Lett. **9**, 366 (1962)

28. Y. Arakawa, H. Sasaki: Multidimensional Quantum Well Laser and Temperature Dependence of Its Output, Appl. Phys. Lett. **40**, 939 (1982)

29. M. Grundmann, F. Heinrichsdorff, N. N. Ledentsov, D. Bimberg: New Semiconductor Lasers Based on Quantum Dots, Laser & Optoelectronics **30**, 70 (1998) (in german)

30. K. J. Ebeling: *Integrated Optoelectronics* (Springer, Berlin, Heidelberg 1992)

31. R. G. Hunsperger: *Integrated Optics: Theory and Technology*, 3rd edn., Springer Ser. Opt. Sci. **33** (Springer, Berlin, Heidelberg 1991)

32. P. Werle: A Review of Recent Advances in Semiconductor Laser Based Gas Monitors, Spectrochim. Acta A **54**, 197–236 (1998)

33. S. Nakamura, M. Senoh, S. Hagahama, N. Iwasi, T. Yamada, T. Matsushita, H. Kiyoku, Y. Sugimoto: InGaN Based Multi-Quantum-Well-Structure Laser Diodes, Jpn. J. Appl. Phys. **135**, 74 (1996)

34. S. Nakamura, G. Fasol: *The Blue Diode Laser*, 2nd edn. (Springer, Berlin, Heidelberg 2000) in press

35. C. Becher, E. Gehrig, K.-J. Boller: Spectrally Asymmetric Mode Correlation and Intensity Noise in Pump-Noise-Suppressed Laser Diodes, Phys. Rev. A **57**, 3952 (1998)

36. K. Liu, M. G. Littman: Novel Geometry for Single-Mode Scanning of Tunable Lasers, Opt. Lett. **6**, 117 (1981)

37. P. McNicholl, H. J. Metcalf: Synchronous Cavity Mode and Feedback Scanning in a Dye Laser Oscillator with Gratings, Appl. Opt. **24**, 2757 (1985)

38. Y. Shevy, H. Deng: Frequency-Stable and Ultra-Narrow Linewidth Semiconductor Laser Locked Directly to an Atomic-Cesium Transition, Opt. Lett. **23**, 472 (1998)

39. D. Botez, D. R. Scifres: *Diode Laser Arrays* (Cambridge Univ. Press, Cambridge 1994)

40. N. W. Carlson: In A. L. Schawlow (ed.): *Monolithic Diode-Laser Arrays*, Springer Ser. Electron. Photon. **33** (Springer, Berlin, Heidelberg 1994)

41. E. Kapon, J. Katz, A. Yariv: Supermode Analysis of Phase-Locked Arrays of Semiconductor Lasers, Opt. Lett. **10**, 125 (1984)

42. G. R. Hadley, J. P. Hohimer, A. Owyoung: High-Order ($v > 10$) Eigenmodes in Ten-Stripe Gain-Guided Diode Laser Arrays, Appl. Phys. Lett. **49**, 684 (1986)

43. J. M. Verdiell, R. Frey: A Broad-Area Mode-Coupling Model for Multiple-Stripe Semiconductor Lasers, IEEE J. QE **26**, 270 (1990)

44. H. Adachihara, O. Hess, E. Abraham, P. Ru, J. V. Moloney: Spatiotemporal Choas in Broad-Area Semiconductor Lasers, J. Opt. Soc. Am. B **10**, 658 (1993)

45. R. Waarts, D. Mehuys, D. Nam, D. Welch, W. Streifer, D. Scifres: 900 mW, CW Nearly Diffraction Limited Output from a GaAlAs Semiconductor Laser Array in an External Talbot Cavity , Int. Conf. Lasers Electro-Optics, CLEO (1991) paper CWE7

46. J. E. Epler, N. Holonyak Jr., R. D. Burnham, T. L. Paoli, W. Streifer: Supermodes of Multiple-Stripe Quantum-Well Heterostructure Laser Diodes Operated (CW, 300 K) in an External-Grating Cavity, J. Appl. Phys. **57**, 1489 (1985)
47. S. MacCormack, R. W. Eason: Near-Diffraction-Limited Single-Lobe Emission from a High-Power Diode Laser Array Coupled to a Photorefractive Self-Pumped Phase-Conjugate Mirror, Opt. Lett. **16**, 705 (1991)
48. A. Larsson, M. Mittelstein, Y. Arakawa, A. Yariv: High-Efficiency Broad-Area Single Quantum-Well Lasers with Narrow Single-Lobed Far-Field Patterns Prepared by Molecular Beam Epitaxy, Electron. Lett. **22**, 79 (1986)
49. L. J. Mawst, D. Botez, M. Jansen, T. J. Roth, J. Rozenbergs: 1.5 W Diffraction-Limited Operation from Resonant-Optical Waveguide, Electron. Lett. **27**, 369 (1991)
50. L. J. Mawst, D. Botez, M. Jansen, T. J. Roth, C. Zmudzinski: 0.5 W CW Diffraction-Limited-Beam Operation from High-Efficiency Resonant-Optical-Waveguide Diode Laser Arrays, Electron. Lett. **27**, 1586 (1991)
51. S. D. DeMars, K. M. Dzurko, R. J. Lang, D. F. Welch, D. R. Scifres, A. Hardy: Angled Grating Distributed Feedback Laser with 1 W CW Single-Mode Diffraction-Limited Output At 980 nm, Proc. Int. Conf. Lasers Electro-Optics, CLEO, OSA Tech. Dig. Ser. **9**, Optical Society of America, Washington, DC (1996) p. 77, paper CTuC2
52. L. Goldberg, H. F. Taylor, J. F. Weller: Injection-Locking of Coupled-Stripe Diode Laser Arrays, Appl. Phys. Lett. **46**, 236 (1985)
53. J. Hohimer, D. R. Myers, T. M. Brennan, B. E. Hammons: Injection-Locking Characteristics of Gain-Guided Diode Laser Arrays with an "On-Chip" Master Laser, Appl. Phys. Lett. **56**, 1521 (1990)
54. L. Y. Pang, E. S. Kintzner, J. G. Fujimoto: Two-Stage Injection Locking of High-Power Semiconductor Laser Arrays, Opt. Lett. **15**, 728 (1990)
55. L. Goldberg, D. Mehuys, M. R. Surette, D. C. Hall: High-Power near Diffraction-Limited Large-Area Travelling-Wave Semiconductor Amplifiers, IEEE J. Quantum Electron. **29**, 2028 (1993)
56. L. Goldberg, D. Mehuys: High-Power Semiconductor Amplifiers, Int. Conf. Lasers Electro-Optics, CLEO (1993) p. 105, paper CTuI1
57. E. Gehrig, B. Beier, K.-J. Boller, R. Wallenstein: Experimental Characterization and Numerical Modelling of an AlGaAs Broad-Area Oscillator Amplifier System, Appl. Phys. B **66**, 287–293 (1998)
58. G. Bedelli, K. Komori, S. Arai, Y. Suematsu: A New Structure for High-Power TW-SLA (Travelling Wave Semiconductor Laser Amplifier), IEEE Photon. Technol. Lett. **3**, 42 (1991)
59. J. Walpole, E. Kintzner, S. Chinn, C. Wang, L. Missagia: High-Power Strained-Laser InGaAs/AlGaAs Tapered Travelling Wave Amplifier, Appl Phys. Lett. **61**, 740 (1992)
60. D. Mehuys, D. Welch, L. Goldberg: 2 W CW Diffraction-Limited Tapered Amplifier with Diode Injection, Electron. Lett. **28**, 1944 (1992)
61. D. Mehuys, L. Goldberg, D. F. Welch: 5.25 W CW Near-Diffraction-Limited Tapered-Stripe Semiconductor Optical Amplifier, IEEE Photon. Technol. Lett. **5**, 1179 (1993)
62. L. Goldberg, D. A. V. Kliner: Tunable UV Generation at 286 nm by Frequency Tripling of a High-Power Mode-Locked Semiconductor Laser, Opt. Lett. **20**, 1640 (1995)

63. A. Robertson, R. Knappe, R. Wallenstein: Kerr-Lens Mode-Locked Cr:LiSAF Femtosecond Laser Pumped by the Diffraction Limited Output of a 672 nm Diode-Laser Master-Oscillator Power-Amplifier System, J. Opt. Soc. Am. B **14**, 672–675 (1997)
64. C. Zimmermann, V. Vuletic, A. Hemmerich, T. W. Hänsch: All Solid State Laser Source for Tunable Blue and Ultraviolet Radiation, Appl. Phys. Lett. **66**, 2318 (1995)
65. B. Beier, D. Woll, M. Scheidt, K.-J. Boller, R. Wallenstein: Second Harmonic Generation of the Output of an Algaas Diode Oscillator Amplifier System in Critically Phase-Matched LiB_3O_5 and β-BaB_2O_4, Appl. Phys. Lett. **71**, 315–317 (1997)
66. D. Woll, B. Beier, K.-J. Boller, R. Wallenstein, M. Hagberg, S. O'Brian: 1 Watt of Blue 465 nm Radiation Generated by Frequency Doubling the Output of a High-Power Diode Laser in Critically Phase-Matched LiB_3O_5, Opt. Lett. **24**, 691 (1999)
67. T. W. Hänsch, B. Couillaud: Laser Frequency Stabilization by Polarization Spectroscopy of a Reflecting Reference Cavity, Opt. Commun. **35**, 441 (1980)
68. A. Yariv: *Quantum Electronics,* 3rd edn. (Wiley, New York 1989)
69. R. L. Sutherland: *The Handbook of Nonlinear Optics* (Dekker, New York 1996)
70. G. M. Gibson, M. H. Dunn, M. J. Padgett: Application of a Continuously Tunable CW Optical Parametric Oscillator for High Resolution Spectroscopy, Opt. Lett. **23**, 40 (1998)
71. G. M. Gibson, M. Ebrahimzadeh, M. J. Padgett, M. H. Dunn: Continuous-Wave Optical Parametric Oscillator Based on Periodically Poled $KTiOPO_4$ and Its Application to Spectroscopy, Opt. Lett. **24**, 397 (1999)
72. R. Al-Tahtamouni, K. Bencheikh, R. Storz, K. Schneider, M. Lang, J. Mlynek, S. Schiller: Long-Term Stable and Absolute Frequency Stabilization of Doubly Resonant Parametric Oscillators, Appl. Phys. B **66**, 733 (1998)
73. T. Ikegami, A. Slyusarev, S. Ohshima, E. Sakuma: Accuracy of an Optical Parametric Oscillator as an Optical Frequency Divider, Opt. Commun. **127**, 69 (1997)
74. N. C. Wong: Optical Frequency Division Using an Optical Parametric Oscillator, Opt. Lett. **15**, 1129 (1990)
75. D.-H. Lee, M. E. Klein, P. Groß, J.-P. Meyn, R. Wallenstein, K.-J. Boller: Self-Injection-Locking of a CW-OPO by Intracavity Frequency Doubling of the Idler Wave, Opt. Express **5**, 114 (1999)
76. R. C. Eckardt, C. D. Nabors, W. J. Kozlovsky, R. L. Byer: Optical Parametric Oscillator Frequency Tuning and Control, J. Opt. Soc. Am. B **8**, 646 (1991)
77. M. Scheidt, B. Beier, R. Knappe, K.-J. Boller, R. Wallenstein: Diode-Laser-Pumped Continuous-Wave KTP Optical Parametric Oscillator, J. Opt. Soc. Am. B **12**, 2087–2094 (1995)
78. K.-J. Boller, M. Scheidt, B. Beier, C. Becher, M. E. Klein, D.-H. Lee: Diode-Pumped Optical Parametric Oscillators, Quantum Semiclass. Opt. **9**, 173–189 (1997)
79. M. Scheidt, B. Beier, K.-J. Boller, R. Wallenstein: Frequency-Stable Operation of a Diode-Pumped Continuous-Wave $RbTiOAsO_4$ Optical Parametric Oscillator, Opt. Lett. **22**, 1287–1289 (1997)
80. D.-H. Lee, M. E. Klein, K.-J. Boller: Intensity-Noise of Pump-Enhanced Continuous-Wave Optical Parametric Oscillators, Appl. Phys. B **66**, 747–753 (1998)

81. M. M. Fejer, G. A. Magel, D. H. Jundt, R. L. Byer: Quasi-Phase-Matched Second Harmonic Generation: Tuning and Tolerances, IEEE J. Quantum Electron. **28**, 2631 (1992)

82. R. G. Batchko, M. M. Fejer, R. L. Byer, D. Woll, R. Wallenstein, V. Y. Shur, L. Erman: Continuous-Wave Quasi-Phase-Matched Generation of 60 mW at 465 nm by Single-Pass Frequency Doubling of a Laser Diode in Backswitch-Poled Lithium Niobate, Opt. Lett. **24**, 1293 (1999)

83. J.-P. Meyn, M. M. Fejer: Tunable Ultraviolet Radiation by Second Harmonic Generation in Periodically Poled Lithium Tantalate, Opt. Lett. **22**, 1214 (1997)

84. A. Arie, G. Roseman, V. Mahal, A. Skliar, M. Oron, M. Katz, D. Eger: Green and Ultraviolet Quasi-Phase-Matched Second Harmonic Generation in Bulk Periodically Poled KTiOPO$_4$, Opt. Commun. **142**, 265 (1997)

85. J.-P. Meyn, M. E. Klein, D. Woll, R. Wallenstein, D. Rytz: Periodically Poled Potassium Niobate for Second-Harmonic Generation at 463 nm, Opt. Lett. **24**, 1154 (1999)

86. W. R. Bosenberg, A. Drobshoff, J. I. Alexander, L. E. Myers, R. L. Byer: Continuous Wave Singly Resonant Optical Parametric Oscillator Based on Periodically Poled LiNbO$_3$, Opt. Lett. **21**, 713 (1996)

87. M. E. Klein, D.-H. Lee, J.-P. Meyn, B. Beier, K.-J. Boller, R. Wallenstein: Diode-Pumped Continuous-Wave Widely Tunable Optical Parametric Oscillator Based on Periodically Poled LiTaO$_4$, Opt. Lett. **23**, 831–833 (1998)

88. M. E. Klein, D.-H. Lee, J.-P. Meyn, K.-J. Boller, R. Wallenstein: Singly Resonant Continuous-Wave Optical Parametric Oscillator Pumped by a Diode Laser, Opt. Lett. **24**, 1142 (1999)

89. M. E. Klein, C. K. Laue, D.-H. Lee, K.-J. Boller, R. Wallenstein: Diode-Pumped Singly-Resonant CW Optical Parametric Oscillator with Wide Continuous Tuning of the Near-Infrared Idler Wave, Opt. Lett. **25**, 490 (2000)

90. A. Robertson, M. E. Klein, M.A. Tremont, K.-J. Boller, R. Wallenstein: 2.5 GHz Repetition Rate Singly Resonant Optical Parametric Oscillator Synchronously Pumped by a Mode-Locked Diode Oscillator Amplifier System, Opt. Lett. **25**, in print (May issue, 2000)

Tapered High-Power, High-Brightness
Diode Lasers: Design and Performance

Michael Mikulla

Fraunhofer-Institut für Angewandte Festkörperphysik,
Tullastrasse 72, D-79108 Freiburg, Germany
mikulla@iaf.fhg.de

Abstract. The development of high-power, high-brightness tapered diode lasers and laser amplifiers is reported in this review. Experimental and theoretical work is described that aims to improve the beam quality of these devices at output power levels well above 1 W. Special emphasis is laid on the dependence of the beam quality on the design of the epitaxial layer structure. For tapered diode lasers as well as for tapered laser amplifiers, the introduction of layer structures with a reduced modal gain leads to an improvement in beam quality by an order of magnitude. At output power levels of 2 W and CW operation beam-quality parameters of $M^2 = 2$ are achieved. The corresponding far-field widths are near the diffraction limit. Furthermore, a tunable diode-laser system is presented that consists of a tapered amplifier chip and a diffraction grating in an external cavity. This high-power high-brightness diode-laser system provides a narrow-bandwidth emission spectrum together with a wide-range tunability.

A new type of tapered laser array is described comprising 25 individual tapered emitters. 25 W CW output power is obtained from a single 1 cm-wide array with an averaged beam-quality parameter of $M^2 = 2.6$ at this power level. Compared to conventional broad-area arrays, the beam quality of the tapered array is higher by an order of magnitude. Due to their beam quality and their low-cost manufacturability, this new type of high-power diode laser array is the most promising device for the direct application of diode-lasers in materials processing.

High-power, high-brightness diode lasers are gaining more and more interest for applications previously dominated by expensive and inefficient solid-state lasers due to their better efficiency, compactness, and reliability. Among others, frequency doubling [1], free-space communication [2], and direct materials processing [3] are fast-growing fields where semiconductor-diode lasers are going to play a major role and will take a considerable part of the world laser market. In all of these applications, high output power together with a nearly diffraction-limited beam quality are either key requirements or strongly improve the system performance.

R. Diehl (Ed.): High-Power Diode Lasers, Topics Appl. Phys. **78**, 265–288 (2000)
© Springer-Verlag Berlin Heidelberg 2000

1 Introduction

The optical mode volume of high-power semiconductor-laser sources has to be enlarged in order to reduce both the junction temperature of the diode laser and the optical power density at the laser facet. These demands lead to the design of broad-area devices with LOC (Large Optical Cavity) epitaxial layer structures that have become the standard layer structure for high-power devices during the last few years [4]. The main drawback of these gain-guided broad-area devices is the loss of a lateral mode confinement that results in a rather poor beam quality due to self-focusing and filamentation processes [5].

A lot of different solutions have been proposed in the last few years to overcome these problems and achieve high output power together with high beam quality. The main effort has been directed to develop broad area structures that support only one lateral mode. Tapered devices [6,7,8], Distributed-FeedBack (DFB) lasers [9], Anti-Resonant Ridge Optical Waveguide (AR-ROW) lasers [10], and monolithically integrated Master-Oscillator Power-Amplifiers (MOPAs) [11] have been demonstrated and all of them are able to produce output powers well above 1 W together with a high beam quality.

Among these, devices based on tapered gain sections seem to be the most-promising candidates when a reproducible and low-cost fabrication is a further requirement.

Although tapered laser oscillators and amplifiers have already demonstrated high beam quality at power levels well above 1 W [6,7,8], beam filamentation remains the problem that limits the device performance [12]. Due to the interaction between the optical power and the carrier density in the active region of broad-area devices, spatial hole-burning leads to a spatially inhomogeneous refractive index that causes the degradation of the optical-beam profile.

This review is focused on the design and the performance of high-power tapered diode-laser oscillators, diode-laser amplifiers, and tunable high-brightness diode-laser systems. It gives an overview of theoretical and experimental work that has been carried out in order to improve the beam quality as well as to enhance the output power of these devices. As an important result, it is shown that the epitaxial layer structure has a large impact on the beam quality of tapered devices, especially at high output power levels. Layer sequences with a low optical confinement have been found to be much less sensitive to beam filamentation because of their reduced differential gain. These structures with confinement factors well below 2% are called LMG (Low Modal Gain) structures in contrast to HMG (High Modal Gain) structures with confinement factors higher than 2%. Experimentally we have shown that the beam quality of tapered laser oscillators and amplifiers can be improved by an order of magnitude when epitaxial layer structures with reduced modal gain are used for the device fabrication.

This review is organized as follows. First, the beam filamentation process is reviewed that leads to the decrease of beam quality in broad-area devices. Then, from the dependence of the refractive index on the carrier density, the concept of LMG structures is explained that results in the reduction of the beam filamentation and increases the beam quality. Second, high and low modal gain structures are investigated theoretically by BPM (Beam-Propagation Method) calculations. From these results, epitaxial layer structures are derived that enabled the experimental comparison of fabricated devices with modal gain structures similar to the modeled designs. The following paragraphs describe the performance of broad-area devices that were fabricated from the different layer structures and some details of the facet coating and the mounting technology are presented. After this, an extensive description of experimental data from both tapered laser oscillators and tapered amplifiers follows. A comparison of the beam quality of LMG devices and devices with a conventional HMG design is performed. Throughout this review, the beam-quality factor M^2 is used to qualify and to compare the experimentally achieved beam profiles.

Tunable high-power, high-brightness diode-laser oscillators in an external grating configuration are introduced in Sect. 7.3. Experimental results, including tuning range, side-mode suppression, and beam quality are presented.

Tapered diode-laser arrays comprising 25 tapered laser oscillators are described in Sect. 7.4. These devices are able to deliver 25 W of CW output power together with a high beam quality of each single emitter.

The chapter concludes with a short discussion of the advantages and disadvantages of tapered LMG devices compared to other approaches for high-brightness and high-power diode-laser sources.

2   Theoretical Background

In broad-area semiconductor-diode lasers, the main reason for the deterioration of the beam quality is the interaction between the amplified optical wave and the carrier density in the active region of the device. High optical power densities lead to spatial hole-burning, increase of the refraction index, and self-focusing of the propagating wave. As a result of these processes, beam filamentation occurs and the beam quality strongly degenerates [5].

The complex refractive index $\overline{n}$ of the diode-laser waveguide as a function of the carrier density n can be written as [13]

$$\overline{n}(n) = n_0 - \frac{1}{2k}g_{\mathrm{m}}(n)\alpha + \frac{\mathrm{i}}{2k}[g_{\mathrm{m}}(n) - \alpha_{\mathrm{i}}] \ . \tag{1}$$

In this equation $g_{\mathrm{m}}(n) = \Gamma g(n)$ is the modal optical gain given by the product of the material gain $g(n)$ and the optical confinement factor Γ. The optical confinement is determined by the overlap between the vertical mode profile in the waveguide structure and the active material. k is the vacuum wavenumber and n_0 is the carrier-independent part of the optical index;

$\alpha(\alpha > 0)$ and α_i are the linewidth enhancement factor and the internal optical losses, respectively. The differentiation of (1) leads to

$$\delta\bar{n}(n) = \frac{1}{2k}\delta g_{\mathrm{m}}(n)(i - \alpha) , \qquad (2)$$

when a constant α-factor can be assumed. From (2), it can be seen that the differential refractive index is determined by the differential modal gain. It follows that a reduced differential modal gain also reduces the variation of the refractive index due to spatial hole-burning, thereby leading to a reduced sensitivity to self-focussing and filamentation. Obviously, a low linewidth enhancement factor α is a further demand in order to achieve a low index variation. The relationship between refractive index variations and the differential gain opens the possibility to increase the beam quality in broad-area devices by proper design of the epitaxial layer sequence. An epitaxial layer sequence with a reduced differential gain will automatically help to suppress the beam filamentation and improve beam quality. These layer structures are called LMG (Low Modal Gain) structures and they can be realized by reducing the optical confinement factor of the waveguiding layers in a diode laser.

As an example, Fig. 1 shows the calculated modal gain (after [14]) of a 8 nm-wide InGaAs quantum well embedded in AlGaAs waveguide layers with different optical confinement factors of $\Gamma = 1.35\%$ and $\Gamma = 2.7\%$. The transparency current density in this material system is around $60\,\mathrm{A/cm^2}$ and the dependence of the modal gain on the current density can be approximated by a logarithmic function.

If, for example, a diode laser needs a modal gain of $20\,\mathrm{cm^{-1}}$ at threshold, the differential modal gain of a device with $\Gamma = 1.35\%$ is by a factor of four lower compared to a device with $\Gamma = 2.7\%$. According to (2) the reduced

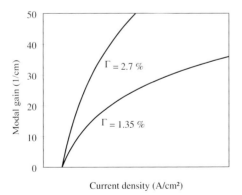

Fig. 1. Calculated modal gain of a 8 nm-wide InGaAs quantum well embedded in AlGaAs-waveguide layers with different optical confinement factors of $\Gamma = 1.35\%$ and $\Gamma = 2.7\%$ ([14])

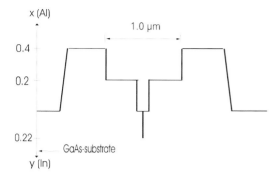

Fig. 4. Composition of the LMG step-index structure

the internal losses corresponds to the amount of optical power that is guided in the doped cladding layers of the two layer structures. A linewidth enhancement factor of $\alpha = 2$ was obtained from the emission spectra of single-mode Fabry–Perot diode lasers and this value was found to be nearly independent of the design of the optical-confinement layers. Of course, the threshold-current density of the LMG devices has to be higher due to the reduced confinement factor. Almost-identical characteristic temperatures around 140 K were found for both layer structures.

Table 1. Basic data of LMG- and HMG-layer structures

Layer structure	LMG	HMG
Optical confinement	$\Gamma = 1.35\%$	$\Gamma = 2.70\%$
Internal efficiency	$\eta_i = 0.9$	$\eta_i = 0.9$
Internal losses	$\alpha_i = 1\,\mathrm{cm}^{-1}$	$\alpha_i = 3\,\mathrm{cm}^{-1}$
Linewidth enhancement	$\alpha = 2.1$	$\alpha = 2.0$
Characteristic temperature	140 K	146 K
Transparency current density	$120\,\mathrm{A\,cm}^{-2}$	$80\,\mathrm{A\,cm}^{-2}$

4.1   Comparison of LMG and LOC Structures

It has to be emphasized that LMG structures have some basic differences compared to the well-known LOC structures. In the LOC structures, the vertical beam dimension in the optical waveguide of the diode laser is enlarged by expanded core layers in order to reduce the optical power density at the laser facets and to raise the Catastrophic Optical Mirror Damage (COMD) level of the device. A possible increase of the threshold current due to a lower optical confinement and a decrease of the differential modal gain is often prevented by the use of two quantum wells [4]. These devices are generally designed for high wall-plug efficiency and high COMD levels.

In contrast, LMG structures, as described in this chapter, take advantage of the reduced optical confinement which suppresses the beam filamentation and improves the beam quality due to the reduction of the differential modal gain as described in Sect. 2. A simple way to reduce the optical confinement is the use of extended waveguiding layers and a single quantum well as the active region. The lower optical confinement is not compensated by using a higher number of quantum wells. As has been shown in this chapter, this results in a near-diffraction-limited beam quality of tapered broad area devices at high power levels.

The LMG structures further benefit from the higher COMD level that is achieved by using extended vertical beam dimensions, similar to LOC structures. Of course, the advantages of LMG structures with respect to beam quality have to be paid for by a higher threshold current and a reduced conversion efficiency due to their lower optical gain.

5 Broad-Area Diode Lasers with LMG Layer Structures

80 μm-wide broad-area diode lasers were fabricated from the LMG layer structures in order to examine the high-power capability of the epitaxial material, the facet coating and the mounting technology. After thinning and cleaving, the front facets of the devices were anti-reflection coated to a residual reflectivity of 1% by a single layer of SiN by reactive magnetron sputtering. The back facets were high-reflection coated to a reflectivity of 90% by a stack of two SiO_2/SiN layers with the same technology. The devices were mounted p-side down on standard copper heat sinks with indium solder.

Figure 5 shows the L–I characteristic and the conversion efficiency of a 2 mm-long device at a heat-sink temperature of $10°$ C [16]. A maximum

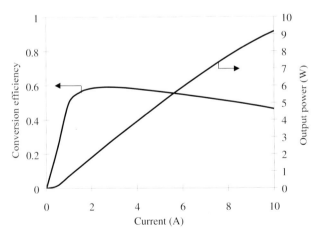

Fig. 5. L–I characteristic and conversion efficiency of a 100 μm × 2000 μm broad-area diode laser fabricated from the LMG layer structure

output power of 9.2 W CW is achieved at a current of 10 A. The threshold current density is $170 \, A/cm^2$ and the external efficiency reaches 82%. The low series resistance of 60 mW and the high internal efficiency result in a high overall efficiency. A maximum efficiency of 59% is achieved at an output power of 3 W. Between 0.7 W and 9.2 W of CW output power, the conversion efficiency is higher than 46%. These results are comparable to the best published data for Metal-Organic Vapor-Phase Epitaxy (MOVPE)-grown and Al-free broad-area diode lasers. Devices driven at 1.5 W CW output power show extrapolated lifetimes in excess of 20 000 h.

These results clearly show the reliable high-power capability of both the MBE-grown LMG epitaxial layer structures and the facet coatings.

6   Fabrication of Tapered Devices

Tapered laser oscillators and laser amplifiers similar to those devices reported in [6] were fabricated from both the HMG and LMG epitaxial layer structures. Figure 6 shows a schematic of the tapered laser oscillator geometry. At the narrow facet dry-etched grooves provide a lateral mode filter in order to improve the beam quality. A taper angle of 6° together with a device length of 2 mm lead to an emitting aperture of about 200 μm width. After thinning and cleaving high-reflection ($R = 90\%$) and anti-reflection ($R = 0.05\%$)-coatings are deposited on the facets in order to achieve maximum power extraction from the broad anti-reflection-coated front facet.

The lateral dimensions of the tapered laser oscillators described here are very similar to the devices published in [6].

The fabrication of tapered laser amplifiers is almost identical to the fabrication of the oscillators. The only difference is the use of anti-reflection coatings on both the input and the output facet. Similar to the broad-area diode lasers, the facets are coated by reactive magnetron sputtering of Si with

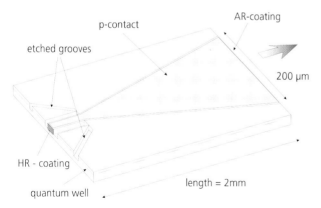

Fig. 6. Tapered laser oscillator

oxygen and nitrogen as reactive gases. The high-reflection coatings consist of two pairs of Si/SiO_2 layers whereas the anti-reflection coating is provided by a single layer of index-matched SiON. The devices are mounted p-side down on copper heat sinks with indium as solder in order to obtain CW operation at high driving currents.

7   Experimental Results

7.1   Tapered Laser Oscillators

Figure 7 shows a typical L–I characteristic of a tapered diode laser based on the LMG layer structure. A maximum output power of more than 3 W is observed under CW operation at a drive current of about 5 A. The wall-plug efficiency reaches a maximum of 36% and remains well above 30% at the highest power level. The threshold current of 540 mA corresponds to a threshold current density of 240 A/cm² for a 2 mm-long device. The external efficiency is 0.65 W/A.

Similar L–I characteristics were obtained from the HMG devices with an optical confinement factor of $\Gamma = 2.7\%$ together with a slightly improved wall-plug efficiency due to a lower threshold. These results are comparable to the data published in [6] and in [7].

In order to investigate the beam quality of tapered laser oscillators, the beam-quality parameter M^2 was measured after ISO 11146 with a commercial beam-analysing system. Figure 8 depicts the experimentally observed beam-quality parameters of typical LMG and HMG tapered diode lasers at different output powers. As expected from the BPM calculations, the beam quality of the HMG device severely degrades at higher power levels. In contrast, the

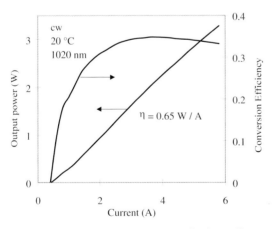

Fig. 7. L–I characteristic and wall-plug efficiency of a tapered diode laser with LMG-layer structure and 2 mm resonator length

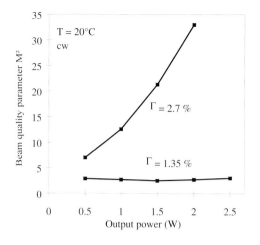

Fig. 8. Dependence of the beam-quality parameter M^2 of tapered diode lasers on the output power for different optical confinements Γ

beam-quality parameter of the LMG device remains below a value of $M^2 = 3$, indicating the high brightness of the emitted optical power.

Far-field profiles of the LMG device were measured after the removal of the quadratic phase front divergence by a cylindrical lens, as described in Sect. 6. An example for the evolution of the beam profiles with increasing output power is given in Fig. 9. A near-diffraction-limited and power-independent far-field angle of $0.24°$ FWHM is obtained up to an output power of 2 W. The measured M^2 parameters range between a value of $M^2 = 1.4$ at 500 mW and a value of $M^2 = 2.2$ at 2 W of output power. The increase in the M^2 parameter is predominantly caused by the rise of small side lobes at high output powers. In the M^2 measurement these side lobes increase the $1/e^2$ width of the fitted Gaussian profile, thereby increasing the calculated beam-

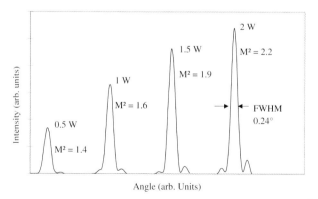

Fig. 9. Far-field profiles and beam-quality parameters M^2 of a tapered LMG diode laser at different output powers

quality parameter M^2, although the far field angle at the Full Width at Half Maximum (FWHM) level remains at a constant value.

The high beam quality of tapered lasers with LMG structures easily allows a highly efficient coupling of the emitted optical power into a multimode fiber. The experimental setup used for the fiber coupling is shown in Fig. 10. The output power of the diode laser is collimated in the vertical dimension by a spherical lens with a focal length of $f = 8$ mm. The strong astigmatism caused by the gain-guiding of the lateral mode in the tapered resonator and the index-guiding of the vertical mode has to be corrected by an additional cylindrical lens. With the $f = 100$ mm cylindrical lens the beam profile is collimated in the horizontal dimension to a beam diameter of approximately 1 cm in both dimensions. The optical power is then focused by a second spherical lens on an optical fiber with 50 µm core diameter and a numerical aperture of NA = 0.2. In order to match the NA of the fiber a focal length of 30 mm is chosen for this lens.

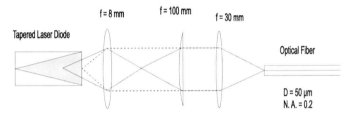

Fig. 10. Experimental setup used for the fiber coupling of a tapered diode laser

The optical power of a 3 mm-long tapered diode laser coupled into the fiber is shown in Fig. 11. At a driving current of 6 A the output power of the diode laser is about 2.8 W and the power in the fiber reaches 1.5 W. The coupling efficiency, measured from the diode-laser facet to the fiber, reaches more than 50%. The corresponding power density of 75 kW/cm^2 is sufficient for cutting and welding polymers.

It has to be emphasized that no optical isolator was used in the experiments. This indicates a high insensitivity of the beam quality against optical feedback from the uncoated fiber.

7.2   Tapered Laser Amplifiers

Tapered amplifiers with LMG and HMG layer structures as described in Sect. 6 were tested in a hybrid master-oscillator power-amplifier configuration. Figure 12 shows a scheme of the experimental setup. A single-mode Fabry–Perot diode laser, that can provide a maximum output power of 50 mW in a near-diffraction-limited beam, serves as the master oscillator. The optical power of this diode laser is collimated by a microscope objec-

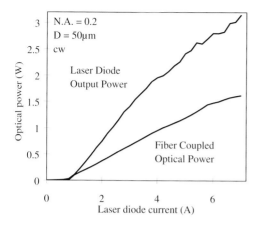

Fig. 11. Fiber coupling of a tapered diode laser

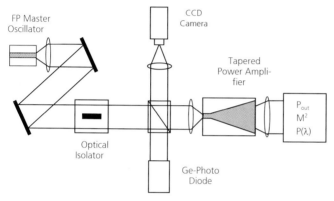

Fig. 12. Experimental setup for the characterization of tapered amplifiers in a hybrid master-oscillator power-amplifier configuration

tive and is coupled into the narrow input waveguide of the tapered power amplifier.

Figure 13 shows the L–I curve of an LMG amplifier at an input power of 32 mW. At a drive current of 4 A, the amplified optical power reaches 2.1 W and the optical gain at this input power is 18 dB. A slope efficiency of 0.7 W/A and a maximum power-conversion efficiency of 35% are achieved. Without input power, the amplifier emits about 250 mW of spontaneous emission. Lasing of the amplifier is completely suppressed by the excellent anti-reflection facet coating. The optical spectrum of the amplified high-power emission is an almost perfect image of the master oscillator's optical spectrum. As shown in Fig. 14, a side-mode-suppression ratio of more than 30 dB can easily be achieved together with a resolution-limited 3 dB bandwidth of less than 0.11 nm. Nearly identical results were obtained from HMG amplifier devices.

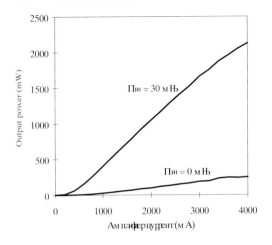

Fig. 13. L–I characteristic of a tapered high-power amplifier of 2 mm length with 32 mW of input power and without input power

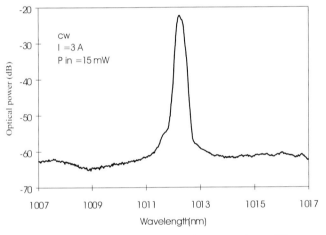

Fig. 14. Optical output spectrum of a tapered amplifier at a driving current of 3 A and 15 mW of master-oscillator input power. The optical output power under these conditions is 1.4 W

The beam quality remains the main difference between these two types of device structures. Similar to the tapered laser oscillators, the beam quality of tapered laser amplifiers with LMG layer structures is by far better than the beam quality of HMG amplifiers. This is shown in Fig. 15 where the beam-quality parameters of amplifiers with different optical confinements are compared. Again, the beam-quality parameter of the LMG devices is by an order of magnitude lower at output powers around 2 W [19,20]. As a best result, a near-diffraction-limited beam-quality parameter of $M^2 = 1.6$ at 2.1 W CW output power could be observed.

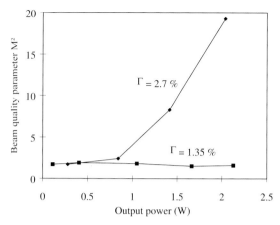

Fig. 15. Dependence of the beam-quality parameter M^2 of tapered laser amplifiers with HMG (*upper curve*) and LMG (*lower curve*) designs on the output power

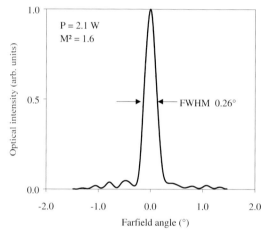

Fig. 16. Far-field pattern of a tapered laser amplifier at 2.1 W of CW output power with a beam-quality parameter of $M^2 = 1.6$

Figure 16 depicts the corrected far-field profile at 2.1 W output power with a near-diffraction-limited far-field angle of $0.26°$ at FWHM.

7.3   Tunable High-Brightness Diode-Laser Systems

Tunable high-power and high-brightness diode-laser systems with a narrow emission spectrum have many possible applications in fields such as spectroscopy, frequency-conversion (e.g. laser TV), pumping of solid-state laser materials, and others. A simple way to build such a laser system is the use of

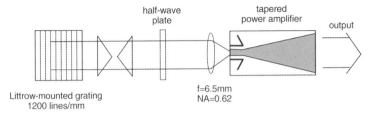

Fig. 17. Schematic experimental setup of the grating-tuned external-cavity high-brightness diode-laser system

a diffraction grating as an external reflector which provides for wavelength-selection and tunability. Figure 17 shows the setup of the laser system containing a tapered amplifier as the active device and a diffraction grating [21]. The amplifier is identical to those described in Sect. 7.2. The optical cavity is built up by the broad facet of the amplifier and the diffraction grating which can be tilted for wavelength-tunability of the laser system. The light emitted by the narrow facet of the amplifier chip is collimated by an aspheric lens of $f = 6.5\,\mathrm{mm}$. The grating is mounted in the Littrow configuration and is oriented so that the grating lines are parallel to the active region of the amplifier in order to disperse the spectrum perpendicular to the epilayers. In this configuration, the only 1 μm-thick vertical waveguide serves as the entrance slit of a monochromator, thereby increasing the side-mode suppression of the optical spectrum. A half-wave plate is inserted into the external cavity in order to rotate the direction of polarization perpendicular to the grating lines. This increases the reflectivity of the grating in the first order by about 13%.

Figure 18 shows the tuning curve of the external-cavity laser for CW operation. An output power of more than 1.0 W is obtained over the wide tuning range of 55 nm from 1030 nm to 1085 nm for a current of 4 A. Also shown in the diagram is the operating current necessary to obtain 1.0 W of output power with a minimum of 2.3 A at 1055 nm. At 1055 nm the maximum output power of 1.6 W is reached at 4 A. At 1083 nm the device emits more than 1.0 W with a drive current of 4 A.

The emission spectra for various grating positions can be seen in Fig. 19. The side-mode suppression over the entire tuning range varies between 40 dB and 50 dB. The highest side-mode suppression is obtained when the wavelength is tuned to the maximum optical gain around 1055 nm wavelength. The measured linewidth (FWHM) is about 0.1 nm, limited by the resolution (0.1 nm) of the optical spectrum analyser. In the spontaneous-emission spectrum a weak modulation with a period of about 2 nm occurs which is attributed to a residual reflectivity at the 80 μm-long input waveguide. This can be seen in operation without the grating or when the lasing wavelength is tuned to the extremes of the tuning range.

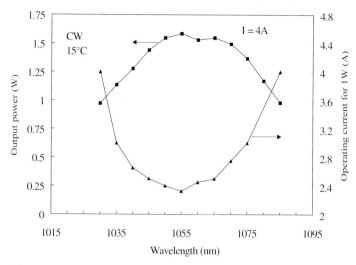

Fig. 18. Tuning curves of the external-cavity diode-laser system

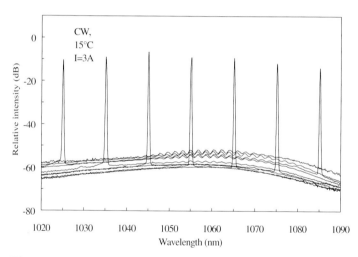

Fig. 19. Tuned emission spectra of the external-cavity diode-laser system between 1025 nm and 1085 nm at a drive current of 3 A

Similar to the tapered laser amplifiers in a hybrid master-oscillator power-amplifier configuration (Sect. 7.2) the tunable laser system provides a high beam quality at high output powers. For 1 W of output power at 3 A the beam-quality factor M^2 is below 2.0 from 1035 nm to 1075 nm.

Figure 20 shows the effective far-field intensity pattern at 1055 nm and at 3 A of driving current. The FWHM is 0.33° and 84% of the 1.3 W total output power is within the main lobe. The measured M^2 value at this operating point

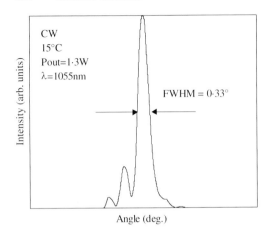

Fig. 20. Corrected far field of a tunable diode-laser system with a tapered laser amplifier in an external-cavity configuration

is 1.7. The FWHM of 0.33° indicates that the output beam would become 1.3 times diffraction-limited by capturing the side lobes with a slit.

7.4   Tapered Diode-Laser Arrays

Depending on heat-sinking and facet preparation the available and reliable output power of a single tapered laser oscillator is restricted to a few watts. Therefore, it is straightforward to fabricate monolithically integrated diode-laser arrays in order to achieve higher output powers and higher optical power densities. A diode-laser bar with tapered laser oscillators instead of broad-area devices comprises 25 single emitters with 200 μm-wide apertures and a pitch of 400 μm from laser to laser.

Figure 21 shows the L–I characteristic of a tapered diode-laser array together with its wall-plug efficiency [22]. The device has a slope efficiency of 0.77 W/ A and a threshold-current of 16.3 A, a threshold current density of 320 A/cm², and a differential resistance of 2.6 mW. The array delivers more than 25 W of optical power under CW operation and a driving current of 50 A which is limited by the current source. The output power is distributed fairly uniformly among the 25 emitters with an average of 1.04 W and a standard deviation of 0.12 W per emitter. No roll-over occurs and the wall-plug efficiency does not reach its maximum in the observed current range; 35% is achieved at the highest output power. All measurements are performed at a heat-sink temperature of 20° C.

The beam-quality factor M^2 of the tapered oscillators was measured with a commercial beam-analysing system. Figure 22 shows the distribution of the M^2 values along the 25 oscillators of the array. At a total power level of 10 W, all oscillators show near-diffraction-limited M^2 values. An average of $M^2 = 2.2$ with a standard deviation of 0.21 is achieved. M^2 values are very similar at 25 W output power with an average value of 2.6. The decrease is

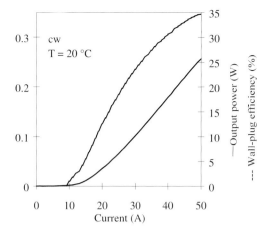

Fig. 21. L–I characteristic and wall-plug efficiency of a tapered-laser array comprising 25 individual emitters

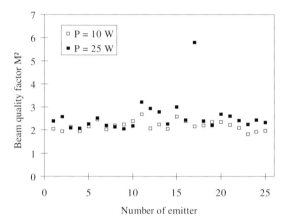

Fig. 22. Dependence of the beam-quality parameter on the number of oscillators and the output power of a tapered-laser array

mainly caused by emitter #17 which shows a significant decrease in the beam quality. This decrease probably indicates thermal problems.

An additional experiment was performed in order to underline the high uniformity of the beam quality. In this experiment the output power of each single emitter was collimated in the slow-axis dimension to a diameter of approximately 6 mm by a spherical lens with $f = 12.5$ mm. After the collimation the beams were focussed by a second lens with $f = 75$ mm, corresponding to a numerical aperture NA $= 0.04$ of the experimental setup. The focus diameter of each emitter was measured using a 5 μm-wide moving slit. The power-dependence of the focus-diameters is depicted in Fig. 23. Similar to the M^2 measurements, the array shows a very uniform distribution at the 10 W

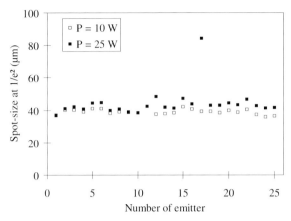

Fig. 23. Dependence of the spot-size diameter on the number of oscillators and the output power of the array. The NA of the experimental setup is 0.04

power level with focus diameters around 40 μm. At the 25 W power level, the diameters still remain below 45 μm, with the exception of emitter #17.

Compared to broad-area emitters, the beam quality of tapered emitters is improved by more than an order of magnitude. Consequently, the power density one can achieve from a tapered diode-laser array is higher by an order of magnitude compared to conventional broad-area arrays.

This power, emitted by 25 individual laser sources, can be superimposed by the well-known techniques for broad-area diode-laser bars. The first and inevitable step that has to be done for the beam focusing of a tapered laser bar is the collimation of each tapered laser which is described in principle in Sect. 7.1. Due to the short distance between the tapered oscillators of a laser bar and the strong astigmatism between the slow and the fast axis of a tapered oscillator, the beam collimation has to be performed by an array of microlenses. In this array each microlens consists of a pair of cross-coupled cylindrical lenses that collimate the slow and the fast axis of a single tapered laser. Figure 24 gives a schematic of a tapered diode-laser bar together with an array of beam-collimating microlenses. For each emitter a pair of crossed microlenses is necessary for the collimation of the astigmatic emission.

8   Manufacturability

From the manufacturability point of view tapered diode lasers compare well with simple broad-area diode lasers. The lateral structuring of tapered devices does not involve more steps than the fabrication of broad area-diode lasers. The coating and cleaving processes are very similar to those of broad-area devices with the exception of the high-quality antireflection coating of the broad output facet.

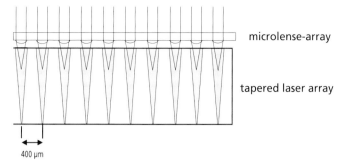

400 μm

Fig. 24. Schematic of a tapered diode-laser bar with a microlens array for beam collimation

Because of their low-cost manufacturability, tapered devices have many advantages compared to other high-brightness diode lasers like monolithic master-oscillator power-amplifiers, a-DFB lasers or ARROW lasers. For all of these counterparts an epitaxial re-growth step or the fabrication of a holographically defined reflection grating is necessary. These additional processing steps are costly and tend to lower the manufacturing yield.

9   Conclusion

In summary, it has been shown theoretically and experimentally that the beam quality of tapered high-power diode-laser oscillators and laser amplifiers can be improved by an order of magnitude using epitaxial layer structures with low modal gain. In these structures, the reduced differential gain leads to a strong suppression of spatial hole-burning, self-focusing and filamentation.

Tapered diode lasers fabricated from LMG layer structures show output powers of more than 3 W CW. Near-diffraction-limited far-field profiles could be observed up to power levels of 2 W. The beam-quality parameter of these devices remains well below $M^2 = 3$ whereas the beam quality of similar devices with high modal gain epitaxial layer structures severely degenerates at moderate output powers. With a LMG tapered oscillator, 1.5 W of optical power can be coupled into a 50 μm-core diameter fiber with more than 50% coupling efficiency. These experimental results have been achieved without the requirement of an optical isolator between the high-power devices and the diagnostic optics. This indicates that the beam quality of LMD devices is inherently insensitive to optical feedback.

Tapered laser amplifiers, fabricated from LMG layer structures, show output powers in excess of 2 W under CW operation. Their far-field profiles remain single-lobed with a near-diffraction-limited beam-quality parameter of $M^2 = 1.6$. The FWHM of the corrected far-field angles is 0.26°. At an input power of 32 mW from an external Fabry–Perot laser as the master oscillator, the devices have a large-signal gain of 18 dB. The emission spectrum of the amplifier is an almost-perfect replica of the master oscillator's optical spectrum with a signal-to-noise ratio of more than 40 dB. In contrast to these results, the beam quality of devices with conventional high modal gain structures severely suffer from beam filamentation and their beam-quality parameter M^2 is reduced by an order of magnitude.

Tunable high-brightness diode-laser systems have been studied that consist of a tapered laser amplifier with an external cavity provided by a diffraction grating at the narrow facet. Tuning of the emission wavelength can be achieved by tilting the grating. An output power of more than 1 W CW can be provided by these systems over a tuning range of 50 nm. With the peak gain of the quantum well centered at 1055 nm the longest emission wavelength that could be observed with 1 W of output power is 1085 nm. The optical spectrum shows a very narrow linewidth that is estimated to have a FWHM well below the measured 0.1 nm which is the resolution limit of the diagnostic system. Over the whole tuning range the beam-quality parameter remains below a value of $M^2 = 2$.

A new type of high-power, high-brightness diode-laser array has been developed. This array comprises tapered oscillators instead of the commonly used broad-area laser oscillators. A single array comprising 25 tapered emitters delivers more than 25 W of output power. A slope efficiency of 0.77 W/A and a conversion efficiency of 35% at the highest output power are observed. At this power level an averaged beam-quality factor of $M^2 = 2.6$ is achieved for each single emitter. Compared to broad-area emitters, the beam quality is higher by more than an order of magnitude. Consequently, the power density one can achieve from a tapered diode-laser array is higher by an order of magnitude compared to conventional broad-area arrays.

Due to their low-cost manufacturability and superior brightness at high power levels, tapered-laser arrays are the most promising candidates for the direct application of diode lasers in materials processing.

Acknowledgements

The author would like to acknowledge the contributions of P. Chazan, A. Schmitt, S. Morgott and A. Wetzel to the theoretical and experimental results. Thanks are also due to G. Bihlmann, J. Schleife, S. Klußmann, R. Moritz, M. Hanel, W. Fehrenbach, and J. Linsenmeier for their excellent technical assistance. J. Braunstein, G. Tränkle, and G. Weimann are gratefully acknowledged for encouraging support and fruitful discussions. Parts of this work were supported by the German Federal Ministry of Education and Research (BMBF) under contract number 13N6378.

List of Symbols

D	Diameter
f	Focal length
g	Optical gain
g_{m}	Modal optical gain
k	Vacuum wavenumber
M^2	Beam-quality parameter
n	Complex refractive index
$\overline{n}$	Carrier density
R	Optical reflectivity
T_0	Characteristic temperature
α	Linewidth-enhancement factor
α_{i}	Internal optical losses
Γ	Optical confinement factor
η	External optical efficiency
h_{i}	Internal quantum efficiency
j	Far-field angle

References

1. B. Beier, D. Woll, K.-J. Boller, R. Wallenstein: Second-harmonic generation of the output of an AlGaAs diode oscillator-amplifier system in critically phase matched LiB3O5 and b-BaB2O4, Appl. Phys. Lett. **71**, 1–3 (1997)
2. S. G. Lambert, W. L. Casey: *Laser Communication in Space* (Artech House, Boston, MA 1995)
3. P. Loosen, H.-G. Treusch, C. R. Haas, U. Gardenier, M. Weck, V. Sinnhoff, St. Kasperowsky, R. vor dem Esche: High-power diode lasers and their direct industrial applications, SPIE Proc. **2382**, 78–88 (1995)
4. A. Al-Muhanna, L. J. Mawst, D. Botez, D. Z. Garbuzov, R. U. Martinalli, J. C. Connolly: 14.3 W quasi-continuous wave front-facet power from broad-waveguide Al-free 970 nm diode lasers, Appl. Phys. Lett. **71**, 1142–1145 (1997)
5. J. R. Marciante, G. P. Agraval: Nonlinear mechanisms of filamentation in broad-area semiconductor lasers, J. QE **32**, 590–596 (1996)
6. E. S. Kintzer, J. N. Walpole, S. R. Chinn, C. A. Wang, L. J. Missaggia: High-power, strained-layer amplifiers and lasers with tapered gain regions, IEEE Photon. Technol. Lett. **5**, 605–608 (1993)
7. D. Mehuys, S. O'Brian, R. J. Lang, A. Hardy, D. F. Welch: 5 W, diffration-limited, tapered-stripe unstable resonator semiconductor laser, Electron. Lett. **30**, 1855–1856 (1994)
8. S. O'Brian, A. Schönfelder, R. J. Lang: 5 W CW diffraction-limited InGaAs broad-area flared amplifier at 970 nm, IEEE Photon. Technol. Lett. **9**, 1217–1219 (1997)
9. S. D. de Mars, K. M. Dzurko, R. J. Lang, D. F. Welch, D. R. Scifres, A. Hardy: Angled grating distributed-feedback laser with 1 W single-mode, diffraction-limited output at 980 nm, Techn. Digest CLEO '96 (1996) paper CTuC2, pp. 77–78

10. C. Smudzinski, D. Botez, L. J. Mawst, A. Bhattacharya, M. Nesnidal, R. F. Nabiev: Three-core arrow-type diode laser: novel high-power, single-mode device, and effective master oscillator for flared antiguided MOPAs, IEEE J. Select. Topics Quant. Electron. **1** (1995)

11. S. O'Brien, D. F. Welch, R. A. Parke, D. Mehuys, K. Dzurko, R. J. Lang, R. Waarts, D. Scifres: Operating characteristics of a high-power monolithically integrated flared amplifier master oscillator power amplifier; IEEE J. QE **29**, 2052–2057 (1993)

12. M. Mikulla, A. Schmitt, P. Chazan, A. Wetzel, M. Walther, R. Kiefer, W. Pletschen, J. Braunstein, G. Weimann: Improved beam quality for high-power tapered diode lasers with LMG (low modal gain) epitaxial layer structures, SPIE Proc. **3284**, 72–79 (1998)

13. R. J. Lang, A. Hardy, R. Park, D. Mehuys, S. O'Brien, J. Major, D. Welch: Numerical analysis of flared semiconductor laser amplifiers, IEEE J. QE **29**, 2044–2051 (1993)

14. L. A. Coldren, S. W. Corzine: *Diode Lasers and Photonic Integrated Circuits* (Wiley, New York 1995)

15. P. Chazan, J. D. Ralston: Beam-propagation model of tapered amplifiers including non-linear gain and carrier diffusion, LEOS Topical Meeting on Semiconductor Lasers, Advanced Devices and Applications, Keystone, CO (1995)

16. M. Mikulla, A. Schmitt, M. Walther, R. Kiefer, R. Moritz, S. Müller, R. E. Sah, J. Braunstein, G. Weimann: High-power InAlGaAs laser diodes with high efficiency at 980 nm, SPIE Proc. **3628**, 80–85 (1999)

17. L. Goldberg, D. Mehuys, M. R. Surette, D. C. Hall: High-power, near-diffraction-limited large-area traveling-wave semiconductor amplifiers, IEEE J. QE **29**, 2028–2043 (1993)

18. L. Goldberg, M. R. Surette, D. Mehuys: Filament formation in a tapered GaAlAs optical amplifier, Appl. Phys. Lett. **62**, 2304–2306 (1993)

19. M. Mikulla, P. Chazan, A. Schmitt, S. Morgott, A. Wetzel, M. Walther, R. Kiefer, W. Pletschen, J. Braunstein, G. Weimann: High-brightness tapered semiconductor laser-oscillators and -amplifiers with low modal gain epilayer-structures, IEEE Photon. Technol. Lett. **10**, 654 (1998)

20. P. Chazan, S. Morgott, M. Mikulla, R. Kiefer, G. Bihlmann, R. Moritz, J. Daleiden, J. Braunstein, G. Weimann: Influence of the epitaxial layer structure on the beam-quality factor of tapered semiconductor amplifiers, Proc. LEOS '97, San Francisco, CA (1997)

21. S. Morgott, P. Chazan, M. Mikulla, M. Walther, R. Kiefer, J. Braunstein, G. Weimann: High-power near-diffraction-limited external-cavity laser tunable from 1030 to 1085 nm, Electron. Lett. **34**, 558 (1998)

22. M. Mikulla, A. Schmitt, M. Walther, R. Kiefer, W. Pletschen, J. Braunstein, G. Weimann: 25 W CW high-brightness tapered semiconductor laser-array, IEEE Photon. Technol. Lett. **11**, 412–414 (1999)

Cooling and Packaging
of High-Power Diode Lasers

Peter Loosen

Fraunhofer-Institut für Lasertechnik,
Steinbachstrasse 15, D-52074 Aachen, Germany
loosen@ilt.fhg.de

Abstract. An overview of cooling and packaging of high-power diode lasers is given. The discussion concentrates on diode lasers in bar geometry, typically 10 mm-wide, which are soldered on actively cooled micro-channel heat sinks, made from copper.

Cooling and packaging of diode-laser chips are among the most essential processes in the production of high-power diode lasers. The discussion in this chapter concentrates on high-power diode lasers with a diode-laser chip in a bar geometry, mounted on an active cooler. This is a combination which at present is frequently used at high power.

The basic setup of these devices is shown in Fig. 1. The diode-laser bar with a typical footprint of 10 mm × 1 mm is soldered on a mechanical mount, which serves simultaneously as a stable carrier, as an electrode and as a heat sink. In order to keep the heat-diffusion paths short, high-power diode-laser bars are usually soldered 'upside down', which means that the epitaxially processed side of the bar with the p-n junction is put directly on the heat-sink surface.

Efficient heat-sinking is essential to keep the temperature of the active zone of the diode-laser bar low, which is an important requisite for long life-time, low threshold currents and low thermal shift of the emission wavelength.

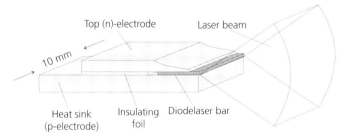

Fig. 1. Schematic setup of a high-power diode laser (not to scale): the diode-laser bar is mounted on a heat sink, serving as the positive electrode, and supplied with a top electrode as the negative contact

R. Diehl (Ed.): High-Power Diode Lasers, Topics Appl. Phys. **78**, 289–301 (2000)

The efficiency of diode lasers is very high compared to classical lasers – typical values range up to approximately 50% – thus creating only a relatively small amount of waste heat. However what makes heat-sinking so demanding is that the excess power is released as heat over a very small area. At, say, 40 W laser power and an efficiency η of 45%, a heat-flow density of roughly $1\,\mathrm{kW/cm^2}$ has to be managed for a diode-laser bar of $10\,\mathrm{mm^2}$ footprint. Low temperatures at these high heat-flow densities require high-performance heat sinks.

The relations in (1) below describe a simplified one-dimensional case, where the waste heat created in the p-n junction of the diode-laser bar is transported solely in the x (cf. Fig. 1) direction by heat conduction, and the thermal properties of the heat sink can simply be described by a constant thermal resistance R_{th}.

$$j = \rho \frac{\partial T}{\partial x}$$

$$P_{\mathrm{H}} = jA = A\rho \frac{\Delta T}{L} = \frac{1}{R_{\mathrm{th}}}\Delta T$$

$$P_{\mathrm{H}} = P_{\mathrm{L}}\left(\frac{1}{\eta} - 1\right)$$

$$\Delta T = T_{\mathrm{j}} - T_{\mathrm{c}}$$

$$R_{\mathrm{th}} = \frac{\Delta T}{P_{\mathrm{H}}} = \frac{L}{\rho}\frac{1}{A}$$

$R_{\mathrm{th}}\left[\dfrac{\mathrm{K}}{\mathrm{W}}\right]$ thermal impedance

$j\left[\dfrac{\mathrm{W}}{\mathrm{cm^2}}\right]$ heat-flow density

$\rho\left[\dfrac{\mathrm{W}}{\mathrm{cm\,K}}\right]$ thermal conductivity

$\dfrac{\partial T}{\partial x}\left[\dfrac{\mathrm{W}}{\mathrm{K}}\right]$ temperature gradient

$P_{\mathrm{H}}\,[\mathrm{W}]$ diode-laser waste heat .   (1)

Figure 2 illustrates this heat resistance for typical cases. Passive heat sinks, e.g. simple copper blocks attached to a water-cooled plate, are characterized by considerably high thermal resistance; in Fig. 2 a typical figure of $1\,\mathrm{K/W}$ for a diode-laser footprint of $10\,\mathrm{mm^2}$ is given. This resistance gives rise to junction temperatures of already $70°\,\mathrm{C}$ at output powers of 25 W and coolant at room temperature. In contrast, active coolers, as discussed in the following, create a tight contact between the active coolant (usually water) and the diode-laser bar. As shown in the diagram and discussed in the following sections, thermal resistances and junction temperatures can be reduced by a factor of 4, thus significantly extending the range of the available laser

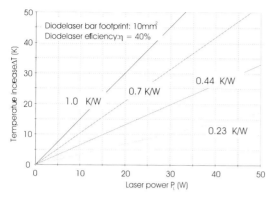

Fig. 2. Illustration of the one-dimensional heat resistance R_{th} of a heat sink. In the diagram the temperature increase ΔT of the p-n junction of the diode-laser bar due to the dissipated heat is given as a function of the output power of the diode laser P_L, for typical values R_{th} of passive heat sinks (high numbers) and active heat sinks (low numbers)

power. In the following sections 1 and 2, principles and techniques of advantageous active coolers are discussed, which are presently very common: namely copper micro-channel heat sinks.

Apart from providing low thermal-resistance coolers, the technique of how to solder the diode-laser bar on the heat sink is a prerequisite to produce powerful and reliable components. In contrast to conventional opto-electronic chips, which usually have small footprints, diode-laser bars are strongly extended in the bar direction, putting some extra demands on the packaging process, such as compensation of the mismatch between the thermal expansion coefficients between the chip (GaAs) and the cooler material. Objectives of this packaging process are:

- to provide a long-term and thermally stable connection between the chip and the heat sink,
- to ensure a precise alignment between chip and heat sink,
- high yields, reproducibility and process stability.

Section 3 discusses aspects of soldering techniques and equipment to perform such packaging processes.

1 Basic Properties of Micro-Channel Coolers for High-Power Diode Lasers

In micro-channel coolers the active coolant (usually water) is fed through small water channels only some 100 μm below the heat-sink surface, thus creating short heat diffusion lengths and a large inner surface for heat transfer. This cooling technique has first been demonstrated for heat sinking of high-power diode lasers and silicon as cooler material [1]. The following discussion concentrates on copper as cooler material, however, because copper yields high thermal and electrical conductivity, both being advantageous for the present application.

In order to obtain some insight into the basic mechanisms and properties of such coolers, a simplified one-dimensional analytical model is used as explained in Fig. 3. In this model the heat solely flows in the x direction; no lateral heat spreading is considered. In this case the total heat resistance can be described as the sum of three series resistances as indicated in the equations in (2) (the symbols refer to Fig. 3). The waste heat of the diode-laser bar (P_H), which is released at the top surface of the cooler, goes via three stages:

- the conductive term P_{cond} describes the conduction of the heat through the material of the top layers to the cooling fins (ρ: thermal conductivity of the material),
- the convective term P_{conv} describes the heat transition from the solid material of the fins to the liquid coolant,
- the capacitive term P_{cap} describes the temperature increase of the coolant while traversing the cooler (m: mass flow rate of the coolant, c: heat capacity of the coolant).

$$P_{cond} = \frac{\rho A}{h}(T_D - T') \qquad\qquad \text{Conductive term}$$

$$P_{conv} = \alpha A'(T'' - T') \qquad\qquad \text{Convective term}$$

$$P_{cap} = 2\,\dot{m}\,c(T'' - T_0) \qquad\qquad \text{Capacitive term}$$

$$P_{cond} = P_{conv} = P_{cap} = P$$

$$\frac{\Delta T}{P} = \frac{T_D - T_0}{P} = R_{cond} + R_{conv} + R_{cap}. \qquad\qquad (2)$$

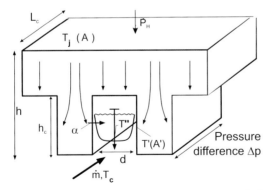

Fig. 3. Schematic view of the top layer and the micro-channel layer of a micro-channel heat sink: the waste heat, being released by the diode-laser bar, flows by heat conduction through the top layer and is transferred to the liquid coolant via the micro-channel fins in the micro-channel layer

The three constituents P_{cond}, P_{conv} and P_{cap}, contributing to the total heat resistance, are discussed in more detail in (3) with regard to their dependence on the widths of the micro-channels (channel width = fin width = d, N: number of channels).

$$R_{\text{cond}} = \varphi_1 \frac{h}{\rho} \frac{1}{A}$$

$$R_{\text{conv}} \cong \frac{1}{\alpha \, \varphi_2 \, h_c/d} \frac{1}{A} \quad \text{with} \quad \alpha_{\text{lam}} \propto \frac{1}{d} \qquad \alpha_{\text{turb}} \propto \sqrt{d}$$

$$R_{\text{conv}} \propto d^{0.7-2}$$

$$R_{\text{cap}} \propto \frac{1}{\dot{m}\,c} \qquad \text{with} \quad \dot{m} = \dot{m}_c(\Delta \rho)N \quad N = \frac{A}{2dL_c} \quad \dot{m} \propto d^{0.7-2}$$

$$R_{\text{cap}} \propto \frac{1}{A} \frac{1}{d^{0.7-2}} \, . \tag{3}$$

The conductive part of the total thermal resistance (R_{cond}) corresponds to (1), disregarding a shape factor φ_1, which describes the effective thickness of the heat-conducting compound, consisting of the top layer and the micro-channel layer of the cooler. The equation shows that under the approximations chosen here, the conductive resistance is not dependent on the channel width.

The approximation for the convective part (R_{conv}) also includes a shape factor φ_2, which reflects the geometry of the fins, at the surface of which the heat is transported into the liquid coolant. The smaller the channel width (factor $1/d$ in the denominator), the higher the total inner surface of the total fin structure and the higher the heat transfer. The second factor in the denominator, α, the heat-transfer coefficient, goes with d^{-1} or $d^{1/2}$, depending on whether the flow is laminar or turbulent [2]. Taking the factors together results in a convective resistance, increasing with a more or less large power of the channel width d.

The capacitive heat resistance is inversely proportional to the heat capacity of the liquid coolant (c) and to the total mass flow through the cooler. This total mass flow is split into N individual flow channels, all being driven by the pressure drop Δp. Introducing the relations for N and $m_c(\Delta$p) for laminar or turbulent flow, respectively [2], results in a relation for R_{cap}, which is roughly proportional to the inverse square of the channel width.

The three contributions to the total heat resistance have been put together in Fig. 4 for typical values of the discussed parameters. At small channel widths the capacitive heat resistance dominates due to the low amount of coolant being transported through the cooler at the given pressure drop; at large channel widths the convective heat resistance strongly increases, because the inner surface available for the heat transfer to the liquid is reduced. An optimum value for the channel width with a minimum heat resistance is achieved at a channel width of approximately 50 μm in the specific case being studied.

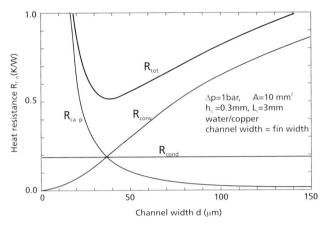

Fig. 4. Contribution of the three constituents R_{cond}, R_{conv} and R_{cap} to the total heat resistance R_{tot}, as a function of the width of the micro-channels

For the design of channel geometry and other parameters of real devices a more detailed numerical analysis is necessary and practical and production considerations have to be taken into account. Although small channel widths in the sub-100 µm range may be advantageous from the point of view of low heat resistance, such channels are hard or expensive to manufacture or may reduce reliability due to potential channel blocking by particles. These are reasons why channel widths in practical devices are more in the range of some 100 µm than below 100 µm.

It is exemplified with the numerical simulation illustrated in Fig. 5 that the heat in real devices is of course not solely flowing one-dimensionally in the x direction. The heat which is assumed to be uniformly produced at the bottom of the diode-laser bar is spread in the $-z$ direction in the top layer of the cooler. Efficient heat spreading, as it occurs in Cu due to the high thermal conductivity of this material, helps to reduce the heat-flow density, thus reducing the thermal resistance. In the design of the top-layer thickness a tradeoff has to be found because increasing the top-layer thickness increases the heat-spreading effect but increases the conductive resistance of the top layer as well. The optimum strongly depends on the material being used.

If a constant thermal resistance R_{th} as introduced above is applied for the characterization and comparison of coolers, it has to be taken into account that this is only a more or less accurate approximation for real cases. In the general three-dimensional case of heat conduction, (4) below can be given for the heat resistance R_{th} [$f(x, t)$: spatial and temporal distribution of the averaged dissipated heat $P_{H,Avg}$, $G(x, t)$: Green's function, Bd: boundary of the heat source]. Similar to the simple one-dimensional case in (1) this heat resistance is a quotient of the temperature T and the dissipated power P_H, but the temperature at the bottom of the diode-laser bar is now varying more or less strongly as can be seen in Fig. 5. The heat resistance is therefore not a

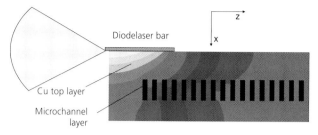

Fig. 5. Two-dimensional numerical simulation of the heat flow in a typical micro-channel cooler. The gray scales in the cooler cross section qualitatively indicate the local temperature

constant any more, it is a function of z (in the two-dimensional case) and of y (in the three-dimensional case). In many practical cases, however, for the sake of simplicity, averaged values are taken and given in data sheets.

$$R_{\text{th}}(x,y,z) = \frac{T(x,y,z)}{P_{\text{H,Avg}}} = \int_{t=0}^{\infty} \int_{\text{Bd}} f(\boldsymbol{x},t)G(\boldsymbol{x} - \boldsymbol{x}', t - t)\mathrm{d}\boldsymbol{x}'\mathrm{d}t$$

$$\boldsymbol{x} = \{x,y,z\}\,. \tag{4}$$

2   Manufacturing and Flow Dynamics of Cu Micro-Channel Coolers

The manufacturing of Cu micro-channel coolers is presently in most cases based on a 2-D layer technique, where thin copper sheets are structured and put on top of each other, forming an inner geometry with the micro-channels and manifolds (Fig. 6). The total production process consists of a sequence of different steps, the most important of which are:

- laser cutting or etching of the copper sheets,
- high-temperature bonding of the sheets,
- conventional milling of the cooler surfaces,
- ultra-precision milling of the surfaces and edges, where the bar will be mounted.

In an industrial environment, the coolers are produced in batches of $5-20$ and the individual elements have to be separated before the milling process.

All machining processes, especially the final ultra-precision milling, have to be performed very precisely and with machining parameters adapted to the sensitivity and necessary accuracy of the components. In many applications, e.g. in diode-laser systems for direct applications, the diode-laser bar has to be mounted extremely flat on the cooler with maximal deviations in the micron or even sub-micron range. This can only be ensured if the surface of the cooler has been precisely machined and if the component is handled carefully during the whole production process.

Quality demands are high not only to the mounting surface but to the front edge of the cooler as well, which has to be well defined along the width with μm-accuracy in order to provide an appropriate reference line for the micro-positioning of the bar during the soldering process (cf. Sect. 3).

The discussion in the foregoing section elucidates that high throughput of the liquid coolant through the micro-channel cooler at the given input pressure is essential for low thermal resistance. If the fluid-dynamical properties of a micro-channel cooler with an inner geometry as exemplified in Fig. 6 are analysed, it turns out that the biggest pressure losses are not friction losses arising in the micro-channels, but are losses produced by hard-edged changes in the flow direction within a layer or between different layers, or are caused by large and discontinuous changes of flow cross section (Fig. 7).

Table 1. Improvements in flow resistance and thermal resistance of micro-channel coolers by proper design of the flow system and the height of the channels [3]

	Pressure loss (at 500 ml/min) (bar)	Thermal resistance (at 4 bar) (K/W)
Original structure	4.5	0.4
Flow-optimized structure, 300 μm channel height	1.5	0.32
Flow-optimized structure, 600 μm channel height	0.5	0.29

In flow-optimized coolers, changes in the flow cross section are kept small, and all changes in cross section and flow direction have to be designed to be continuous and smooth (lower Fig. 7). A special advantage of the copper-layer setup is that this technology can be easily and flexibly adapted to these

Fig. 6. Micro-channel coolers, made from Cu sheets: Layer technique (*left*), micro-channel layer (*middle*, with a matchstick to show the proportions), final element (*right*). The width of the cooler is approximately 11 mm

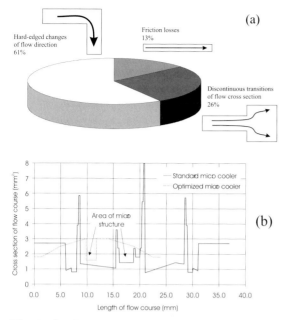

Fig. 7a,b. Pressure losses in a non-flow-optimized ('standard') micro-channel cooler: most of the pressure is lost at hard-edged changes in the flow direction or in discontinuous transitions of the cross section [4]

requirements. The following table shows the results of such optimizations for some selected designs: in the best case the pressure loss could be reduced by a factor of approximately 8. As illustrated in Table 1, this achievement can be invested in a twofold way: either profit is taken from the lower heat resistance at high input pressures and high water throughput or the cooler is operated at moderate water throughput with the advantage that simple and low-cost water pumps can be utilized.

3   Packaging of Diode-Laser Bars

In Fig. 8 the principal setup of a completed diode laser is shown as well as a picture of a typical component. The diode-laser bar is soldered on the Cu cooler, which usually acts as the p-electrode. With a second soldering process, the top (n)-electrode, which is insulated against the cooler with an isolating foil, is fixed to the bar. As already mentioned in the forgoing section, the emitting front of the bar has to be aligned precisely with respect to the front side of the cooler. The required protrusion of the bar with respect to the front edge of the cooler amounts to some μm: if the protrusion is too low or if the bar even stands back behind the cooler front side, solder may be squeezed out in front of the facets, resulting in damage of this most sensitive part of the

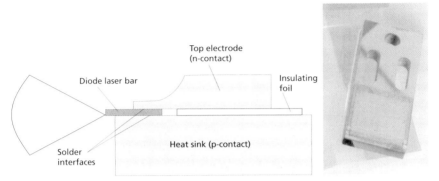

Fig. 8. Structure of a packaged diode-laser bar: schematics (*left*) and technical example (*right*)

bar. If, on the other hand, the protrusion is too high, cooling of the highly loaded laser facets is low thus increasing possible facet damage and reducing lifetime.

The different packaging steps succeeding cooler production are summarized in the following list and will be detailed hereunder:

- galvanizing the Cu coolers: here diffusion barriers (for instance Ni) as well as appropriate solder interfaces (for instance Au) are deposited on the cooler,
- solder deposition is usually performed by physical vapor deposition of the solder material (for instance In),
- the bar and top-electrode soldering process is carried out by precise, optically controlled setups, ensuring the high positional accuracy required,
- after the soldering processes the component is electrically and optically characterized and undergoes a burn-in procedure.

In contrast to packaging of single-emitter chips, which have an extension of only some $100\,\mu\mathrm{m}$, packaging of $10\,\mathrm{mm}$ extended bars requires more consideration of the differences in thermal expansion of chip and cooler material. When a GaAs chip with a thermal expansion coefficient of $6.5 \times 10^{-6}\,1/\mathrm{K}$ is soldered on a copper mount with, approximately, $16 \times 10^{-6}\,1/\mathrm{K}$, a considerable amount of mechanical stress may be introduced into the semiconductor material due to the soldering process and other thermal cycles. In order to compensate for this, In as a comparatively soft material is a common solder material.

In Fig. 9 it is illustrated schematically how the diode-laser bar is gripped by a vacuum gripping tool and subsequently aligned with respect to the heat sink, which is heated by a heating plate during the soldering process. Usually, the alignment process is controlled optically with a microscope setup as shown in the photograph.

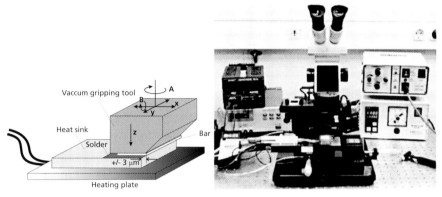

Fig. 9. Soldering of a diode-laser bar on a heat sink: schematic setup of the equipment (*left*) and experimental realization (*right*)

The last steps in the production process are characterization and burn-in. At the time of writing it is not possible to predict precisely the final properties of the diode-laser component without further measurements. It is therefore required to characterize the diode laser after the packaging process. In this characterization procedure diode-laser properties such as output power, current–voltage characteristics and optical parameters (divergence angle, state of polarization, etc.) are either measured at each laser or statistically at some elements of a lot.

Especially important is the subsequent burn-in procedure, during which each laser is individually operated during a period of several hours at specific burn-in parameters (current, heat-sink temperature, etc.). During burn-in output power and, if required, further parameters are monitored. In Fig. 10 the mechanical setup of such burn-in equipment is shown in an overview as well as details of the snap-in holders, in which the diode lasers are mounted

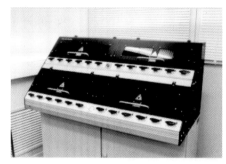

Fig. 10. Equipment for the burn-in of high-power diode lasers: overview (*left*), details of the holders (*right*)

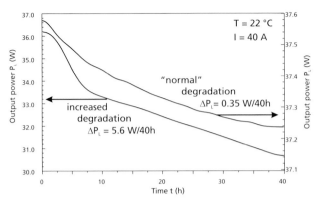

Fig. 11. Typical burn-in result for a diode laser with "normal" and with increased degradation

and which supply the components with cooling water and current. The laser emission is fed into beam dumps and measured by appropriate photodiodes.

A typical example of a burn-in sheet, measured with this device, is given in Fig. 11 [5]. The diagram shows on the one hand a diode laser with increased degradation, which would be sorted out in the production process. On the other hand, a normally operating component as given on the right-hand side of the diagram in many cases shows in the first tens of hours a slightly increased degradation rate, which decreases in further operation.

List of Symbols

A	Footprint of the diode-laser bar
A'	Area for heat transfer from liquid to solid
c	Heat capacity
d	Channel width
h, h_c	Cooler height, channel height
j	Heat-flow density
L	Thickness of the heat sink
L_c	Cavity length of the diode-laser bar
m	Total mass-flow rate
m_c	Mass-flow rate of one individual channel
N	Number of channels
Δp	Pressure drop
P_L	Laser power
P_H	Waste heat
R_{th}	Thermal impedance
T	Temperature
T_j	Temperature of the p-n junction

T_c	Coolant temperature
x, y, z	Orthogonal room directions
α	Heat-transfer coefficient (liquid–solid)
η	Laser efficiency
ρ	Thermal conductivity
$\varphi_{1,2}$	Shape factors

References

1. R. Beach, W. J. Benett, B. L. Freitas, D. Mundinger, B. J. Comaskey, R. W. Solarz, M. A. Emanuel: Modular microchannel-cooled heat sink for high average power laser-diode array, IEEE J. QE **28**, 966–976 (1992)
2. Verein Deutscher Ingenieure (ed.): *VDI-Wärmeatlas* (VDI, Düsseldorf 1984)
3. T. Ebert: Optimierung von Mikrokanalkühlern zur Steigerung der Ausgangsleistung von Hochleistungs-Diodenlasern, Dissertation RWTH Aachen, Aachen (1999)
4. T. Ebert, G. Treusch, P. Loosen, R. Poprawe: Photonics West 98, SPIE Proc. **3285**, 25–29 (1998)
5. J. Jandeleit, N. Wiedmann, P. Loosen, R. Poprawe: Photonics West 99, SPIE Proc. **3626**, 217–221 (1999)

High-Power Diode Lasers
for Direct Applications

Uwe Brauch[1], Peter Loosen[2], and Hans Opower[1]

[1] DLR, Institut für Technische Physik,
Pfaffenwaldring 38-40, D-70569 Stuttgart, Germany
u.brauch@dlr.de

[2] Fraunhofer Gesellschaft, Institut für Lasertechnik,
Steinbachstraße 15, D-52074 Aachen, Germany
loosen@ilt.fhg.de

Abstract. Diode-laser systems have a good chance to be used as light sources for direct applications. For this purpose a large number of relatively weak light bundles originating from individual oscillators have to be coupled to a single powerful beam. The coupling may be done in a completely incoherent way in cases where only a moderate beam quality is appropriate or with increasing degrees of coherence corresponding to applications where high demands on the quality are necessary. In this article all current methods of incoherent as well as of coherent beam combining are described and judged with respect to their present and future potential.

1 Introduction

Hans Opower

The outstanding feature of semiconductor lasers is their exceedingly high efficiency, which surpasses the most prominent types of lasers by nearly an order of magnitude. The spectrum of optical wavelengths attainable under advantageous conditions reaches from the visible red regime to the near infrared until about 1.5 μm and therefore comprises the spectral range which is commonly available with the most powerful solid-state lasers. But apart from this general behavior there are fundamental differences to the conventional solid-state lasers. These differences refer to the design of individual elements as well as to the way by which powerful systems may be arranged: diode lasers function like forward biased p-n diodes, with ultrathin active layers between the p and n material. In the case of so-called edge emitters the optical resonator extends along the active area, whereas Vertical Cavity Surface-Emitting Lasers (VCSEL) are operated with the beam direction perpendicular to the area of the active layers, whose extremely small thickness determines the optical gain length [1]. Presently, the edge-emitter concept is commonly adopted for power applications despite the difficulties of beam shaping. Due to the relatively high gain there exists a practical limitation of the size of the active zone belonging to one individual laser resonator. This fact leads to the important consequence that really high optical powers can only be attained by adding the beams of a lot of individual oscillators, each

R. Diehl (Ed.): High-Power Diode Lasers, Topics Appl. Phys. **78**, 303–367 (2000)

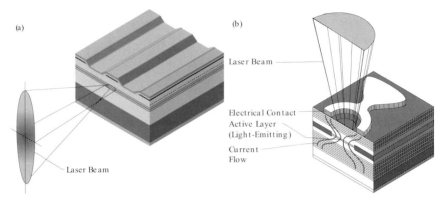

Fig. 1. Schematic structure of an edge emitter (**a**) and of a vertical cavity surface emitter (**b**)

of them delivering some tenths of a watt depending on the specific mode structure. This situation remains essentially unchanged, while VCSELs will possibly play a large role in future. Figure 1 shows the principal design of a single edge emitter (Fig. 1a) and of a vertical cavity structure (Fig. 1b).

It is evident that the addition of many sources may cause serious problems which increase with the degree of beam quality being required. Not least for this reason powerful diode-laser systems have been primarily considered as pumping sources for solid-state lasers because in this case the spectral width is uncritical and a moderate degree of spatial coherence is already sufficient. With these relatively simple requirements a remarkable improvement of the emission characteristics of solid-state lasers will be obtained compared to the situation with the present gas-discharge pump lamps.

Nevertheless there are good arguments to draw attention to applications where the diode lasers act immediately as the exclusive sources of light to perform the processing tasks. The advantage is that by doing this the overall efficiency will be enhanced by nearly a factor of two as the excitation losses of the second laser, namely the solid-state laser, can be avoided.

In order to get an impression of applications being most favorable for the employment of diode-laser systems, it is advisable to classify the different kinds of processing according to the specifications the light beams have to meet. Looking at different applications in the field of materials processing in agreement with the present state of the art, a subdivision into three classes may be found to be adequate:

- Processes which need only a moderate power density (about $10^4\,\mathrm{W/cm^2}$) within a relatively large focal area, for instance surface hardening or soldering,
- Processes requiring high power densities (about $10^6\,\mathrm{W/cm^2}$) not within extremely small focal areas; examples are welding and surface alloying of metals,

- Precise cutting and drilling under the condition of the highest possible beam quality (power densities on the order of $10^7 \, \mathrm{W/cm^2}$).

Not explicitly mentioned because they are not yet frequently in use are applications employing patterns of many distinct and independently controlled light spots [2].

It is self-evident that – besides the distinct patterns – the first kind of processes offers the best opportunity for diode-laser arrays, with or without a fiber-coupling arrangement. Even with an incomplete filling of the space between individual diode elements within an array or between the arrays in a stack, power densities on the order of a few tens of kilowatts per $\mathrm{cm^2}$ are obtainable. These completely incoherent devices may be improved by spatial rotations of the beams in order to get a better overall symmetry of the whole system and by bringing the fill factor close to one. With such measures power densities of at least $10^5 \, \mathrm{W/cm^2}$ should be attainable [3,4]. As long as we persist in working with essentially incoherent sources there remains only one further way of increasing the brightness, namely to combine beams of different wavelengths and/or different polarizations. This so-called wavelength- or polarization-multiplexing principally enables an improvement by nearly a factor of ten depending on the effort being applied.

With these measures the applications summarized under the second point are quite achievable. But it is clear that the above-mentioned power densities are not a reliable indication of an already good beam quality as they are dependent on the numerical apertures of the focusing systems. As a correct measure of the real beam quality we may take the M^2 values (diffraction parameter) or, sometimes more convenient, the wavelength-independent values of the aperture–divergence product. With respect to these generally valid criteria the incoherently coupled diode lasers do not quite reach the performance of conventional solid-state lasers. The development of diode-laser systems with really satisfactory optical-beam quality requires a higher degree of coherence. The specific meaning of this term in the context with beam-combining will be explained in more detail below.

Qualitatively speaking the lowest degree of coherence is characterized by the fact that each individual diode oscillates with the spatial fundamental mode but with an arbitrary longitudinal-mode composition. In this case the optimum beam quality of a light bundle linearly decreases with the number of individual beams lying parallel along one direction, which means the M^2 value increases proportionally to the square root of the total number of beams, when the whole arrangement shows a rotational symmetry [5,6]. With respect to a fixed amount of the total light power the resulting quality of the bundle now strongly depends on the power of each individual beam as this contribution determines the number of beams necessary. If each light beam is generated by just a single oscillator its power will be limited to a few tenths of a watt; if a costly oscillator–amplifier system acts as light generator a power of some watts may be obtained [7]. In the latter case a one-kilowatt laser could be operated with an M^2 factor between 10 and 20, which represents quite a good performance compared to conventional solid-state lasers. A possibility of even better figures may be obtained with the help of the afore mentioned wavelength- or polarization-multiplexing in such cases where a well-defined wavelength or state of polarization are unimportant.

The problem is that oscillator–amplifier systems in the form of integrated devices are extremely sensitive against internal backscattering and therefore very complicated. So one has to look for more appropriate solutions being characterized by phase-controlled beam-combining methods. Such tasks require a higher degree of coherence, because all lasers to be coupled have not only to oscillate in the lateral fundamental mode but also in a longitudinal single mode as this is one prerequisite for an exact frequency adjustment. If all individual lasers contributing to a light bundle can be coupled in a completely coherent way the resulting M^2 factor will remain close to one independent of the total power and the number of elements [2].

Key questions are whether there exists a unique physical possibility of coherently coupling a large number of individual elements and what enabling means are at hand to achieve this. Here one has to keep in mind that not only the frequencies have to be locked but, in addition, the phases must be adjusted. The frequency-locking can be brought about with exclusively optical methods. Two ways have been experimentally investigated:

- all individual diode lasers are allowed to oscillate under equal conditions but their beams are guided in such a way that neighboring elements can mutually interact via evanescent or leaky waves and so the whole system gets a chance of self-organization [8];
- one diode laser is selected to serve as a master oscillator whereas all others, being carefully kept isolated from each other, are controlled by the radiation of this master laser [2,9].

It may easily be understood that the first method based upon the hope of self-organization proves to be successful for a few oscillators but will soon lead to insurmountable difficulties when the number of individual elements grows, as in this case instabilities unavoidably arise and finally bring the whole system into chaotic oscillations.

The situation presents itself far more favorably when the concept of a master oscillator is applied. The slave lasers to be controlled run as independent oscillators with complete resonator configurations. Each of them is seeded by a small part of the master radiation.

The seeding concept may be seen as an extension of the simple amplification scheme with many separated parallel-lying amplifiers. These amplifier arrangements guarantee in general an optimal degree of coherent coupling but they suffer from the limited amplification along a single path. The step to a multipass operation finally leads to the master–slave principle. Such devices need a more careful adjustment between the individual elements but they have the advantage of an effective multiplication of the original radiation. In any case additional means for the final adjustment of the phases must be provided. For this purpose phase-correcting optical elements either fixed once and for all or controlled by electronic circuits are unavoidable.

Besides the exclusive optical methods of frequency locking by amplification or seeding, a possible combination of optical metrology and electronic processing should at least be generally mentioned. It works similarly to the master–slave principle by detecting the interference pattern of a reference master beam with each of the slave beams and by controlling the beam frequency and phase of all individual elements via electronic circuits according to the detected interference signals. Despite the undeniable elegance of such an electronic way of frequency locking and phase control it seems presently to be too costly for laser systems of technical importance.

The technical design of coupling schemes will be influenced by the specific semiconductor structures of the individual elements. Edge emitters can be equipped with amplifiers or arranged as seeded oscillators with the aid of waveguide systems, the fabrication of which is well known including the switches and phase-shifting elements. In the case of VCSELs, one is confronted with a somewhat different situation: the amplifier scheme is hardly useful due to the low amplification per path, whereas seeding arrangements seem to be quite feasible but they have to be designed as three-dimensional devices.

A general perspective concerning all concepts irrespective of the specific peculiarities must thoroughly be considered. The realization of coherent coupling has been proven to be principally feasible. The question remains: are the solutions scalable to an arbitrarily large number of elements? An example will illustrate the problem.

Master oscillators of any kind may be correctly assumed to emit radiation, the power of which lies in the range of ten to a hundred mW. In this case the total power of the corresponding slave system does not surmount the order of a few watts. Really high-power lasers therefore need cascade networks with each slave of the predecessor stage acting as master for a slave system of the successive stage. Arrangements of this kind are – by neglecting the more-efficient seeding technique – similar to conventional many-stage amplifier systems with gradually increasing diameters of the active elements. The essential difference is that in the case of semiconductors the increasing diameter must be thought of as being replaced with an increase of the number of active elements which are geometrically separated from each other but operated as optically parallel elements, meaning that in this case the individual beam propagations are characterized by a multiple beam splitting and combining. Figure 2 illustrates the principal similarities and differences of conventional amplifier systems and comparable diode-laser arrangements.

In this context one point should be emphasized: the final combination of a large number of parallel beams necessary for highly qualitative systems

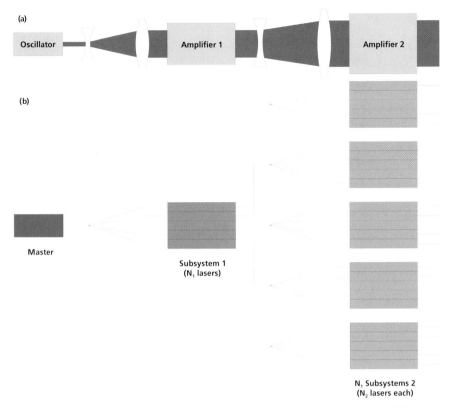

Fig. 2. Comparison between the principal design of conventional 'monolithic' amplifier systems (**a**) and diode-laser arrangements with distributed sources (**b**)

demands a sophisticated adjustment of the phases and powers within each individual beam of the last amplification stage, because otherwise even a coherent addition of a bundle of beams to be focused leads to multiple side maxima of the light-power density in the focal plane [9].

Having the physical features of diode-laser systems in mind, especially the disseminated sources, it seems to be worthwhile to make a technical comparison with the 'monolithic' solid-state lasers. In particular, the size and the complexity of systems as a function of the beam quality are of interest. As a rule of thumb, in all cases we choose a low Fresnel number as the necessary condition to obtain fundamental-mode radiation. Under this assumption the relevant parts of an individual semiconductor device delivering a light power of about 100 mW take a volume – without periphery – on the order of 10^{-6} cm^3, whereas a solid-state oscillator of a similar capacity typically occupies about 1 cm^3. This means that by comparison of the active parts the solid-state oscillators need nearly a million times larger volumes than diode lasers of the same integral output power. When, in addition, the high efficiency of the latter is taken into account there seems, at first glance, to be an overwhelming superiority of the diode lasers. On the other hand, the disadvantages and difficulties of diode-laser systems may not be disregarded: the necessity to deal with a huge number of individual elements causes a remarkable effort and may lead to a considerable increase of the volume of the system as well as to growing expense. The problems can only be mastered with an accessible route to fabricate the systems as integrated devices, which means with the methods of process techniques instead of an assembling technique being mainly applied to conventional lasers whose design resembles that of machines. Process techniques are well established in the field of microelectronics but not yet for optical devices in general. Nevertheless, the worldwide efforts and the current progress of microoptics fabrication techniques will surely offer the prerequisites for an economic production of efficient and powerful diode-laser systems with gradually increasing beam quality for direct applications.

2   Incoherent Beam Combining
Peter Loosen

Incoherent beam combination of individual diode lasers is presently widely used in laser systems for materials processing, medical applications and for the pumping of solid-state lasers. It is a straightforward way to achieve relatively high output powers at high system efficiency, although the potential to achieve high beam quality is limited (Sect. 2.2).

Schematic pictures of present technical setups are given in Fig. 3. At low output power in the range up to approximately 100 W, where only a couple of diode lasers is needed, the techniques of direct coupling into a fiber bundle or the utilization of a beam-transformation system (Sect. 2.3) are common. At higher output power up to several kilowatts another setup is used: the individual diode-lasers are stacked on top of each other or arranged in parallel.

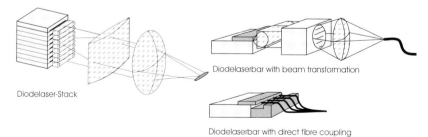

Fig. 3. Schematics of different technical setups for incoherent beam combination: stacking (*left*), fiber coupling of a transformed beam (*upper right*) and direct coupling into a fiber bundle (*lower right*)

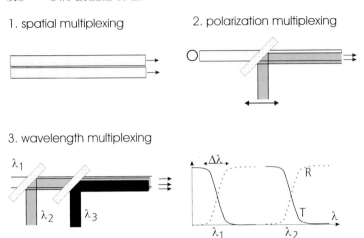

Fig. 4. Basic principles of incoherent beam combination

The basic physical principles used in all systems are illustrated in Fig. 4. The most common arrangement utilizes 'spatial multiplexing' where the individual laser beams are simply set side-by-side in a one- or two-dimensional array, thus increasing output power as well as the size of the beam. Two complementary techniques, which increase output power at constant beam size, are also shown: in polarization-multiplexed systems two mutually perpendicularly polarized beams are coupled via a polarization coupler; in wavelength-multiplexed systems several beams with different wavelengths are coupled with edge filters so that the combined beams are collinear. The discussion in the following sections mainly refers to spatial multiplexing; only where explicitly mentioned, polarization and frequency multiplexing is addressed.

Figure 5 illustrates for the one-dimensional case how N adjacent beams are overlapping in the far-field, e.g. in the focal plane of a focusing lens. In the fully incoherent case the intensity of the total beam is N times the intensity of one individual beam. If the coherence among the beams is increased, the individual beams start to interfere with each other, thus producing an intensity distribution with a much smaller width having a peak intensity proportional to the square of the number of beams N. This is the case of a 'classical' laser with a well-defined phase relation across the full cross-section, whereas laser systems with incoherent beam superposition are in between a classical laser and a pure incoherent light source. However, well-designed systems with incoherent beam combination are comparatively simple to realize and have the perspective of achieving beam properties and output powers which cover nearly all commercially relevant applications, as will be discussed in the next sections.

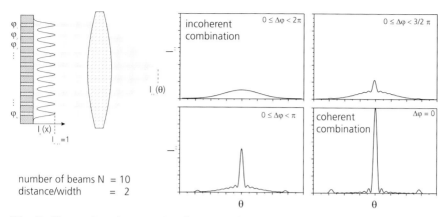

Fig. 5. Comparison between incoherent and coherent beam combination. In a numerical simulation several coherent beams with similar intensity distributions are superimposed in the focal plane of a lens. The coherence between the beams is increased from the *top left* to the *lower right* by reducing the statistical phase differences $\Delta \varphi$ among the beams

2.1   Properties of High-Power Diode Lasers for Direct Applications

High-power diode-laser systems in most cases are based on diode-laser bars as shown in Fig. 6. In order to achieve high output powers, a number of individual diode-laser structures are arranged in parallel on a diode-laser bar, which is typically 10 mm wide.

The individual diode-laser structures on the bar may consist of an array of stripe emitters or are built up as broad-area emitters. In the direction perpendicular to the plane of the p-n junction (x dimension in Fig. 6) the beam quality in general is close to diffraction-limited, because the beam is emitted from a region with only an x extension in the μm range. The corresponding divergence angle is quite large and may achieve figures of up to NA 0.8 (approximately 100° full angle). Since the beam diverges quite fast in this dimension, the x axis in Fig. 6 is usually called the 'fast axis'.

In contrast to the fast axis, the emitting width in the y direction of an individual diode-laser structure (Fig. 6) is usually much larger, up to some 100 μm. The beam quality is much lower compared to the fast axis and typically amounts to some times up to some ten times the diffraction limit. Typical divergence angles are thus in the range of some degrees: this slowly diverging behavior is the reason why this axis is usually called the 'slow direction'. In most present diode-laser bars, adjacent diode-laser structures are completely incoherent to each other. A coupling between the individual structures is often suppressed by v-grooves as shown in Fig. 6 in order to suppress 'super modes' and lasing in crosswise directions.

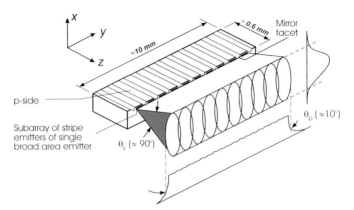

Fig. 6. Diode-laser bar as typically used in diode-laser systems

In all discussions in the current section the coordinate system is labeled as given in Fig. 6: the propagation coordinate of the beam is the z direction; x and y denote the transverse coordinates.

In order to achieve the high output powers from diode bars needed in diode-laser systems, proper cooling of the components has to be ensured. This is done very efficiently by means of micro-channeled heat sinks, which are used as diode-laser mounts [10]. Figure 7 illustrates a specific technique of this kind, based on micro-structured Cu sheets [11].

By means of laser cutting, fine channel structures are created, through which water as an active coolant is guided, thus creating a large inner surface and a close inner contact between the heat-conducting solid and the coolant. The coarse structures, which can be seen in the picture, serve as water manifolds, guiding the water from the inlet openings to the micro-

Fig. 7. (*Left*) Micro-channel coolers, built up from thin Cu sheets (typical thickness 0.3 mm). Fine channel structures down to a width of 100 μm are laser-cut into the sheets which are subsequently diffusion-bonded. (*Right*) Final cooler with the water inlets and outlets going through the cooler in order to enable stacking of several components

channels and back to the outlet openings. Typical water-flow rate is around 500 ml/min per cooler at feeding pressures in the range 1–4 bar.

On the right-hand side of Fig. 7 a completed diode laser is given, where the diode-laser bar has been soldered on the front edge of the cooler and has been supplied with a top electrode as n-contact.

In many diode-laser systems, for instance in stacks as shown in Fig. 3, the diode lasers are put on top of each other in order to increase the output power to the desired level. The diode lasers are put in parallel with regard to the cooling circuit and in series with respect to the electrical supply. The water connections have to go through the cooler as shown in Fig. 7. In the interface between each two diode lasers in a stack, means for water sealing have to be supplied as well as proper electrical contacts in order to conduct without undue losses the quite large driving currents, which run typically up to around 50 A.

2.2   Beam Quality of Incoherently Combined Beams

Beam quality is a measure of how tightly a beam can be focused. The higher the beam quality, the smaller the spot size and the higher the laser intensity. According to the ISO standard [12] this property can be characterized by the beam-parameter product (Q), multiplying the waist radius (w_0) and the far-field divergence (θ_0) of the beam. In the best case, i.e. without any aberrations, this beam-parameter product remains constant, if the beam is transformed by passive optical components such as lenses or mirrors [13].

$$Q = w_0\,\theta_0/2 = \text{const.} \tag{1}$$

For a given Numerical Aperture of the focusing system (NA_f) and a rotationally symmetric beam with a total power of P, the average laser intensity in the focal plane (I_f) is given by:

$$
\begin{aligned}
I_f &= \frac{P NA_f^2}{\pi}\frac{1}{Q^2}\;,\\
&= \pi NA_f^2 B\;,\\
B &= \frac{P}{\pi^2 Q^2}\quad\left(\frac{W}{cm^2\,str}\right)\;.
\end{aligned}
\tag{2}
$$

B denotes the brightness of the beam and is introduced into (2) as an abbreviation; the physical meaning will be discussed later in this section. Relation (2) is widely used in many applications of conventional lasers with a single coherent and circular symmetric beam.

In contrast to conventional lasers, diode-laser systems with incoherent beam combination are built up from one- or two-dimensional arrays of individual beams as discussed in section 2. For the calculation of the beam-parameter product of such beams it is necessary to determine the far-field divergence angle (θ_{tot}) as well as the radius of the beam in the waist (w_{tot})

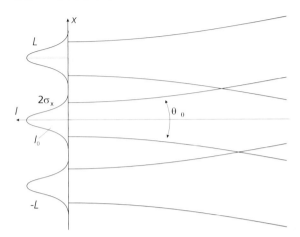

Fig. 8. Example for a one-dimensional array of individual, similar emitters, propagating from the beam waist at the *left* to the *right*

for the total intensity distribution of the emitter array. In Fig. 8 a typical geometry is shown for the simplified one-dimensional case, where $2N + 1$ similar intensity distributions $I_0(x)$ are equally spaced with a distance of x_0, each distribution having a radius of w_0.

Both x_0 and w_0 are defined according to the ISO standard [12] as first- or second-order moments of the intensity distribution respectively (I_0: one-dimensional intensity distribution, σ_x: second-order moment of I_0, P_0: total power of I_0):

$$w_0 = 2\sigma_x \,,$$

$$\sigma_x^2 = \frac{1}{P_0} \int x^2 I_0(x)\mathrm{d}x \,,$$

$$x_0 = \frac{1}{P_0} \int x I_0(x)\mathrm{d}x \,. \tag{3}$$

The intensity distribution (I_tot) and the power (P_tot) of the total beam are given by

$$I_\text{tot} = \sum_{n=-N}^{N} I_0(x + nx_0) \,,$$

$$P_\text{tot} = (2N + 1) \int I_0(x)\mathrm{d}x = (2N + 1)P_0 \,. \tag{4}$$

From Fig. 8 it can easily be concluded, that the far-field divergence angle of the total beam (θ_tot) equals that of an individual beam:

$$\theta_\text{tot} = \theta_0 \,. \tag{5}$$

The beam radius of the total beam (w_{tot}) can be calculated according to the ISO standard from the second-order moment (σ_x) of the total intensity distribution ($I_{\text{tot}}(x)$) as:

$$w_{\text{tot}}^2 = 4\sigma_x^2 = \frac{4}{P_{\text{tot}}} \int x^2 I_{\text{tot}}(x)\mathrm{d}x . \tag{6}$$

Insertion of (4) into (6), coordinate transformation ($\tilde{x} = x + nx_0$) and minor rearrangement gives the following relation for the radius of the total beam, consisting of M individual beams ($M = 2N + 1$):

$$w_{\text{tot}}^2 = 4\sigma_x^2 + \frac{1}{3}(M^2 - 1)x_0^2$$

$$w_{\text{tot}} = 1.14 M w_0 F \quad (\text{for } x_0 \gg \sigma_x, M > 2), \quad F = \frac{x_0}{2w_0}(\text{filling factor}) . \tag{7}$$

As a rule of thumb it can be stated that a beam consisting of M individual beams has a radius which is roughly M times larger than the radius of an individual beam, if the beams are densely packed (filling factor $F \approx 1$). The beam-parameter product of the total beam is then M times the beam-parameter product of the individual beam:

$$Q_{\text{tot}} = w_{\text{tot}}\theta_{\text{tot}}/2 \cong M w_0 \theta_0/2 = M Q_0 . \tag{8}$$

Real beams are always two-dimensional. In many practical devices (Fig. 3), arrays as shown in Fig. 9 are used. With (8) the beam quality of the total beam can easily be calculated separately for each dimension.

The brightness of the total beam $B[\text{W}/\text{cm}^2/\text{str}]$ is the laser power per emitter area and solid angle and for the case of the rectangular symmetry in Fig. 9 is given by

$$B_{\text{tot}} = \frac{P_{\text{tot}}}{16Q_{x,\text{tot}}} = \frac{MNP_0}{16MQ_{x,0}NQ_{y,0}} = \frac{P_0}{16Q_{x,0}Q_{y,0}} = B_0 , \tag{9}$$

which is the well-known brightness theorem [14], which states that in the best case, i.e. without any aberrations, the brightness of the total beam is

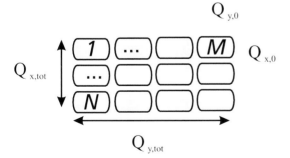

Fig. 9. Two-dimensional array of emitters, as used in the devices shown in Fig. 3

the same as for each individual beam. Real systems however suffer from several degradation mechanisms, which reduce brightness and are approximately summarized in the following factors η_i ($\eta_i \leq 1$):

$$B_{\text{tot}} = \eta_{\text{FAC}}\eta_{\text{PL}}\eta_{\text{StQ}}\eta_{\text{FF}} B_0$$

with:

η_{FAC} optical quality of the fast-axis collimation, measured as the relation of beam-parameter products (BPPs)

$$\eta_{FAC} = \frac{Q_x}{Q_{x,0}}, \text{ where}$$

Q_x : BPP behind the micro-optical lens

$Q_{x,0}$: BPP before the micro-optical lens,

η_{PL} power losses at the micro-optical lens;

$$\eta_{\text{PL}} = P_{\text{tot}}/(MNP_0).$$

η_{StQ} 'Stacking quality'

$$\eta_{\text{StQ}} = \frac{\theta_{x,\text{tot}}}{\theta_x}, \text{ where}$$

$\theta_{x,\text{tot}}$: fast-axis divergence of the total beam

θ_x : fast-axis divergence of an individual beam,

η_{FF} filling-factor quality;

$$\eta_{\text{FF}} = F_x F_y, \text{ where}$$

$F_{x,y}$: filling-factors in x and y directions.   (10)

Typical values for the different factors and measures to minimize losses in beam quality and laser power will be discussed in the next sections.

2.3   Technical Devices

In Fig. 10 a practical setup is shown for a low-power fiber-coupling scheme according to Fig. 3 (upper right), utilizing a beam-transformation device. Beam transformation is needed in many diode-laser systems, as the beam of individual diode lasers and stacked devices is mostly asymmetric in terms of size and beam quality and often requires shaping, for example in the case of fiber coupling. A collimated beam from a diode-laser bar, for instance, has a typical dimension of $1\,\text{mm} \times 10\,\text{mm}$ and a beam quality $Q_{x,0} \times Q_{y,0}$ of $1\,\text{mm mrad} \times 2000\,\text{mm mrad}$.

The basic purpose of transformation techniques is to change beam size and divergence angle in the two lateral dimensions, while maintaining the brightness of the system, i.e. the product $Q_{x,0}Q_{y,0}$. This objective can be achieved with several optical arrangements, e.g. [4,15,16]. In Fig. 10 a simple and efficient solution with a step-shaped mirror is illustrated [17]. In all systems the line-shaped beam, e.g. from a diode-laser bar after fast-axis

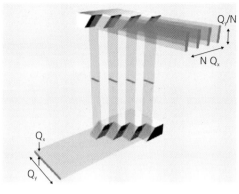

matching size and quality
of the beam by means of step-mirrors

$N_x \cdot Q_x \approx Q_y / N_x$

principle   technical set-up

Fig. 10. Transformation of beam qualities. Schematics (*left*), technical setup with two step-shaped Cu mirrors (*right*). With this technique the coupling of 30 W diode-laser power into a fiber of 200 mm core diameter (NA = 0.2) has been demonstrated

collimation, is cut or divided into N parts, which are spatially rearranged. After rearrangement the beam quality in x and y directions has changed into $Q_{x,1} Q_{y,1}$. With (8) the values for these modified beam qualities can be easily calculated as well as the number of beam parts, M, needed to produce a symmetric beam according to

$$Q_{x,1} = N Q_{x,0} \,,$$
$$Q_{y,1} = \frac{Q_{y,0}}{N} \,,$$
$$M = \sqrt{\frac{Q_{y,0}}{Q_{x,0}}} \qquad (\text{for } Q_{x,1} = Q_{y,1}) \,. \tag{11}$$

In order to extend output powers into the kilowatt range, stacking techniques are commonly used [18,19,20,21]. The photograph in Fig. 11 shows two stacks, each incorporating approximately 25 diode lasers, every diode laser being individually collimated by a microlens. One of the stacks operates at 808 nm, the other at 980 nm; the beams of both are combined by the edge filter in front of the stacks.

Conventional lasers for industrial applications such as CO_2- and lamp-pumped solid-state lasers are usually characterized by their output power and their beam-parameter product, which is taken for the rotationally symmetric case. Whether a laser is fit for the different fields of application, discussed in Sect. 2.6, depends mainly on these parameters. In order to compare diode-laser systems with conventional lasers, the beam-parameter products (Q_x, Q_y) of commercial diode-laser systems have been converted to a single number $Q_{x,y}$, characterizing the equivalent rectangular-symmetric beam. This conversion is performed on the basis of (11) according to the following

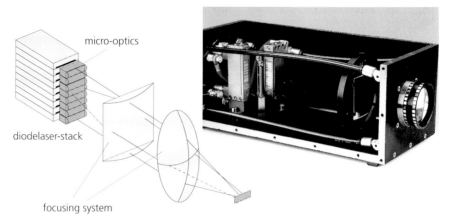

micro-optics

diodelaser-stack

focusing system

Fig. 11. Wavelength-multiplexed diode-laser stacks. The total output power is approximately 1200 W; the beam can be focused down to a spot of $2\,\mathrm{mm} \times 4\,\mathrm{mm}$ (NA approx. 0.2 in both directions)

relation

$$Q_{x,y} = Q_{x,0}M = Q_{y,0}M$$
$$= Q_{x,0}\sqrt{\frac{Q_{y,0}}{Q_{x,0}}}$$
$$= \sqrt{Q_{x,0}Q_{y,0}}\,. \tag{12}$$

These calculated data have been put together with data of commercial CO_2- and lamp-pumped solid-state lasers in one diagram (Fig. 12).

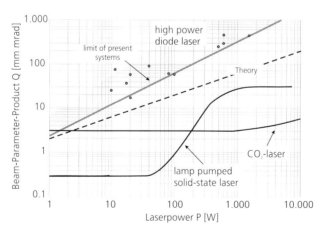

Fig. 12. Rated power and beam-parameter product of CO_2-, lamp-pumped Nd:YAG- and diode-laser systems. The parameters of commercially available systems are all located above the indicated lines

The lower limit for the beam-parameter product or the highest beam quality is always given by the diffraction-limited Gaussian beam of wavelength λ (GB: Gaussian Beam):

$$
\begin{aligned}
Q_{\mathrm{GB}} &= \frac{\lambda}{\pi} \\
&= 0.3\,\mathrm{mm\,mrad} \qquad \text{for a Nd:YAG laser with } \lambda = 1.06\,\mu\mathrm{m} \\
&= 3.0\,\mathrm{mm\,mrad} \qquad \text{for a CO}_2 \text{ laser with } \lambda = 10.6\,\mu\mathrm{m} \\
&= 0.24\text{–}0.28\,\mathrm{mm\,mrad} \\
&\qquad\qquad\qquad \text{for a diode laser with } \lambda = 0.8\text{–}1.0\,\mu\mathrm{m}\,.
\end{aligned}
\tag{13}
$$

In principle Q_{GB} does not depend on the output power, from which it follows that the brightness of a laser is unlimited in theory. In practical lasers, however, the maximum beam quality (or minimum beam-parameter product) can only be maintained up to a certain output power level. At present, diffraction-limited CO$_2$ lasers are available up to approximately $2\,\mathrm{kW}$, and diffraction-limited lamp-pumped Nd:YAG lasers up to some $10\,\mathrm{W}$. Above these power levels beam quality decreases due to several physical and technical shortcomings such as wave aberrations in the active medium.

Of course, in diode-laser systems a lower limit for the beam-parameter product exists as well. However, the method of power scaling by spatial multiplexing discussed in the previous sections manifests itself in a different shape of the curve. According to (9) the one-dimensional beam-parameter product scales with the square root of the laser power. The starting point for the line describing this scaling behavior is given by the power and beam quality of one elementary emitter of the two-dimensional array. For the dotted curve, labeled 'Theory' in Fig. 12, present data for a typical broad-area emitter, which emits $1\,\mathrm{W}$ at $2\,\mathrm{mm\,mrad}$, have been taken. At low output powers, real diode-laser systems are available with a beam quality which closely matches this theoretical limit. At higher powers, where many individual emitters have to be combined in order to achieve the required power, actual systems have not come to the limiting curve, labeled 'Theory' in Fig. 12, as yet.

The main reasons for this discrepancy have already been summarized in (10). Firstly, beam degradation and power losses can occur at the micro-optical collimation lens if nonoptimized components are used or if alignment tolerances are excessive. These aspects will be discussed in detail in Sect. 2.4.

A second source for beam degradation can be aperture-underfilling in either direction of the two-dimensional array of diode-laser emitters. Figure 13 illustrates this effect for the dimension in which the diode lasers are stacked. If the stacking geometry is not designed properly, the beams have considerable gaps and the filling factor F is well below one, thus decreasing beam quality according to (10). Methods to overcome this limitation will be discussed in Sect. 2.5 as well as techniques to solve the problem of stacking tolerances, shown in the right-hand part of Fig. 13.

Fig. 13. Reasons for beam degradation in diode-laser stacks. Aperture-underfilling (*left*) and angular tolerances of the individual diode lasers (*right*)

In conventional stacks, where a number of diode lasers are put on top of each other, the unavoidable mechanical tolerances due to water sealings, spacers, electrical contacts, etc. may sum up to quite large angular tolerances (α) of the individual diode lasers. If α is on the order of the divergence angle of the collimated beam, which is typically in the millirad range, the divergence angle of the total beam will exceed that of an individual beam. In this case (5) is violated and the beam quality is decreased. This problem, however, can be overcome by properly modified stacking techniques, as will be discussed in Sect. 2.5.

2.4   Beam Collimation

A typical property of diode lasers is the highly divergent emission in the direction perpendicular to the p-n junction ('x' in Fig. 6), the 'fast axis'. Numerical apertures of up to 0.8, i.e. full divergence angles of approximately 100°, are not unusual. Therefore, in most cases where diode lasers are directly used, a collimation of this fast axis is necessary. A simple and low-cost solution for beam collimation is a cylindrical lens as shown in Fig. 14. An important requirement for this cylindrical lens is that as much beam quality and laser power as possible must be conserved, i.e. the beam-parameter product in the fast axis behind the lens should be close to the diffraction limit as is the case right at the emission facet of the diode laser.

In order to compare different lens types and to assess the lens performance in relation to the physical limit, ray-tracing calculations have been carried

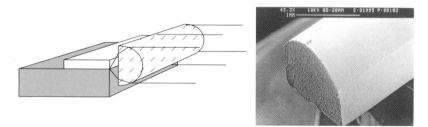

Fig. 14. Principle of fast-axis collimation with a cylindrical lens (*left*) and SEM picture of a sample (*right*), which has been manufactured from high-index glass with an ultra-precision grinding technique

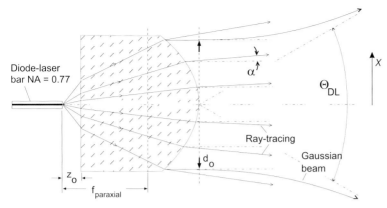

Fig. 15. Ray-tracing calculation for the collimation of the diode-laser emission by a cylindrical lens

out as explained in Fig. 15 [22]. For the calculations, a diode laser with a diffraction-limited beam, i.e. with Gaussian intensity profile in the x direction, has been used as a source. The numerical aperture (NA = 0.77 in all examples discussed below) is the divergence angle where the intensity is reduced to $1/e^2$ of the center value. For the one-dimensional case approximately 95% of the total power falls within this numerical aperture. In addition to the fast-axis divergence, a numerical aperture in the y direction ('slow axis') of 0.084 (approximately 5°) has been taken into consideration in all the calculations given below. This is a typical divergence angle for many diode-laser emitters.

From the source point a fan of equally spaced rays has been traced through the lens, each ray carrying the local laser intensity. Behind the lens the intensities of the rays are summed and plotted against the angle α (Fig. 16). These curves are compared with the case of a diffraction-limited Gaussian beam with the same diameter d_0 as the fan of rays which leaves the lens (dotted line in Fig. 16). If the curves calculated by ray-tracing are in the range of this dotted line, the lens is assumed to be sufficient for a high-quality collimation of the beam.

Figure 16 clearly shows that beams with high numerical aperture as discussed here cannot be collimated adequately with standard plano-spherical or bi-convex lenses: even if lens material with high refractive indices is used, the numerical aperture of the collimated beam (defined as the angle with 95% power enclosure) is more than 10 times bigger that for the ideal case of a diffraction-limited beam (θ_{DL}). This is the reason why easy-to-manufacture plastic lenses cannot be applied here, because plastics typically have low refractive indices. The lens with the highest performance in the diagram is a single lens made from high refractive index glass which has an aspherical surface, optimized for the emission characteristics of the diode-laser bar.

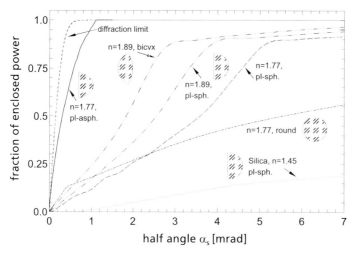

Fig. 16. Comparison of the quality of diode-laser beams after collimation by different types of cylindrical lenses (diode laser $NA_{fast} = 0.77, NA_{slow} = 0.084$)

However, even with such an optimized lens the diffraction limit cannot be achieved under all conditions. The reason for this is the coupling of the fast-axis and the slow-axis divergence [23] due to Snell's law.

At the time of writing several manufacturing techniques are available to produce aspheric glass microlenses, as shown in Fig. 14, with the necessary accuracy in the sub-micrometer range: ultra-precision grinding [24] or ultra-sonic shaping with a hard-metal mould [25]. With such commercially available components a power throughput in excess of 90% and a beam quality behind the lens of approximately two times the diffraction limit can be achieved in the best cases [22].

The requirements for positional accuracy of the microlenses are as high as those for the shape of the lens. The most sensitive processes, leading to beam degradation due to misalignment, are axial defocusing and lateral beam twist. Axial defocusing, shown schematically in Fig. 17, means that the cylinder lens and the emitter facet of the bar are not perfectly arranged in parallel. While the emitters in the front part of the picture are assumed to be perfectly aligned in the back focal position of the lens (z_0), thus creating a beam with minimal divergence, the emitters in the rear part of the picture are axially out of focus, resulting in a beam with an enlarged divergence angle. If several such collimated diode lasers with statistically varying lens misalignment angles are combined, relation (5) no longer holds: the divergence angle of the total beam is larger than that of an individual beam and the beam quality of the total beam is reduced.

With the ray-tracing modeling discussed above, this process has been studied quantitatively, the result being summarized in Fig. 18. The upper-most emitters in the picture (labeled with '1') are located in the back focal

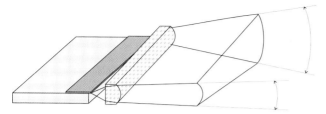

Fig. 17. Beam degradation due to axial misalignment between microlens and diode-laser bar

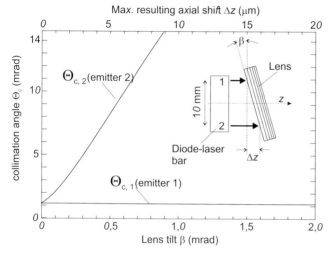

Fig. 18. Increase of the collimation angle due to a misalignment angle β between the microlens and the emitting facet of the diode-laser bar ($\mathrm{NA_{fast}} = 0.77$, $\mathrm{NA_{slow}} = 0.084$, paraxial focal length of the lens $f = 0.88\,\mathrm{mm}$)

position of the lens ($z = 0$); the lowest emitters (labeled with '2') are shifted in the z direction due to lens tilt as indicated by β. A local tilt of the lens without any defocus only has a negligible impact on the divergence angle of the emitters. This can be seen for the emitter '1', where the divergence angle remains constant over the whole displayed range of tilt angles. In contrast, the divergence angle of the emitter '2' strongly increases approximately linearly, due to the tilt-induced axial shift into a defocus position. If a variation of the divergence angle along the bar of some 10% is accepted at maximum, the axial position of the lens has to be maintained with tolerances well below 1 μm in the case studied in Fig. 18. This imposes high demands on alignment accuracy and on the long-term mechanical and thermal stability of the fixtures of the lenses.

Apart from axial deviations, the back focal point of the lens may of course also deviate laterally (in the x direction) from its nominal position, giving rise to a strong twist of the collimated beam in the x direction. This process,

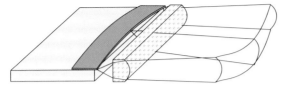

Fig. 19. Beam degradation in the case that the diode-laser bar is not perfectly straight. The exit beam looks like a 'smile'

however, can be compensated by an appropriate tilt of the diode laser as a whole, including the lens. Another difficult to handle phenomenon, which cannot be compensated for, however, is the so-called 'smile', being explained in Fig. 19. If either the diode-laser bar or the micro-cylinder lens is not perfectly straight, a part of the beam is shifted laterally upwards or downwards, with respect to its nominal position. The example in Fig. 19 shows as a result that the beam behind the lens then looks like a 'smile'.

The upper part of Fig. 20 illustrates a 'smile' in a typical experimental case. The picture has been taken by applying an appropriate imaging system, which strongly enlarges the x dimension. If a couple of such beams with statistically varying smiles are combined, the outgoing beam angle (or far-field image) of the total beam is enlarged as shown in the lower part of Fig. 20.

In order to obtain quantitative data on the sensitivity of lateral bar position and geometry, ray-tracing calculations for a beam slightly shifted in the x direction have been performed with the model and data discussed in

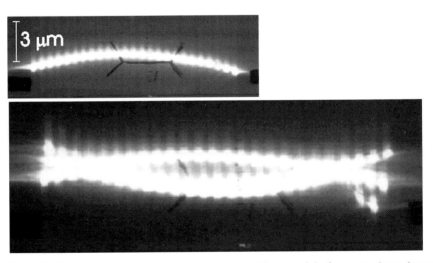

Fig. 20. Enlarged picture of a diode-laser bar with a 'smile' of approx. 2 μm (*upper picture*); if a couple of these bars with statistically varying 'smiles' are superimposed, the far-field picture is enlarged on average (*lower picture*)

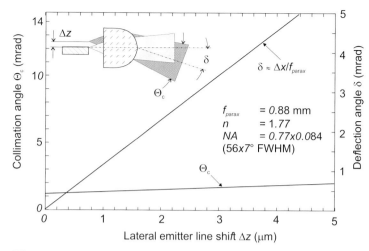

Fig. 21. Dependence of the collimation angle and the beam tilt on the lateral shift of the diode-laser emitter in the x direction

the foregoing section. Figure 21 shows the small effect of this shift on the divergence angle Θ_c, which increases approximately 10%, and the much larger effect on the total tilt of the beam δ as it leaves the lens. This large tilt angle is caused by the short focal length of the lens as is indicated in the figure. For a discussion of the smile effect, the tilt angle in Fig. 21 is considered to apply to the center part of the bar, while the wings are assumed not tilted. In order to keep the tilt of the beam δ well below the divergence angle Θ_c, a maximum smile of the bar below 0.5 µm has to be ensured.

2.5 Aperture Filling and Stacking Accuracy

If the emission of a diode-laser bar is used directly without any further optical processing of the slow axis, the beam quality in this direction is given by the product of the bar width (divided by 2) and the divergence angle of the individual emitter or the bar, respectively. Depending on the filling factor of the bar, this beam quality may be considerably lower than theoretically possible due to aperture underfilling as illustrated in Fig. 22. If the emitters shown were not spaced by the pitch of x_0, but were located densely side by side with $x_0 = 2w_0$, the total beam width and thus the beam-parameter product could be reduced by a factor of $F = 2w_0/x_0$.

An alternative way to achieve this improvement is shown in Fig. 22 where, by means of a lens array, each individual emitter is individually collimated and the divergence angle thus reduced. The lenses are put at an axial position shortly before the beams start to overlap. The focal length of the lenses is chosen such as to transform the local phasefront curvature of the individual beam into a plane phasefront. The diameter of the individual beam is in-

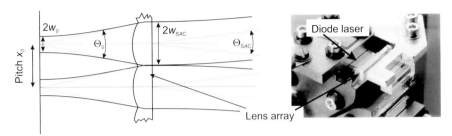

Fig. 22. Principle and experimental setup for beam-quality improvement by slow-axis collimation

creased by a factor of $1/F$ and according to (1) the divergence angle reduced by F, leading to a decrease of Q by a factor of F at best.

Table 1 summarizes the results which have been achieved with the experimental device of Fig. 22: the beam quality could be improved by a factor of approximately 3, corresponding to the figure expected for the filling factor of 0.3. However, the table shows also that considerable power was also lost due to reflection losses at the uncoated lens surfaces and due to the fact that in the specific case shown here the lens-array geometry was nonperfectly matched to the structure of the emitter array. Under optimized conditions a power throughput of more than 90% and thus a brightness increase slightly below a factor of 3 are achievable.

Table 1. Improvement of the beam quality by slow-axis collimation illustrated with experimental results obtained with the setup of Fig. 22 (bar width: 10 mm, emitter width: 150 μm, emitter spacing: 500 μm)

	Before lens array	Behind lens array
Divergence angle (slow axis)	9° (178 mrad)	3° (52 mrad)
Beam-parameter product (slow axis)	395 mm mrad	130 mm mrad
Laser power	P_0	$0.7P_0$
Brightness	B_0	$2.2B_0$

Beam-quality reduction due to aperture-underfilling has to be taken into consideration not only in the slow direction but also in the case of stacked diode-laser bars in the fast direction. In many cases a dense package of the diode lasers in the fast direction, such as shown in Fig. 22, is hard to achieve in practice or has serious technical drawbacks. In order to obtain high output power from compact systems, thin heat sinks in the range 1-2 mm are often preferred. However, with such thin and mechanically sensitive heat sinks the beam may be seriously degraded due to mechanical deformations which, in turn, lead to smile (Fig. 19) or angular tolerances (Fig. 13). Thick heat sinks on the other hand are mechanically stable but the stack height and thus the output power is limited by the size and handling capability of the optical systems needed to focus the beam on the workpiece.

This restriction can be overcome by beam-staggering techniques as explained in Fig. 23. In the specific case illustrated here, three stacks are used. The beams from the individual diode-laser bars are each collimated in the fast axis and have a spacing which is nearly three times the height of the individual beams. By means of comb-shaped mirrors the emissions of the three stacks are staggered together into one outgoing beam, which is now densely packed. Filling factors in the fast direction up to about 90% have been proven to be achievable by applying such techniques.

Even with structures described above, the problem remains, that small mechanical tolerances from water sealings and electrical contacts, located between each pair of diode lasers, may sum up to considerable numbers, which are beyond acceptable tolerances for high beam quality systems (Fig. 13). The stacking technique, shown in Fig. 24 ('optical' stacking), yields a solution for this problem as well as for the problem of aperture-underfilling while using thick heat sinks. The individual diode lasers are located side by side instead of on top of each other, on a staircase-shaped mechanical holder. Each beam from a diode laser is folded by an angle of 90° by means of individually aligned folding mirrors. This mechanical arrangement ensures high filling factors, even if thick, mechanically stable heat sinks are utilized. Due to the adjustable folding mirrors, mechanical tolerances of each diode laser can be compensated for individually, thus ensuring highly parallel emission of the beams.

In Table 2 the results achieved with such an optical stack comprising 8 diode lasers and a total power of 240 W are compared with a stack, built from 8 diode lasers as well but according to conventional design. Due to the more-precise alignment and the dense package in the fast direction, the brightness of the optical stack was increased by a factor of 3 over conventional technology.

Fig. 23. Beam staggering for increasing the filling factor in the fast direction of a diode-laser stack. Principle (*left*) and experimental setup (*right*)

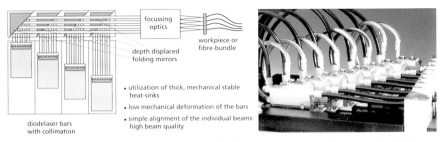

Fig. 24. 'Optical stacking'. Schematics (*left*) and experimental set up (*right*)

Table 2. Improvement of the beam quality by 'optical' stacking as shown in Fig. 24

	'Conventional' stack (Fig. 11)	'Optical' stack (Fig. 24)
Beam aperture (fast axis)	14.5 mm	10.4 mm
Divergence angle (fast axis)	8.2 mrad	3.6 mrad
Beam-parameter product (fast axis)	58.0 mm mrad	18.7 mm mrad
Laser power	240 W	240 W
Brightness	B_0	$3.1 B_0$

2.6   Applications

Two of the most important parameters for laser materials processing are laser power and laser intensity. Each application is characterized by a specific range for these two parameters. With relation (2) the process intensity can be linked to the beam-parameter product Q of a given laser, if a numerical aperture of the focusing system is assumed. In Fig. 25 this has been carried out for a typical numerical aperture of $NA = 0.12$, corresponding to a $F\#4$ focusing system. By comparing Fig. 25 with Fig. 12 it can easily be determined what type of laser is most suitable for a specific application, at least as far as laser power and beam quality are concerned.

The present high-volume markets for materials processing are marking and cutting/welding of sheet metal. All three require relatively high beam quality and, for cutting and welding, high output powers above approximately 1 kW. Typical process intensities are in the range above approximately 10^5 W/cm^2. Apart from these high-power and high-intensity applications a broad range of 'low-intensity applications' exists, where the beam merely has to be concentrated to moderate spot sizes, and lasers with lower beam quality are adequate. The comparison of Fig. 25 with Fig. 12 elucidates why present diode-laser systems are mainly used in this regime.

Two examples of low-power and low-intensity applications are shown in Fig. 26: welding of plastics and soldering.

Plastics welding with lasers [26] in most cases requires that the two parts to be welded together are chosen from differently pigmented plastics in order to create a geometry where the laser penetrates through the transparent top layer and is fully absorbed at the surface of the layer underneath. In the

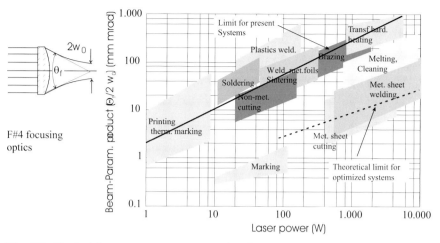

Fig. 25. Typical beam-parameter ranges for the most common laser applications in materials processing

Fig. 26. Examples of low-power and low-intensity laser applications, using diode-laser systems. Welding of the plastic casing of an electronic car key (*left*) and soldering of an electrical connecting braid (*right*)

example shown in Fig. 26 (left), a plastic cover has to be welded to the plastic casing. The cover is colored with a pigment which appears black in the visible spectral range, but is transparent for the IR radiation of the diode laser. The requirements on laser power are quite moderate, because the process power is applied very effectively, it only heats the joining zone and the melting energy for plastics is quite low. At typical widths of the joining zone, and thus spot size, of about 1 mm, a welding speed of 1 m/min is typically achieved with a laser power of a few tens of watts.

Advantages of welding by laser instead of conventional methods such as ultrasonic welding are, e.g., flexibility, high quality of the weld (important in the case of visible seams), controllability of the power input and the possibility of integration into automated quality control.

Most of these advantages also hold for the second example shown in Fig. 26, the soldering of electronic or electrical components. Geometries being soldered range from millimeters for electric parts to tenths of a millimeter for electronic components. Required spot sizes and laser powers, respectively, vary from $1/10$ mm and tens of watts up to hundreds of watts for the soldering of the connecting braid in Fig. 26. For soldering as well as for plastics welding, mostly fiber-coupled diode lasers are utilized, for instance according to the principle discussed in [26].

Considerably higher output powers are required in the two further examples illustrated in Fig. 27. The right-hand side picture shows a side-cutter, the cutting edges of which have been hardened with diode lasers. The principle of laser transformation hardening involves fast surface heating of steel to the austenitic temperature and subsequent rapid self-quenching by heat conduction to the cold base material. Due to the relatively high temperatures required (approximately 750–$1000°$ C) and the macroscopic geometry of typical parts (Fig. 27) the laser power needed is typically in the range exceeding 1 kW. Diode-laser stacks, which have recently been employed for these applications, are distinguished from classical lasers by the property of having a rectangular beam, which usually is the most desirable geometry. In the example of Fig. 27 the beam of the diode-laser stack was shaped such that the full length of the edge could be hardened in one step without any movement of beam or workpiece.

The other example of Fig. 27 refers to metal welding with diode lasers: a stainless-steel coin, 1.5 mm thick, has been welded with a speed of 1.8 m/min, using a diode-laser stack of 0.7 kW power. The diode-laser beam has been focused to a spot size around 1.5 mm $\times$ 3.8 mm, resulting in a processing intensity of approximately 10^4 W/cm^2. At these intensities heat-conduction welding at relatively slow speed is achievable. While deep penetration or key-hole welding [27] allows much higher speeds, albeit at simultaneously

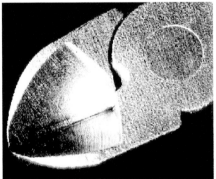

Fig. 27. Transformation hardening of the edges of a side-cutter (*left*) and heat conduction welding of a stainless-steel coin (*right*)

much higher power densities, an advantage of the heat-conduction welding with diode lasers is, that the weld seams are very smooth and have a very high surface quality. Recently, deep penetration welding of stainless steel (welding depth up to 6 mm) could be demonstrated for the first time by applying a compact high-power diode-laser stack delivering some 10^5 W/cm^2 of intensity [27].

2.7   Discussion and Perspectives

The discussion in the foregoing section showed that present diode lasers are excellent tools for a wide range of laser applications, which require only moderate beam qualities or focusabilities. Even-larger application fields would be opened however, if output power and simultaneously beam quality could be raised in order to achieve figures comparable to present lamp-pumped solid-state lasers, thus enabling routine cutting and welding of metal parts.

The sum of all laser power and beam-quality degradation processes, summarized in Sect. 2.3, is the reason why the beam-parameter product of present systems is above the theoretically achievable one indicated in Fig. 12: the more individual diode lasers are combined, the bigger is the difference between theory and technical devices due to the increasing effect of degradation processes. Introduction and combination of all the technical means discussed in Sect. 2.5 can considerably enhance the beam quality. In addition to these means, several wavelengths and the two states of polarization may be combined in future high-power systems. If one takes supplementarily into account that output power as well as beam quality of diode lasers have constantly been raised over the past years, it seems feasible to build diode-laser systems in the future with an output power in the kilowatt range and a beam quality comparable to present lamp-pumped solid-state lasers.

3   Coherent Beam Combining

Uwe Brauch

The output power of diode lasers with diffraction-limited beams is limited mainly by two effects: Catastrophic Optical Damage (COD) of the laser facets [28,29] and instabilities in broad-area lasers [30,31,32]. Nonlinearities in the semiconductor material cause spatial and temporal inhomogeneities in the gain and index of refraction which make it difficult to stabilize the large fundamental mode of a broad-area laser or to get undistorted amplification across the aperture. On the other hand, the emitting surface cannot be reduced arbitrarily because the power density on the facets must not exceed values on the order of magnitude of 10 MW/cm^2. This means that the power from small-area oscillators is limited to approximately 1 W [33], and the power from broad-area amplifiers to approximately 10 W [7]. Despite the ongoing effort to increase the diffraction-limited output power through special

resonator designs like the α-DFB (Distributed FeedBack) laser [34] or lasers
with vertical output coupling through gratings [35], there has been limited
success to develop a commercial product. But independent of the actually
realized output, for high-power applications, the beams of a corresponding
number N of emitters have to be combined.

There is a fundamental limit on the maximum brightness (power density
per solid angle) which can be achieved when combining laser beams or, in
general, when manipulating any radiation field: there is no way to increase the
brightness beyond that of the original radiation source, i.e., applied to laser
beams, the brightness of the combined beams cannot exceed the brightness
of a single beam. This follows from Liouville's law[1] which states that the
volume of an element $\Delta \Omega$ in the phase space remains constant:

$$\frac{\mathrm{d}}{\mathrm{d}t}\Delta \Omega = \frac{\mathrm{d}}{\mathrm{d}t}(\Delta p_1 \ldots \Delta p_{\mathrm{f}}\Delta q_1 \ldots \Delta q_{\mathrm{f}}) = 0 . \tag{15}$$

Taking into account the well-known relation between momentum p and
wavenumber k

$$\Delta p = \hbar \Delta k \tag{16}$$

and the geometrical relation between the beam half-angle $\theta_i/2$ and a wavevector component Δk_i perpendicular to the beam propagation, it follows that

$$\frac{\theta_i}{2} \approx \tan\frac{\theta_i}{2} = \frac{\Delta k_i}{k} = \frac{\lambda}{2\pi}\Delta k_i . \tag{17}$$

This means that the product of beam diameter – proportional to Δq_i – and
beam divergence angle θ_i and hence the (inverse) brightness, remains at best
constant.

The absolute minimum is given by Heisenberg's uncertainty relation

$$\Delta p_i \Delta q_i \geq \frac{\hbar}{2} , \tag{18}$$

Δp_{i} and Δq_{i} being the square root of the average quadratic deviation from
the mean value (RMS)

$$\Delta p_{\mathrm{i}} = \sqrt{(p_i - \overline{p_i})^2}$$
$$\Delta q_{\mathrm{i}} = \sqrt{(q_i - \overline{q_i})^2} . \tag{19}$$

[1] A similar result can be deduced from the second law of thermodynamics by
considering an (ideal) concentrator (surface area and acceptance angle of the
entrance and exit apertures $A_1, \Theta_1, A_2, \Theta_2$, indices of refraction n_1, n_2) in the
radiation field of a black-body source. Since there cannot be any heat transfer
between bodies of the same temperature, it can be shown that the maximum
(two-dimensional) concentration is [36,37]

$$C = \frac{A_1}{A_2} = \frac{n_2^2 \sin^2 \Theta_2}{n_1^2 \sin^2 \Theta_1} = \frac{N A_2^2}{N A_1^2} . \tag{14}$$

In the case of a beam with Gaussian power-density and angle distribution the half width measured at $1/e^2$ of the maximum corresponds to two times the RMS values so that the Heisenberg relation reads for a Gaussian beam

$$w_{0i}\frac{\theta_i}{2} \geq \frac{\lambda}{\pi} \quad \text{or}$$
$$w_{0i}\frac{\theta_i}{2} = M_i^2 \frac{\lambda}{\pi}, \qquad M_i^2 \geq 1 \tag{20}$$

with w_{0i} being the waist radius, θ_i being the full angle (measured at $1/e^2$), and M_i^2 being the beam-parameter product. The maximum brightness (for a given power) is that of the so-called diffraction-limited beam with $M_i^2 \equiv 1$.

When combining N (incoherent) beams of identical polarization and spectrum the power density for a given NA at best remains constant, i.e. the beam parameter product $M_x^2 M_y^2$ is increased at least by the factor N, the brightness reduced by the factor $1/N$. Methods for power scaling with constant brightness or power density are position- or angle-multiplexing (Sect. 3.2). These methods principally allow us to realize any power by simply adding more lasers.

The limitation of the brightness to that of a single laser means that for applications requiring beams with high-brightness single-mode lasers ought to be used. Despite their limited power they have the highest brightness which compares favorably with standard fiber-coupled high-power Nd:YAG lasers used for material processing: 3 kW out of a 600 μm fiber (NA = 0.2) corresponds to a power density of 1.1 MW/cm². Focusing the 200 mW beam out of a single-mode fiber ($\lambda = 980$ nm) with the same NA results in twice the average power density. That means that typical power densities and powers needed for material processing can already be realized with the available single-mode oscillators in combination with angle- or position-multiplexing.

If beams of different polarization states or spectral distribution are to be combined, polarization and frequency become additional generalized coordinates so that the brightness can be increased at the cost of spectral purity and degree of polarization.

The limitation of the brightness does not apply to mutually coherent beams. They occupy the same elements in phase space and behave as if they came from one coherent source. Therefore only coherent superpositioning allows truly scalable output powers and diffraction-limited quality of the combined beams.

There are three basic approaches to get mutually coherent beams, viz.

- independent oscillators with external electronic phase control,
- self-organizing of the oscillators by coupling through evanescent/leaky waves or common resonators,
- external optical phase control by a master oscillator through injection-locking or amplification.

Methods for combining coherent beams are the phased-array technique with a setup similar to the angle-multiplexing of incoherent beams and the direct superpositioning (position and angle) with beam splitters or single-mode fiber couplers.

In Sect. 3.1 the coherence properties of individual diode lasers will be summarized. The presentation of methods for incoherent combining of a large number of coherent lasers (Sect. 3.2) will show the limitations of this approach and will also be a basis for the discussion of methods for coherent beam combining of single-frequency lasers (Sect. 3.3). Finally, the limitations and prospects of the method of coherent beam combining will be discussed in Sect. 3.4.

3.1   Coherence Properties of Diode Lasers

The outstanding property of lasers compared to thermal-radiation sources is their spatial and temporal coherence due to the dominating stimulated-emission process in a resonant cavity. For the following, the definitions given below appear helpful [38,39].

The *coherence* is defined as the degree of correlation of the complex amplitudes $E_i(t_i)$ of the radiation fields between two points in space and time. It is mathematically described by the *cross-correlation function* Γ_{ij} (or autocorrelation function if $i = j$):

$$\Gamma_{ij}(\tau) = \langle E_i(t + \tau)E_j(t)\rangle , \tag{21}$$

where the brackets stand for time averaging over many periods of the light waves and $\tau = t_i - t_j$. The degree of coherence is directly related to the measured *contrast* K in an interferometric superposition

$$K = \frac{I_{\max} - I_{\min}}{I_{\max} + I_{\min}} = \frac{2\sqrt{I_i I_j}}{I_i I_j}|\gamma_{ij}| \tag{22}$$

with $I_i = \langle E_i E_j\rangle$ being the power densities and

$$\gamma_{ij}(\tau) = \frac{\Gamma_{ij}(\tau)}{\Gamma_{ii}(0)\Gamma_{jj}(0)} \tag{23}$$

being the normalized cross- (auto-) correlation function or the complex degree of coherence.

The *spatial coherence*, i.e. the correlation between two points in space at the same time, is important for the maximum achievable concentration (brightness). Single-transverse-mode lasers have a very high degree of coherence between the instantaneous amplitudes and phase angles at any two points across the output beam and therefore can be focused to the minimum diffraction-limited spot size and to the maximum brightness for a given output power [39].

The *temporal coherence*, i.e. the correlation between the radiation of one source at two points in time, can also be characterized by the coherence time τ_c

$$\tau_c = \int_{-\infty}^{\infty} |\gamma_{ii}(\tau)|^2 d\tau \tag{24}$$

or, alternatively, by the *coherence length* l_c

$$l_c = c\tau_c . \tag{25}$$

In the case of a Lorentzian lineshape

$$I_L(\nu) = \frac{I_0}{1 + \frac{4(\nu - \nu_0)^2}{\Delta \nu_L^2}} \tag{26}$$

which is typical for homogeneously broadened laser radiation (centered at the frequency ν_0 with a full linewidth measured at half maximum $\Delta \nu_L$); the *coherence time* τ_c gives the time difference for which the self-coherence is reduced to $1/e$.

The relationship between spectral linewidth $\Delta \nu$ (FWHM) and coherence time τ_c is

$$\tau_c^L = \frac{1}{\pi \Delta \nu_l} \qquad \text{Lorentzian lineshape},$$

$$\tau_c^G = \sqrt{\frac{2 \ln 2}{\pi}} \frac{1}{\Delta \nu_G} \qquad \text{Gaussian lineshape}. \tag{27}$$

The stability of single- (fundamental-) mode operation (wavelength λ), i.e. the *spatial coherence*, of diode lasers depends on gain and loss which the different transverse modes experience in the active resonator. In classical Fabry–Pérot resonators with mirror radius a and length L, the losses due to diffraction effects depend on the Fresnel number [39,40]

$$N_F = \frac{a^2}{\lambda L} . \tag{28}$$

N_F can be thought of as the ratio of the acceptance angle a/L of one mirror as viewed from the center of the opposing mirror to the diffraction angle λ/a of the beam. Resonators with smaller Fresnel numbers ($N_F < 1$) have higher diffraction losses, because only a portion of the beam will be intercepted by the mirrors. Since the losses of the higher-order modes are growing faster than the losses of the TEM_{00} mode this mode can be stabilized by using resonators with small Fresnel numbers. A reasonable compromise between loss and mode discrimination are Fresnel numbers $N_F \cong 1$.

Edge-emitting diode lasers have their resonator axis ($L \cong 1\,\text{mm}$) within the p-n junction [1]. The vertical-mode extension (orthogonal to the junction) is limited to approximately $1\,\mu\text{m}$. The index-of-refraction profile is designed to form a single-mode waveguide. The lateral-mode extension (parallel to the junction) is chosen according to power and beam quality. Single-mode oscillators have emitters usually not wider than 4λ; the wave guiding is realized either by gain guiding (through the injected current) or by index guiding. Due to nonlinearities and inhomogeneities it is difficult to operate broad-area emitters in the fundamental mode.

In vertical-cavity lasers, the cavity is orthogonal to the p-n junction. The output power and the number of transverse modes depend on the (rotational) beam-cross-section area.

The *temporal coherence* [41,42] depends on the number of longitudinal modes that are oscillating quasi-simultaneously and their linewidths. In homogeneously broadened lasers theoretically only the mode with the highest net gain can oscillate in steady state, all other modes being below threshold. In edge-emitting lasers, the mode separation $\Delta\nu_{\mathrm{mode}}$

$$\Delta\nu_{\mathrm{mode}} = \frac{c}{2nL} \tag{29}$$

with approximately 100 GHz is between one and two orders of magnitude smaller than the gain linewidth $\Delta\nu_{\mathrm{gain}}$, so that already small external parameter changes like current or temperature fluctuations, optical feedback etc. prevent the laser from single-longitudinal-mode (single-frequency) operation. A grating within the cavity (DBR, Distributed Bragg Reflector or DFB, Distributed FeedBack) strongly increases the longitudinal-mode selectivity. In VCSELs the resonator is much shorter so that only one longitudinal mode can possibly fall into the gain region.

The linewidth of a single longitudinal mode is determined by internal and external effects. Intrinsic noise sources like amplitude and phase noise due to spontaneous emission and shot noise cause a homogeneous, Lorentzian line broadening. External perturbations like vibrations, current or temperature fluctuations are the reason for inhomogeneous, Gaussian broadening. Typical linewidths for lasers with a Fabry–Perot cavity are of the order of 1 MHz – 100 MHz, corresponding to a coherence length of 100 m – 1 m. DFB or DBR lasers have similar linewidths. External resonators allow us to reduce the linewidth by approximately one order of magnitude. Since the frequencies of the transversal modes belonging to a given longitudinal mode are slightly different, a strong side-mode suppression is additionally necessary.

3.2   Combining of Diffraction-Limited Beams

In this section methods will be presented to combine diffraction-limited (spatially coherent), but mutually incoherent, beams. Some of these techniques can also be used for coherent beam combining, which requires a set of diode lasers with mutual spatial and temporal coherence. Both combination techniques are facilitated by single-mode fiber coupling.

The output beam of an edge-emitting laser is given by the properties of the internal optical waveguide. In order to achieve a high lasing efficiency and fundamental-mode operation the typical size of the waveguide, or more precisely, the mode field diameter $2w_0$ (measured at the $1/e^2$ points of the power density at the laser facet) is at most $1\,\mu m$ orthogonal to the p-n junction ('fast axis') and 4λ laterally ('slow axis'). Assuming an ideal diffraction-limited Gaussian beam the corresponding far-field divergence angle θ (full width at $1/e^2$) and the numerical aperture NA are given by [39,40]

$$\theta = \frac{2\lambda}{\pi w_0}, \tag{30}$$

$$\mathrm{NA} \equiv \sin\frac{\theta}{2} = \sin\frac{\lambda}{\pi w_0}. \tag{31}$$

Typical numbers for high-power oscillators ($\lambda = 980\,nm$, $P_{\mathrm{out}} = 250\,mW$) are $\theta = 54°$, $\mathrm{NA} = 0.45$ for the fast axis and $\theta = 14°$, $\mathrm{NA} = 0.12$ for the slow axis [32]. The propagation of a Gaussian beam near the waist is shown in Fig. 28. A convenient property of Gaussian beams is that in paraxial approximation (small angles θ) the power-density distribution always remains Gaussian. With (ideal) lenses the waist can be transformed to arbitrary sizes while the product $w_0\theta$ according to (20, 30) remains constant.

The strong divergence requiring diffraction-limited collimating optics with NA values between 0.5 and 0.6 and the high aspect ratio of almost 4 requiring cylindrical optics suggest using Single-Mode (SM) fiber-coupled lasers. In this way a diffraction-limited, symmetrical Gaussian beam with a NA of 0.1 is provided which can be easily transformed by standard optics.

SM-fiber coupling with high efficiency requires an optimal adaptation of the laser beam to the fiber mode. The electric and magnetic fields that can propagate in waveguides with rotational symmetry can be described by (modified) Bessel functions [43,44,45]. In the case of SM fibers with an index

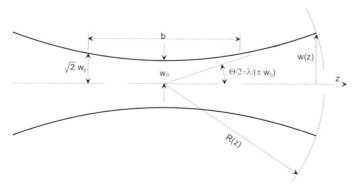

Fig. 28. Hyperbolic contour of a Gaussian beam near the waist defining the waist radius w_0, the Rayleigh range $b/2$ and the far-field divergence angle θ ([40])

difference

$$\Delta n = n_{\mathrm{core}} - n_{\mathrm{cladding}} \ll 1 \,, \tag{32}$$

the longitudinal field components can be neglected and the modes are called LP_{lm} (linearly polarized modes). The lowest-order mode is the rotationally symmetric LP_{01} mode. The appearance of higher-order modes depends on the normalized frequency V

$$V = a \frac{2\pi}{\lambda} \sqrt{n_{\mathrm{core}}^2 - n_{\mathrm{cladding}}^2} \,. \tag{33}$$

For $0 < V \leq V_{\mathrm{c}}$, with $V_{\mathrm{c}} = 2.405$, only the x- or y-polarized LP_{01} exists. Hence, for a given fiber (core radius a) the operational wavelength λ has to be larger than the cut-off wavelength λ_{c} with

$$\lambda_{\mathrm{c}} = \frac{2\pi a}{2.405} \sqrt{n_{\mathrm{core}}^2 - n_{\mathrm{cladding}}^2} \,. \tag{34}$$

Usually, the cut-off wavelength is chosen to be 5–10% below the operational wavelength. In this case the power-density distribution of the LP_{01} mode can be approximated by a Gaussian distribution. The following approximation holds for the mode field diameter mfd (error $< 0.5\%$ for $2.1 < V < 2.3$) [44]:

$$\mathrm{mfd} \equiv 2w_0 = 2a \left[0.65 + 0.434 \left(\frac{\lambda}{\lambda_{\mathrm{c}}} \right)^{3/2} + 0.0149 \left(\frac{\lambda}{\lambda_{\mathrm{c}}} \right)^6 \right] \,. \tag{35}$$

Using the relationship between waist radius w_0 and divergence angle θ of a Gaussian beam (20, 30) the NA of the beam leaving or entering the SM fiber can be calculated according to

$$\mathrm{NA}_{\mathrm{beam}} \equiv \sin \frac{\theta}{2} = \sin \frac{2\lambda}{\pi \mathrm{mfd}} \,. \tag{36}$$

This NA is different from the NA given by the fiber manufacturers. They define the maximum NA by the maximum angle of total reflection $\theta_{\mathrm{totrefl}}$ at the core–cladding interface which is meaningful for multimode fibers only but can formally also be applied to SM fibers:

$$\mathrm{NA}_{\mathrm{fiber}} \equiv \sin \theta_{\mathrm{totrefl}} = \sqrt{n_{\mathrm{core}}^2 - n_{\mathrm{cladding}}^2} \,. \tag{37}$$

$\mathrm{NA}_{\mathrm{beam}}$ is approximately 20% smaller than $\mathrm{NA}_{\mathrm{fiber}}$.

With optimized NA and neglected reflection losses the theoretical coupling efficiencies are $> 99\%$. In reality, tolerances in lateral, longitudinal and angular fiber positioning and in NA reduce the efficiency. Table 3 gives the maximal tolerances for these parameters for a maximal reduction of the coupling efficiency of 10% each.

A further reduction of the coupling efficiency comes from the nonideal collimated laser beam with a typical M^2 of 1.2 containing contributions from

Table 3. Tolerances for -10% $(-0.46\,\text{dB})$ efficiency when coupling a diffraction-limited Gaussian beam $(\lambda = 980\,\text{nm})$ into a SM fiber $(\lambda_\text{c} = 938\,\text{nm},\ a = 3.14\,\mu\text{m},\ n_\text{core} = 1.4551,\ n_\text{clad} = 1.4506)$ [6]

Parameter	Tolerances
Displacement perpendicular to the optical axis	$\pm 1.1\,\mu\text{m}$
Displacement along the optical axis	$\pm 19\,\mu\text{m}$
Tilt	$\pm 1.6°$
NA	± 0.03

the diode laser (index inhomogeneities, higher-order modes), from the collimating lens (fabrication and positioning tolerances, diffraction effects), and from reflections from the various antireflection-coated glass–air interfaces.

A macroscopic setup for SM-fiber coupling of a high-power oscillator is shown in Fig. 29. The optically relevant parts are the diode laser $(\lambda = 980\,\text{nm},\ P_\text{out} = 250\,\text{mW},\ \text{NA}_\perp = 0.45,\ \text{NA}_\parallel = 0.12)$, a biasphere $(f = 4.5\,\text{mm}, \text{NA} = 0.55)$ for collimation of both axes, a cylindrical Galilean telescope to reduce the fast-axis diameter by a factor of 3.7 to the slow-axis diameter, and a second biasphere for focusing the now symmetrical beam onto the SM-fiber tip (data see Table 3). The best experimental results with the above setup are also indicated in Fig. 29. The optical-to-optical coupling efficiency was 85%, the electrical-to-optical efficiency 30%, and the maximal output power of the SM-fiber-coupled laser 225 mW. Typical values, realized with a larger number of lasers, were 10–20% lower.

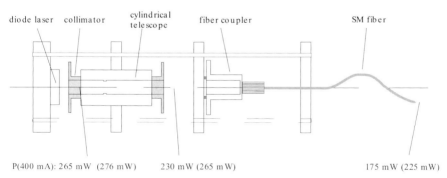

P(400 mA): 265 mW (276 mW)   230 mW (265 mW)   175 mW (225 mW)

Fig. 29. Cross section of an SM-fiber-coupled diode-laser module. Indicated are typical (in *brackets*, optimized) laser powers in the collimated beam and behind the SM fiber from a Tutcore/Jenoptik 980 nm oscillator ([6])

3.2.1   Multiplexing Techniques

Having a number of fiber-coupled lasers with well-defined beam properties or, alternatively, one- or two- dimensional arrays of edge or surface emitters the next step is to find the appropriate superpositioning method for the highest possible brightness. As mentioned above, Liouville's law does not allow us to exceed the brightness of a single radiation source when combining a number N of sources unless these sources have different polarization, spectrum, or are mutually coherent. So, depending on the beam properties, different multiplexing methods have to be applied [2,5]:

- *Angle/position multiplexing.* Any set of laser beamlets can be combined by angle or position multiplexing. In angle multiplexing the beams are focused from different angles onto the same spot while in position multiplexing the beams are focused from one direction onto different spots. In a first approximation the brightness, i.e. the power per area and per solid angle, remains constant assuming an optimized geometry.
- *Polarization multiplexing.* Beams with different polarization states can be directly superimposed by using a polarization beam splitter in opposite directions. Since there are only two independent polarization states, the brightness can maximally be increased by a factor of two (in the case of equal powers).
- *Wavelength multiplexing.* Beams with emission lines centered at different wavelengths (with linewidths smaller than the separation of the emission lines) can be directly superimposed by reversing the beams through dispersive elements like prisms, gratings, or interferometers. The increase in brightness depends on how many beams with different wavelengths (that are separable by the dispersive element) can be accommodated in a certain wavelength range. In optical communication systems wavelength-division multiplexing (WDM) is beginning to become a widely used technique to increase the long-distance transmission capacity without increasing the number of fibers. In integrated optics, for example, add-drop filters can be realized with directional couplers (X couplers) having Bragg gratings inside the coupling region [46].
- *Coherent superpositioning.* The combination of coherent sources will be discussed in the next section.

Concepts and experimental results of angle and position multiplexing of SM-fiber-coupled lasers will be discussed in the following in more detail. The angle-multiplexing technique will also be useful for coherent superpositioning.

Angle Multiplexing (AMP). When using the beam bundles as they come from the fiber or diode-laser array directly, the average brightness is dramatically reduced because the relation between radiating area total area is very unfavorable. In order to optimize the fill factor F, the divergent beams have to be collimated by a lens array in such a way that the neighboring beams just touch each other. A cross section of such an arrangement is shown in Fig. 30. For circular beams a hexagonal arrangement of the fiber ends and collimating lenses gives the highest fill factor. The total beam is then focused

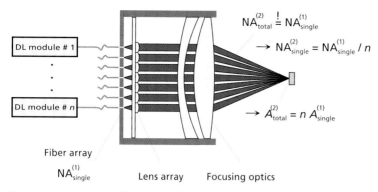

Fig. 30. Angle MultiPlexing (AMP). Shown is a cross section along the optical axis of a system with $N = 37$ channels. The number of channels n along the diagonal is approximately $\sqrt{N}$. The comparison of the numerical apertures and beam cross sectional area A in the fiber-array plane (1) and the image/focal plane (2) show the power scalability with constant power density. $(P/A)_{\text{total}} = (NP_{\text{single}})/(NA_{\text{single}}) = P_{\text{single}}/A_{\text{single}}$. Neglected are the aperture effects of the lens array (see text)

by a common lens onto the target. If the NA of the total (convergent) beam is chosen to be equal to the NA of the divergent individual beams the NA of the convergent individual beams is reduced by a factor corresponding to the number of fibers in the sectional plane. Therefore, the diameter of the focal spot of the individual lasers – which is in the ideal case equivalent to the diameter of the total spot – is increased by that same factor leaving the power density constant.

The optimum fill factor is determined by contrary effects: an increase of the fill factor increases the NA of the focused individual beams and therefore reduces the spot sizes and increases the power density. On the other hand, the individual beams have a Gaussian power-density distribution, which is truncated by the apertures of the lens array resulting in a reduced transmission and stronger diffraction effects. The dependence of average power density and transmitted power on the fill factor can be calculated using Kirchhoff's integral [38]:

$$E(x', y', z') = \frac{1}{i\lambda} \int_{-\infty}^{\infty} \int_{-\infty}^{\infty} E(x, y) \frac{e^{ikr}}{r} \cos(\hat{k}, \hat{r}) dx dy , \tag{38}$$

where

$$\boldsymbol{r} \equiv r\hat{r} \equiv \begin{pmatrix} x' \\ y' \\ z' \end{pmatrix} - \begin{pmatrix} x \\ y \\ 0 \end{pmatrix} \tag{39}$$

is the vectorial distance between a point in the image plane (x', y', z') and a point in the 'diffraction plane' $(x, y, 0)$, i.e. the plane where the diffraction takes place. As both amplitude and phase of the electrical field are known in the 'diffraction plane', the complex electrical field in any other plane can be calculated. Kirchhoff's integral can be unterstood as the mathematical description of Huygen's concept of describing the propagating electrical field as a superposition of elementary spherical waves with the initial field distribution $E(x, y)$, the amplitude reduction proportional to $1/r$, the phase factor e^{ikr}, and the direction cosine between wavevector k and image point direction r.

In the far-field or Fraunhofer approximation, i.e. at a distance z' from the diffracting structure, large compared to the size of this structure measured in wavelengths,

$$z' \gg \frac{\pi}{\lambda}(x^2 + y^2) , \tag{40}$$

the cosine function and the quadratic and higher-order phase factors can be neglected and the diffraction image is simply the Fourier transformation of the field-amplitude distribution in the diffraction plane

$$\begin{aligned} E(x', y', z') &= \tilde{E}(\tilde{\nu}_x, \tilde{\nu}_y) \\ &= A(x', y', z') \int_{-\infty}^{\infty} \int_{-\infty}^{\infty} E(x, y) e^{-2\pi i(\tilde{\nu}_x x + \tilde{\nu}_y y)} dx dy \end{aligned} \tag{41}$$

with the spatial frequencies

$$\tilde{\nu}_x \equiv \frac{x'}{\lambda z'} \approx \frac{\theta_x}{\lambda} \quad \text{and}$$

$$\tilde{\nu}_y \equiv \frac{y'}{\lambda z'} \approx \frac{\theta_y}{\lambda} , \tag{42}$$

and the complex phase factor

$$A(x', y', z') = \frac{e^{ikz'}}{i\lambda z'} e^{i\frac{\pi}{\lambda z'}(x'^2 + y'^2)} . \tag{43}$$

With a lens of focal length f the angular distribution $E(\theta_x, \theta_y)$ can be transformed into a spatial distribution in the focal plane $E'(\xi, \eta)$ of the lens using the following relationships

$$\tilde{\nu}_x = \frac{\theta_x}{\lambda} = \frac{\xi}{\lambda f} ,$$

$$\tilde{\nu}_y = \frac{\theta_y}{\lambda} = \frac{\eta}{\lambda f} . \tag{44}$$

The calculated dependence of power density and transmitted power on the (linear) fill factor (F, ratio of beam diameter at $1/e^2$ to center-to-center distance of two adjacent beams) is shown in Fig. 31. The optimum fill factor for highest power density is 0.9; in this case the transmitted power is 90%.

The AMP was demonstrated with 37 fiber-coupled diode lasers each having an average output power of 176 mW which corresponds to an average

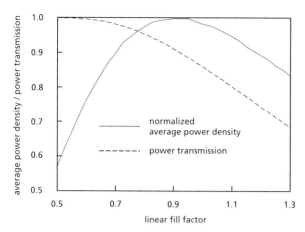

Fig. 31. Power transmission T and average power density as a function of the linear fill factor F (Gaussian-beam diameter divided by the lens diameter) calculated for a single beam ([6]). Assuming an ideal superposition in the common spot this holds also for angle-multiplexed systems unless the total NA is too large

power density of 2.0 MW/cm^2 ($\lambda = 980$ nm, NA $= 0.2$) [6]. The beams from the hexagonally arranged fiber ends were collimated by an array of 4 mm-diameter plano-convex lenses made of the Schott glass LaSFN 31 ($n = 1.86$) cemented on a BK 7 plate ($F = 0.94$) and then focused by an objective (2 lenses, $f = 100$ mm, NA $= 0.2$).

In the common focus the average power density of the nearly Gaussian distribution was 175 kW/cm^2 (NA $= 0.14$ of the focused beam) which corresponds to a power density of 350 kW/cm^2 for the 'standard' NA $= 0.2$ (see the left two columns of Table 4). This is only 18% of the power density of the individual laser. The reasons for this reduction are of principle and of experimental nature, viz.

- the hexagonal arrangement with a linear fill factor of 0.94 – already larger than the power-density optimized fill factor of 0.9 – results in an area fill factor (illuminated area divided by the area of the smallest possible circle) of 0.68 which is equivalent to the reduction factor for the average power density (-1.7 dB),
- diffraction effects at the collimating lens array increase the focus diameter by 36%, thus reducing the power density by a factor 0.54 (-2.7 dB),
- the nonoptimum alignment of the individual beams results in a further power-density reduction by a factor of 0.7 (-1.5 dB),
- the reduction of the total power compared to the sum of the individual powers by the factor of 0.7 is composed of aperture effects (factor 0.89 or -0.5 dB) and nonoptimum antireflection coatings and alignment (factor 0.83 or -0.8 dB).

The power-density reduction is mainly due to the unavoidable geometric and diffraction effects yielding $-(1.7 + 2.7 + 0.5)\,\mathrm{dB} = -4.9\,\mathrm{dB}$, which corresponds to a reduction factor of 0.33. Experimental shortcomings reduce the power density by another factor 0.58 or $-2.4\,\mathrm{dB}$. Hence, the scalability of the total power at constant power density or brightness with AMP is confirmed, but the brightness is at least $5\,\mathrm{dB}$ lower than that of the individual lasers.

Position Multiplexing (PMP). The preparation of the beams to be combined is similar to the case of AMP. The individual (diffraction-limited) beams are collimated in a hexagonal arrangement with a fill factor of 0.9. But instead of focusing all beams into a common focus from different angles as in AMP the power-density distribution behind the collimating-lens array (object plane) is imaged with the necessary reduction factor onto the image plane. In a first-order approximation one would expect a total spot size similar to the focal spot of AMP but with individually resolvable and addressable spots.

A prerequisite is that each beamlet uses the full aperture of the imaging optics. In this way it can be focused to its original size (same NA assumed) so that the total area is N times the area of one element and therefore the power density is preserved. This can be done in two steps: first, an inverted telescope reduces the beam diameter and increases the beam divergence. The remaining reduction is then accomplished by imaging optics in a second step (Fig. 32).

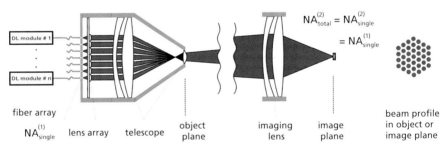

Fig. 32. Position MultiPlexing (PMP). Shown is the cross section along the optical axis of a system realized with $N = 37$ channels. The beam profile on the image plane has a hexagonal symmetry reflecting the power density profile behind the lens array and behind the demagnifying telescope ([6])

An experiment similar to the one on AMP has been executed that allows the direct comparison of the two methods [6]. The collimation of the beamlets was identical to the AMP experiment. The focusing lens ($f = 100\,\mathrm{mm}$) served in the PMP experiment as the first lens of a Newtonian telescope.[2] The second lens of the telescope was a biasphere with $f = 4.5\,\mathrm{mm}$. As imaging

---

[2] The simplest telescope consists of two lenses (objective of focal length f_1 and ocular of focal length f_2) at a distance $f_1 + f_2$. The (angular) magnification

optics a second focusing lens ($f = 100$ mm) at a distance of 3.85 m from the telescope was used. The distance was chosen so that a total NA similar to that of the AMP experiment resulted. The total (linear) reduction factor was $(4.5/100)(100/3850) = 1/850$. Table 4 allows for a comparison of both methods.

Table 4. Comparison of angle and position multiplexing with a system of 37 fiber-coupled lasers [22]. For both a single laser and the system the power on the work-plane, the diameter of the collimated beams, the numerical aperture of the focused beams, the focus diameter (at $1/e^2$), and the average power density (within $1/e^2$) are tabulated

37-channel beam	Angle multiplexing single	total	Position multiplexing single	total
P_{focus} (W)	0.13	4.8	0.13	4.8
$2w_0$ (mm)	3.8	28	28	29
NA	0.02	0.14	0.14	0.14
$2w_{\text{focus}}$ (μm)	45	55	4.8	35
I_{focus} (kW/cm^2)	7	175	620	500

The big advantage of PMP in terms of power density, I_{focus}, is that all the effects that lead to an increase of the (total) spot diameter $2w_{\text{focus}}$ in AMP only affect the individual spot size in PMP thus increasing the total spot size by only a few percent. The two remaining factors reducing the power density are the transmission of the optics and the (area) fill factor. The average power density with PMP is therefore more than 2.5 times the power density with AMP.

The other big advantage of PMP is that it produces a spatially modulated top-hat profile that can be varied without inertia almost arbitrarily by modulation of the diode-laser power. Two examples produced by the 37-channel setup are shown in Fig. 33. A disadvantage is the more complex and space-consuming optics.

3.3 Phase Coupling and Beam Combining of Single-Longitudinal-Mode Lasers

Power scaling while preserving the diffraction-limited beam requires the coherent combining of lasers. This in turn requires that the individual lasers emit spatially coherent beams that have, at the point of superposition, time-independent phase differences. One possible way of realization is the use of phase-coupled diode lasers with diffraction-limited beams and sufficiently large coherence lengths to account for different distances.

---

is given by $-f_1/f_2$. The Newtonian type uses two converging (positive) lenses while the Galilean type uses a converging (positive) objective and a diverging (negative) ocular which reduces the overall length and avoids a real focal point within the telescope.

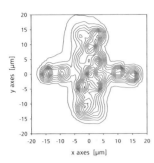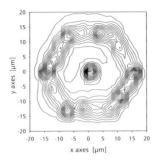

Fig. 33. Contour plot of measured power-density distributions produced with a PMP´setup when activating a part of the 37 fiber-coupled diode lasers such that a cross or ring pattern results. The power-density difference between two contour lines is 7% of the maximum, i.e. the second outermost contour gives the $1/e^2$ size. The maximum power density according to Table 4 is approximately $600\,\text{kW/cm}^2$ ([6])

3.3.1   Methods of Phase Coupling

Electronic Control.   The electronic coupling scheme allows all lasers to emit independently, i.e. mutually incoherently. The phase difference of all oscillators relative to an arbitrarily chosen reference laser is measured interferometrically and corrected by changing the individual drive currents of the oscillators or via an additional phase modulator. The phase control is typically realized with an OPLL (optical phase-locked loop) well known from homodyne receivers in optical communication systems [47]. Problematic is the need of extremely fast electronics or extremely stable lasers (ratio of electronic bandwidth to laser linewidth of 10 or higher).

Self-Organization. Self-organization of independent lasers requires some kind of interaction of the laser fields within the resonators.

An obvious choice is to pack closely the individual emitters in a linear array in the case of edge emitters or in a two-dimensional array in the case of VCSELs [8,48]. There are a number of options depending mainly on the type of waveguide employed, as shown in Table 5.

The simplest structure technology of a single emitter is the gain-guided laser which is preferred for high-power multimode lasers. The laser modes are defined only by the (plane) Fabry–Perot resonator and the gain distribution. Therefore, the laser is sensitive to current and thermally induced index changes and spatial hole burning which makes it difficult to achieve stable single-mode operation. Adjacent lasers of an array couple through 'leaking' of optical power between the two lasers. Their phase is shifted by π ('out of phase') which gives a lower power density in the lossy area between the gain guides. The width of the resulting two-lobed far-field pattern is many times the diffraction limit due to poor intermodal discrimination.

Table 5. Brief history of phase-locked diode-laser array research and development ([8])

Time period	Array type	Preferred array mode	Overall inter-element coupling	Single-mode selectivity	Max. diffraction-limited power
1978 –	gain-guided	leaky in-phase or leaky out-of-phase	series	poor	—
1981	antiguided	leaky in-phase or leaky out-of-phase	series	poor	—
1983–88	positive index-guided	evanescent out-of-phase	series	moderate	0.2 W
1988	antiguided	leaky in-phase or leaky out-of-phase	series	moderate	0.2 W
1989–	in-phase resonant antiguided	leaky in-phase	parallel	excellent	2.0 W

Index-guided emitters with a central higher index-of-refraction section allow us to favor a single Gaussian-like mode similar to an optical fiber (Sect. 3.2). The small index difference makes them still vulnerable to laser-induced index changes. Neglecting absorption or scattering losses the guiding of the eigenmodes of the waveguide is lossless despite the fact that part of the wave is propagating outside the core. These so-called evanescent waves also allow us to couple two adjacent lasers (Fig. 34b,c). But similar to the gain-guided case the strong nearest-neighbor coupling is a series coupling which does not lead to strong overall coupling resulting again in poor intermodal discrimination.

So both the gain- and index-guided SM laser arrays suffer from mode instability of the single lasers when operated at high power levels and from the limitation to nearest-neighbor interaction, a coupling in series which does not allow stable coupling for higher numbers of lasers. Antiguided lasers with a lower index of refraction in the center and a comparatively large index difference solve both problems: the single laser modes are more stable and coupling to the neighbors due to leaky waves is stronger (Fig. 34d,e). A drawback for the single laser is the strong loss through the leaky waves, which can be reduced by placing more antiguides parallel to the laser in such a way that the leaky waves are anti-resonantly coupled back into the resonator (Anti-Resonant Ridge Optical Waveguide, ARROW-type lasers) [49]. In an array the loss due to the leaky waves is compensated by the radiation 'losses'

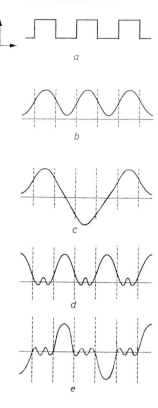

Fig. 34. Modes of arrays of periodic real-index variations. (**a**) index profile; (**b**) in-phase evanescent-wave type; (**c**) out-of-phase evanescent-wave type; (**d**) in-phase leaky-wave type; (**e**) out-of-phase leaky-wave type ([50])

of the neighbors. For the outermost lasers an ARROW-type structure can be used to reduce the losses.

Especially favorable is resonant coupling which corresponds to lateral standing waves, so called resonant optical waveguide (ROW) arrays. In this case the width of the gain and the interelement (high-index) region are an integer number of half-lateral wavelengths – odd integers giving in-phase, even integers out-of-phase modes. Schematic representations of the lowest-order modes are shown in Fig. 35, whereas near- and far-field patterns of an eight-element array are displayed in Fig. 36. The resonant leaky-wave fan-out in a tree-like manner ensures parallel coupling unless the resonance frequencies within the array, e.g. by thermal effects, are slightly different. Figure 35 also shows that with increasing thermal gradient (higher pump power) the coupling is more and more reduced to the central channels, thereby increasing the width of the central peak. Therefore, the highest diffraction-limited CW power demonstrated was several hundred milliwatts (also including the power of the satellite peaks).

A coupling between the individual oscillators can also be achieved by using a common external resonator. This could be, e.g., a Talbot resonator [37,53] or a resonator with a filter in a Fourier plane within a telescope [54]. Talbot

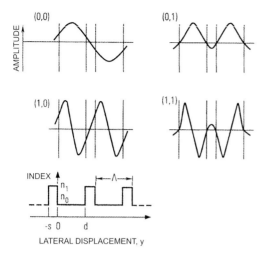

Fig. 35. Leaky-type array modes. In (l, m) l is the (lateral) mode order in the antiguide core region, and m is the number of field-intensity peaks in the interelement regions ([51])

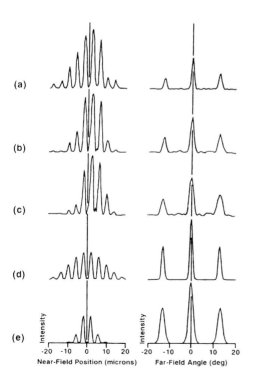

Fig. 36. Near- and far-field patterns for a leaky-mode array under CW operation, (**a**) 1.4× threshold, (**b**) 2.4× threshold, (**c**) 3.4× threshold; (**d, e**) calculated far-field pattern in the absence of thermal effects (**d**) and with an 8 K temperature rise in the center of the array (**e**) ([52])

has discovered that an infinitely extended array of coherent emitters with a lateral period of d and wavelength $\lambda = \lambda_{\text{vac}}/n$ is reimaged with the period

$$Z_{\text{T}} = \frac{2d^2}{\lambda} \ . \tag{45}$$

Hence, for a laser array with a feedback mirror at a distance of $Z_{\text{T}}/2$ maximum feedback occurs if all lasers emit coherently (Talbot resonator, Fig. 37). A small fill factor gives the strongest discrimination. However, the resonator losses in the coherent mode increase as well due to the finite size of the array and deviations from the paraxial propagation. Another problem is the discrimination of the various supermodes for different Talbot distances. For the typically used distances the self- and nearest-neighbor coupling is strongest, and distant neighbors do not couple at all.

The filtering in the Fourier plane of the total radiation field can be realized by inserting a telescope in front of the common plane end mirror (Fig. 38). The losses for the different super modes depend on the spatial transmission of the filter placed in the focal plane. A trade-off has to be found between strong mode discrimination and low additional resonator losses.

External Phase Control (Master Laser). A general problem of the self-organizing schemes may be an increasing instability with a growing number of interacting lasers. This can be overcome by implementing a master laser as external reference. The master laser is protected against any feedback. Its radiation is fed into a tree-like amplifier chain. With a monolithically

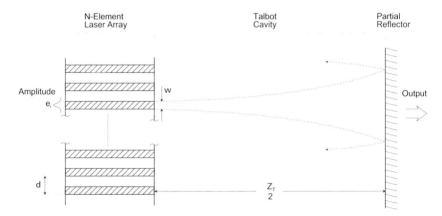

Fig. 37. Talbot resonator for the coherent coupling of an N-element laser array ([53]). The radiation which is diffracted at the aperture of the laser elements is refocused and mixed after propagating twice the Talbot cavity length, i.e. the Talbot distance Z_{T}. The losses for each laser element are minimal if the radiation scattered into the resonator from the other elements adds in phase to the internal laser field, therefore, eventually leading to an array of self-phase-locked lasers

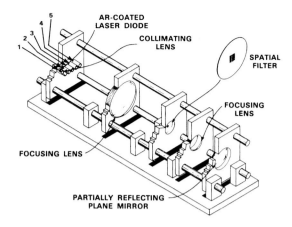

Fig. 38. External cavity laser system with modal filtering ([54]) for the coherent coupling of a 5-element laser array. Within the resonator the radiation of each element is collimated and propagates through a Newtonian telescope (the two focusing lenses) and is reflected back by the partially reflecting outcoupling mirror. A spatial filter (pinhole) in the common focal plane of the telescope (the Fourier plane of the focusing lenses) allows us to favor or suppress certain super modes

integrated array of 4 flared amplifiers a spectrally and spatially coherent pulsed power of more than 5 W has been achieved [55]. Roughly the same coherent power has been realized with an amplifier stack of nine linear 100-emitter arrays fed by one master oscillator [56].

Alternatively, the master-laser beam can be divided into several beams (in a tree-like arrangement, if necessary, after amplification) and fed into so-called slave oscillators. Under certain conditions the slave oscillators emit in phase with the injected master radiation. Since there is no interaction between the slave lasers – each oscillator 'sees' only the master radiation – this method is principally not limited in the number of coherently coupled lasers.

In the following we will concentrate on the master-slave approach because of its power scalability. The self-organizing approaches have been tried for a long time but with success limited to small numbers of coupled lasers. The total power achieved up to now has not been appreciably higher than the power of an individual single-mode laser. The feasibility of the electronic approach has been demonstrated by laser transmission through space with homodyne detection. The required stable lasers and fast electronics render this approach still too complex to be implemented with a larger number of lasers.

Injection Locking of Diode Lasers. If an electromagnetic field (e.g. from a master laser) is injected into a laser resonator this so-called slave laser will oscillate with the frequency and phase of the injected signal if the gain for this signal is sufficiently large to suppress lasing on the free-running frequency. This injection-locking process can be described by the semi-classical laser theory with an external laser field [39]. The frequency range within which locking occurs, the locking range,

$$\Delta\nu_{\text{Lock}} \equiv 2|\nu_{\text{S}} - \nu_{\text{M}}| = 2r_{\text{e}}\sqrt{\frac{\delta P_{\text{M}}}{P_{\text{S}}}} , \tag{46}$$

depends on the ratio of injected master power to slave power $\delta P_{\text{M}}/P_{\text{S}}$ and the loss rate

$$r_{\text{e}} = -\frac{\ln\sqrt{R_1 R_2}}{t_{\text{R}}} \tag{47}$$

with R_1 and R_2 being the mirror reflectivities and $t_{\text{R}} = 2nL/c$ the resonator round trip time. For a typical edge-emitting laser ($L = 500\,\mu\text{m}$, $n = 3.5$, $R_1 = 1$, $R_2 = 0.3$, $\delta P_{\text{M}}/P_{\text{S}} = 1\%$) a locking range of 10 GHz can be expected. Due to light-carrier interaction in the semiconductor, the injection changes the index of refraction and hence the resonator frequency, resulting in a larger, asymmetrical locking range [57,58]

$$\Delta\nu_{\text{Lock}}^{\text{DL}} \equiv 2|\nu_{\text{S}}^{\text{DL}} - \nu_{\text{M}}| = \left(1 + \sqrt{1 + \alpha^2}\right) r_{\text{e}}\sqrt{\frac{\delta P_{\text{M}}}{P_{\text{S}}}} . \tag{48}$$

In addition, instabilities can occur for high injected-power levels at the high-frequency end of the locking range resulting in a smaller and more symmetric locking range approximately as given by (46).

Stable injection-locked operation therefore requires constant operating conditions, in the example above maximum current fluctuations of 3 mA or temperature variations of 0.3 K assuming typical sensitivities of 3 GHz/ mA and 30 GHz/ K. Equally important is to reduce any feedback to levels well below the injected signal and to maintain an optimal injection, i.e. mode-overlap of resonator and injected field.

Phased Arrays. The angular multiplexing scheme discussed above for the incoherent combining can also be used advantageously for combining mutually coherent laser beams leading in the ideal case to an electronically steerable, diffraction-limited beam [59,60,61].

The mathematical description is similar to the incoherent case. The propagation of each beam is described by the Fourier transformation of the electric-field distribution in the lens-array plane (35,36,37). But instead of adding the power densities, in the coherent case the electric-field distributions of the N beams in the far-field (or in the focal plane of the focusing lens) have to be added taking into account the phase factors of each beam E_j [centered at the position (x_j, y_j)] [9]:

$$E_{\text{sys}}(x, y) = \sum_{j=1}^{N} E_j(x - x_j, y - y_j) . \tag{49}$$

Assuming a normalized single-emitter near-field distribution $\hat{E}_0(x, y)$ with a field amplitude $E_j^{\max}$ and constant phase factors ϕ_j for each of the beams j in the lens-array plane,

$$E_j(x - x_j, y - y_j) = E_j^{\max}\, \hat{E}_0(x - x_j, y - y_j)e^{\mathrm{i}\varphi_j}\,, \tag{50}$$

the field distribution of the superimposed beam in the focal plane can be separated into two factors, the interference and the envelope factor

$$
\begin{aligned}
E_{\mathrm{sys}}(\xi, \eta) = A(\xi, \eta) \qquad &\text{phase factor}\\
\times \sum_{j=1}^{N}\left(E_j^{\max}\, \mathrm{e}^{-2\pi\mathrm{i}\left(\frac{\xi}{\lambda f}x_j + \frac{\eta}{\lambda f}y_j\right)}\mathrm{e}^{\mathrm{i}\varphi_j}\right) \qquad &\text{interference factor}\\
\times \iint \hat{E}_0(x, y)\mathrm{e}^{-2\pi\mathrm{i}\left(\frac{\xi}{\lambda f}x_j + \frac{\eta}{\lambda f}y_j\right)}\mathrm{d}x\mathrm{d}y \qquad &\text{envelope factor}\,. \tag{51}
\end{aligned}
$$

The envelope factor [including the complex phase factor $A(\xi, \eta)$] is well known from the incoherent superposition where it describes the field distribution of a single beam (41,42,43). The interference term is caused by the periodic modulation of the amplitude distribution due to the Gaussian power distribution of the beamlets and the apertures of the lenslets of the lens array. The width of the central peak, in the case of equal phases φ_j, is limited by the total numerical aperture as opposed to the incoherent superposition where the sub-apertures (smaller by a factor of $\sqrt{N}$) are responsible for the spot size. The maximum of the coherent superposition is N times the maximum of the incoherent superposition and N^2 times the maximum of a single beam.

The situation is very similar to the diffraction effects behind a grating illuminated by a plane wave. The interference factor is given by the periodicity of the grating (rays emerging from neighboring slits have to have a phase difference of multiples of λ); the envelope factor is given by the (usually constant) power distribution within a single slit leading to an Airy-type modulation of the periodic interference peaks (56). The factor $\sqrt{N}$ corresponds in the case of the grating to the number of illuminated slits in one dimension.

A linear phase gradient over the system aperture corresponding to a phase difference of two neighboring beams

$$\frac{\phi_j - \phi_{j-1}}{x_j - x_{j-1}} = \frac{\partial\phi}{\partial x} \tag{52}$$

gives a shift of the interference pattern and especially of the central peak in the focal plane of

$$\Delta\xi = \frac{\partial\phi}{\partial x}f\,. \tag{53}$$

In this way fast beam steering or focus shifting by electronically shifting the relative phases of the lasers is possible ('phased array'). The maximum

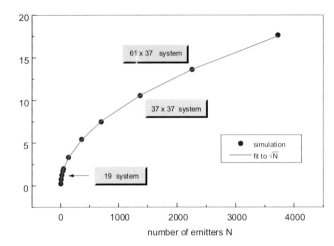

Fig. 39. Maximal useful shift of the central peak of the power-density distribution as function of the number of emitters of a phased array. The criterion for the maximal shift is the decrease of the power density of the central peak by 10% compared to the power density of the unshifted beam ([2,9])

shift of the central peak of a two-dimensional array (given by a 10% decrease of the peak power) normalized to the peak diameter is shown in Fig. 39. It increases approximately with the square root of the number of lasers N.

The considerations above are correct for mutually coherent beams with a coherence degree $\gamma = 1$. In the case of partially coherent emitters with $\gamma < 1$ the electric field can be divided in to a completely coherent part E_{coh} and a completely incoherent part E_{incoh} [9]

$$E(x,y) = \sqrt{\gamma}E_{\mathrm{coh}} + \sqrt{1-\gamma}E_{\mathrm{incoh}} \,. \tag{54}$$

These can be treated separately, which means that the power-density distribution is composed of a broad Gaussian-like baseline (the width of which is given by the NA of a single beam) with the central peak and the associated satellite peaks (due to the diffraction effects) on top. The power ratio is given by γ, which means that the height of the central peak is also reduced by the factor γ.

Experimental Realization of a Phased Array of Coherently Coupled Diode Lasers. A conceptual setup of a system with 19 coherently coupled diode lasers is shown in Fig. 40.

Nineteen Fabry–Perot-type slave lasers (20 mW, 675 nm) have been injection locked each by approximately 30 μW injected power from a master laser (30 mW, 675 nm), followed by two Faraday isolators. The beams are fiber-coupled into polarization-maintaining fibers that are connected to the laser head which consists of a plate holding the fiber ends, a lens array (hexagonal symmetry, fill factor 0.6), a beam splitter (uncoated BK7 plate) in the

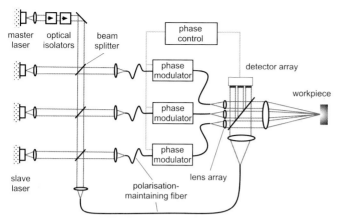

Fig. 40. Concept for the coherent coupling of diode lasers. The output of the master laser is injected via beam splitters and the collimating optics into the slave lasers leading to a coupling of frequency and phase. The laser power of the slaves is transported by polarization-maintaining fibers to the laser head. There, the phase is controlled by a circuit containing a phase-sensitive detector, PI controller and phase modulator for each channel ([6])

collimated beam, a focusing objective for the main beam, and a detector array for the reflected beam. The beam splitter allows us to superimpose the phase-modulated master radiation with each of the slave beams for an interferometric phase control which is necessary because of acoustic and thermal perturbations. The relative points in time of the zero transits are a measure for the phase difference, which is then corrected with a piezo-disk-shaped electric fiber stretcher by an analog control circuit with proportional-integral characteristics. The contrast of each of the interferometric superpositions is used to optimize the degree of coherence between master and slaves by adjusting the slave currents. Since the temperature drift is in the range of minutes the corrections can be done in series by a single PC.

The realized average degree of coherence between slaves and master was 79% with a standard deviation of 8%. With minimized back-reflections the coherence was higher with $(83 \pm 3)\%$. The locking range for $> 50\%$ coherence was 2.8 GHz and 3.2 GHz, respectively, in good agreement with the calculated 2.6 GHz (46). The phase-control loop reduced the phase fluctuations to 20 mrad RMS (1.1°) under standard laboratory conditions and 140 mrad RMS (8°) with strong mechanical perturbations present.

The power-density distribution together with a numerical simulation is shown in Fig. 41. The increase compared to the incoherent superposition (master laser switched off) is a factor of 13, indicating a degree of coherence of the system of $\gamma = 70\%$. The power contained in the central peak is 16% of the total power. From simulations, reasons for this small amount are derived as follows. The diffraction caused by the lens-array apertures with unoptimized

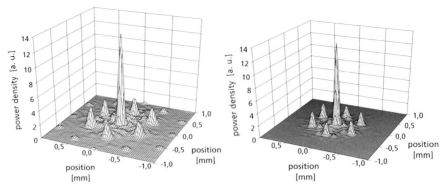

Fig. 41. (a) Measured power-density distribution in the focal plane (20× magnified) of the coherent superposition of 19 slave lasers. The symmetry reflects the hexagonal symmetry of the fiber/lens array. (b) Calculated power-density distribution for the system used in (a) taking into account the nonideal coherence, alignment errors and different output powers ([9])

fill factor is responsible for the satellite peaks and reduces the power in the central peak to 43% for otherwise-ideal conditions. Taking into account the 30% incoherence this number is reduced to 32%, and through additional alignment errors and deviations in output power to 26%. A further reduction can be attributed to the non-diffraction-limited lens array.

Phase-Correcting Plates. Even with an optimized fill factor of 0.9, approximately one third of the power is lost in the satellite peaks. Another 10% of the power is lost at the apertures. To increase the power in the central peak a beam transformation would be desirable that converts the laser field of the coherent array into a diffraction-limited Gaussian laser beam. Fortunately, it is – at least in principle – always possible to convert a well-defined amplitude and phase distribution in a first plane into an arbitrary field distribution in a second plane. For such a transformation two phase-modulating surfaces A and B are necessary: surface A between the two planes rearranges the power distribution in the second plane and surface B located on that plane corrects the phase distribution to get the desired wavefront. The advantage of using pure phase elements for the transformation is that in principle no transformation losses occur that would reduce the system efficiency. A few exemplary transformation schemes will be described in the following.

- Beam shaping of the individual beams [62]

Each beamlet consisting of a slightly truncated Gaussian beam (fill factor $\cong 0.5$) is transformed into a flat-top distribution (by surface A) with plane phasefront (by surface B) so that the total beam cross section is homogeneously filled (Fig. 42). Thus the total beam has a uniform power density

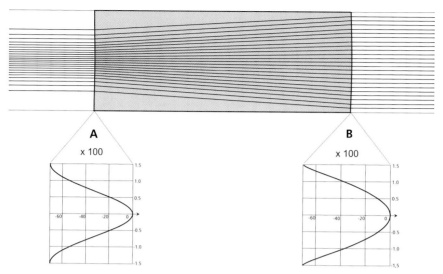

Fig. 42. Ray-trace simulation of a beam transformer from Gaussian to flat-top with a biaspherical phase plate of length $L = 20\,\mathrm{mm}$ [64]. The contour of the surfaces A and B (maximum amplitude $\cong 20\lambda$) is indicated below with $100\times$ magnification. The function can be understood in the frame of geometrical optics. The rays going from A to B are directed by surface A such that ring segments with equal power dP on A are transformed into segments of the same size (dA) and power (dP) and hence constant power density dP/dA on B. Surface B then directs the rays again parallel to the optical axis

and can be focused to a single spot (Airy disk[3], the size given by the total beam aperture) while the total transmission would still be close to 100%. The collimating micro-lens array and phase-plate array A can be combined into a single array. Allowing for an overlap of neighboring beams on surface B by modifying surface A should reduce the the discontinuities of the slope of surface B and by that the sensitivity to alignment errors.

By further transforming this uniform (top-hat, flat-top) profile into a single Gaussian profile with a plane wavefront, the total beam can be focused into a diffraction-limited Gaussian spot without diffraction fringes.

---

[3] The angular power-density distribution of a homogeneous plane wave with $k = 2\pi/\lambda$ diffracted at a circular aperture of radius a is given by the Airy pattern [63]

$$I(\theta) = I(0) \left[\frac{2J_1(ka \sin \theta)}{ka \sin \theta} \right]^2 \tag{55}$$

with $J_1(u)$ being the Bessel function of the first kind of order 1. A strong central maximum is surrounded by circular diffraction fringes with decreasing power density, the radius of the first minimum – when focused with a lens of focal length f – is at $r = 1.22 f\lambda/(2a)$

- Power-density redistribution in the Fourier plane [65]

Instead of individually reshaping the power-density distribution of each beamlet into a flat-top with phase-plate array A two Fourier transformations with a phase plate in the Fourier plane can be used. A phase shift of $\phi = \arccos(2F - 1)/(2F)$, F being the fill factor of the collimated beam, applied to the central peak in the Fourier plane, leads to a nearly constant amplitude distribution after collimation. The phase is corrected by phase plate B (Fig. 43).

- Phase correction in the Fourier plane [66]

Phase plate A applies phase shifts between 0 and 2π on the beams exiting the laser array. These phase shifts are chosen such that the power-density distribution in the Fourier plane is nearly Gaussian and that the phase distribution is not too strongly modulated. The phases are then corrected with phase plate B in the Fourier plane (Fig. 44). After filtering the side lobes with less than 5% power content originating from the modulation of the Gaussian power-density distribution, M^2 is theoretically very close to 1.

Since all transformation schemes are supposed to deliver diffraction-limited beams at least one of the optical surfaces has to be aligned to sub-wavelength tolerances. In addition, the minimal dimensions that have to be realized on the phase plates can be quite small (several micrometers) so that they are difficult to fabricate. Equally important is the efficiency of the transformation optics which is critical if diffractive optics is used.

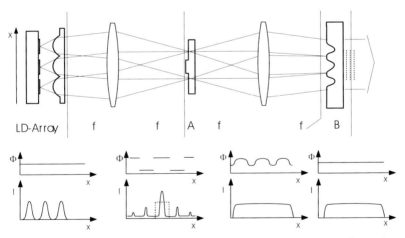

Fig. 43. The *upper part* shows the cross section of a beam-transformation system to convert an array of collimated multi-Gaussian beams into one flat-top beam. Phase plate A in the Fourier plane transforms the power density, phase plate B in the image plane the phase distribution. The *lower part* shows the phase distribution $\Phi(x)$ and the power-density distribution $I(x)$ at different places of the optical path indicated by the *broken line*

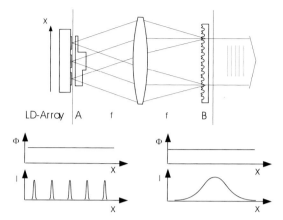

Fig. 44. Cross section of a beam-transformation system to convert an array of Gaussian beams into a single Gaussian beam with phase plate A causing a nearly Gaussian power-density distribution in the Fourier plane and phase plate B for correcting the phase. The phase and power-density distributions before and after the transformation are shown *below*

Direct Superpositioning. For the experimental realization angle multiplexing was used to superimpose the beams in the far field. On the one hand, this has the advantage of electronic beam steering; on the other hand, a non-negligible part of the power is lost in the satellite peaks unless a suitable low-loss beam transformation is added. Another option is to totally superimpose the beams (angle- and position-wise). In this way the superimposed beams have the same beam parameters as a single beam but N times the power and power density. This does not work with mutually incoherent beams because the relative phase is randomly changing with time, but is, at least theoretically, no problem with phase-controlled beams: one can simply use any beam-splitting device in opposite directions. Such a beam splitter could be a dielectrically coated glass plate or cube as well as an X or 3 dB coupler in integrated optics. The splitting ratio has to correspond to the power ratio of the beams to be superimposed. By adjusting the phase difference of the two input beams the power ratio at the two exits can be chosen. By minimizing the power in one channel by a feedback loop the power in the other channel can be maximized.

The combining of N beams can be realized either with $(N-1)$ 1:1 beam splitters in a tree-like arrangement or with one 1:N beam splitter. Figure 45 shows a simulation of a 1:4 beam splitter realized with three 3 dB couplers followed by a 4:1 combiner. Since the beam-splitter section defines the relative phases of the beams entering the combiner section the beams can be almost completely recombined in one waveguide. The problem of combining large numbers of lasers is that $\log_2 N$ steps are required each with a certain attenuation $(1 - \eta_{\mathrm{comb}})$, which leads to an exponential decrease of the

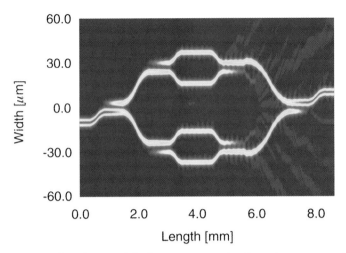

Fig. 45. Simulation of the beam propagation in a planar waveguide structure showing the splitting of the input power equally into four waveguides and the subsequent recombination in one waveguide

efficiency η_{cascade} with the number of combined beams,

$$\eta_{\text{cascade}} = \left(\eta_{\text{comb}}\right)^{\log_2 N} , \tag{56}$$

and the total power, respectively [67]. The 1:N beam splitter has potentially lower losses (unless it is made up of 1:1 beam splitters), but it is more difficult to control the phases because there are no places where the degree of constructive interference can be measured separately for each channel.

3.4   Prospects and Limitations of Coherent Beam Combining

Principally, coherent beam combining allows us to extend the advantages of (low-power) semiconductor-diode lasers – compactness, high efficiency, compatibility with microelectronics, high modulation frequency, nearly diffraction-limited beam – into the medium- and high-power regions.

Most attempts have been undertaken to couple a number of emitters on a chip via self-organization of the resonators by neighbor–neighbor coupling. It turned out that nearest-neighbor coupling is not sufficient to get a stable locking of the resonators. Some success has been achieved with a 'parallel-coupling' scheme (ROW), but the total power is still limited to that of high-power single emitters. This could be a problem of homogeneity of the larger-size wafer or the more general problem of self-organization of a large number of more or less independent emitters.

A more complex scheme is to use external resonators which gives a stronger parallel coupling but relatively high resonator losses and hence low efficiency if good mode discrimination is needed. The problem of self-organization remains.

Starting with radiation from one master laser which is either amplified in a tree-like amplifier chain or fed in parallel into slave oscillators basically solves the problem of phase locking and guarantees stable mode quality which is given by the design of the slave oscillators or amplifiers. The amplifier solution might suffer from Amplified Spontaneous Emission (ASE) if several amplification steps are required. Also, the feedback into the master laser may become quite strong. From the point of view of scalability, the injection-locking scheme seems best suited since there is almost no coupling between the slave lasers. In addition, only a very small master power is necessary for an adequate locking which requires no or only slight amplification. Standard, highly efficient oscillators can be used, and the optical system efficiency is only marginally below that of the single oscillator.

The need for an additional master laser as well as a distribution optics/network complicates the systems and requires system realization in mass-producible hybrid or integrated-optics technology. One possibility would be, e.g., a linear array of edge-emitting single-mode lasers the beams of which are coupled into a planar waveguide array that combines the beams with a series of X couplers. The distribution and feeding of the master radiation can be done in the same element in the opposite direction (Fig. 46). Another approach uses a two-dimensional array of vertical-cavity lasers with an array of micro-optics for collimation, fill-factor optimization and injecting the master radiation (Fig. 47). The design is very similar – except for the size – to the phased fiber array discussed above.

The macroscopically realized injection-locked arrays required a phase and coherence control because of thermal and acoustic coupling to the outer world

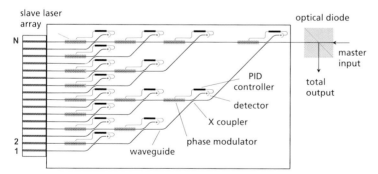

Fig. 46. Schematic layout of a planar waveguide structure to combine the output of $N = 16$ diode lasers by a tree-like arrangement of 15 directional couplers each with a phase-control circuit. The 15th coupler allows us to modulate the output power of the array. The master radiation is injected from the exit; the optical diode (e.g. Faraday isolator) separates the output power and protects the master laser against feedback

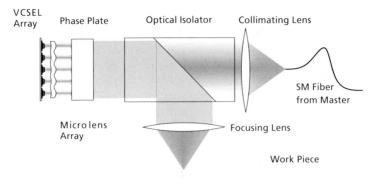

Fig. 47. Design for the coherent coupling of a two-dimensional VCSEL array. The lens array collimates the individual beams and improves the fill factor, the phase plate transforms the spatially modulated beam into a homogeneous beam. The master radiation is injected in the reverse direction via the optical isolator. Here it is assumed that no dynamic phase control is necessary [68]

which is different for each channel. First experiments with a miniaturized version consisting of a 5×5 VCSEL array showed that very stable and highly coherent operation is possible without any current or temperature control provided all the lasers including the master laser are on the same chip (and heat sink) and all the lasers are operated from one power supply [68].

Common to all coupling schemes is the task to combine the phase-coupled beams into one beam of diffraction-limited beam quality without losing the potential high efficiency of the diode lasers. The phased-array approach has the advantage of parallel processing which guarantees scalability, but necessitates beam-transformation arrays of high quality and efficiency.

List of Symbols

a	Resonator mirror radius; fiber-core radius
A	Area of aperture, focal spot, etc.
$A(x, y, z)$	Complex phase factor
B	Brightness
c	Speed of light
C	Concentration factor (of radiation)
$E(t), E(x, y, z)$	Complex amplitude of electrical field
$E_{\text{sys}}(x, y), E_{\text{sys}}(\xi, \eta)$	Complex electrical-field distribution of the system in the far field/focal plane
$E^{\max}, \hat{E}$	Normalized/peak amplitude of electrical field
$\tilde{E}(\tilde{\nu}_x, \tilde{\nu}_y)$	Fourier-transformed E-field in spatial frequencies
f	Focal length
F	(linear) fill factor

$\hbar = h/2\pi$	Reduced Planck constant
I	Power density, laser intensity
I_0	One-dimensional intensity distribution
I_{tot}	Intensity distribution of total beam
k	Wavenumber
Δk_i	Deviation (RMS) of wavevector component
K	Contrast
l_c	Coherence length
$J_1(u)$	bessel function of the first kind of order one
L	Resonator length; thickness of phase plate
M^2, M_x^2, M_y^2	Times-diffraction-limit factor (in x or y direction)
mfd	Mode field diameter
n	Index of refraction
$n_{\text{core}}, n_{\text{cladding}}$	Index of refraction of core/cladding
N	Number of lasers, beams, lenses, etc. of an array
NA	Numerical aperture
$\text{NA}_{\text{beam}}, \text{NA}_{\text{fiber}}$	NA of beam/fiber as defined by manufacturers
N_{F}	Fresnel number
OPD	Optical path difference
P	(optical) power
δP_{M}	Injected master-laser power
P_0	Total power of I_0
P_{tot}	Power of total beam
δP_{S}	Injected slave-laser power
$P^{\text{enc,norm}}(r_{\text{A}})$	Normalized encircled power within radius r_{A}
p	Momentum
p_i	Generalized momentum coordinate
Δp_i	Deviation (RMS, root mean square) of p_i
q_i	Generalized room coordinate
Δq_i	Deviation (RMS, root mean square) of q_i
Q	Beam-parameter product
$r, \boldsymbol{r}$	Distance/unit vector between two room points
R	Reflectance
r_{e}	Resonator loss rate
T	Transmittance
w_0	Beam-waist radius (measured at $1/\text{e}^2$)
V, V_{c}	Normalized frequency (at cut-off)
x, y, z	Room coordinates
$z_{\text{A}}(r_{\text{A}}), z_{\text{B}}(r_{\text{B}})$	Profile of entrance/exit surface of phase plate
α	α factor
α	Half divergence angle of a diode-laser beam after collimation reduced by the diffraction-limited part
β	Misalignment angle of collimating lens
γ	Degree of coherence

γ_{ii}/γ_{ij}	Normalized auto-/cross-correlation function
$\Gamma_{ii}(\tau), \Gamma_{ij}(\tau)$	Auto-/cross-correlation function
φ_j	Phase angle of beam j
η_{comb}	Efficiency of combining two beams
η_{cascade}	Efficiency of combining all N beams of the cascade
η_{FAC}	Quality of fast-axis collimation
η_{FF}	Fill-factor quality
$\eta_P L$	Power transmission of micro-lens array
η_{StQ}	Stacking quality
λ	Wavelength
λ_{i}	Filter cut-off/cut-on wavelength
λ_{c}	Cut-off wavelength of a single-mode fiber
$\Delta \lambda$	Difference between cut-on and cut-off wavelengths
ν, ν_0	Frequency, center frequency
$\nu_{\mathrm{M}}, \nu_{\mathrm{S}}$	Master/slave frequency
$\Delta \nu$	Full linewidth measured at half maximum (FWHM)
$\Delta \nu_{\mathrm{G}}, \Delta \nu_{\mathrm{L}}$	fWHM of Gaussian/Lorentzian line
$\Delta \nu_{\mathrm{mode}}$	Frequency difference of two modes
$\Delta_{\mathrm{Lock}}, \Delta_{\mathrm{Lock}}^{\mathrm{DL}}$	Locking range (of diode laser)
$\tilde{\nu}_x, \tilde{\nu}_y$	Spatial frequencies
θ	Full-beam divergence angle (measured at $1/e^2$)
$\theta, \theta_x, \theta_y$	Angle (far field)
$\theta_{\mathrm{s}}, \theta_{\mathrm{p}}$	Far-field divergence angle perpendicular/parallel to p-n junction
Θ_{c}	Collimation angle
$\Theta_{\mathrm{totrefl}}$	Angle of total reflection
Θ_1, Θ_2	Max. acceptance angle of entrance/exit aperture
σ_x	Second-order moment of one-dimensional power-density distribution
τ	Time difference; lifetime
τ_{c}	Coherence time
τ_{R}	Photon lifetime ($1/e$) in resonator
$\Omega, \Delta \Omega$	Phase space, volume element in phase space
ξ, η	Room coordinates of the focal plane
$\Delta \xi$	Shift of central peak in the focal plane

References

1. K. J. Ebeling: *Integrierte Optoelektronik* (Springer, Berlin, Heidelberg 1992)
2. L. Bartelt-Berger, U. Becker, U. Brauch, C. Fleig, A. Giesen, B. Luecke, H. Opower, C. Schomburg, M. Schubert, R. Springer: Systems of fiber-coupled diode lasers as versatile sources of high-brightness radiation, SPIE Proc. **3110**, 310 (1997)
3. P. Albers, H. J. Heimbeck, E. Langenbach: Focusing of Diode Lasers for High Beam Quality in High-Power Applications, SPIE Proc. **1780**, 533 (1993)

4. A. Clarkson, D. C. Hanna: Two-Mirror Beam Shaping Technique for High-Power Diode Bars, Opt. Lett. **21**, 375 (1996)
5. W. Chen, C. S. Roychoudrhuri, C. M. Banas: Design approaches for laser-diode material-processing systems using fibers and micro-optics, Opt. Eng. **33**, 3662 (1994)
6. M. Schubert: Leistungsskalierbares Lasersystem aus fasergekoppelten Grundmode-Diodenlasern. Dissertation, Universität Stuttgart, Stuttgart (1999)
7. S. O'Brien, A. Schoenfelder, R. J. Lang: 5 W CW diffraction-limited InGaAs broad - area flared amplifier at 970 nm, IEEE Photon. Techn. Lett. **9**, 1217 (1997)
8. D. Botez: Monolithic phase-locked semiconductor laser arrays, in D. Botez, D. R. Scifres (eds.): *Diode Laser Arrays*, (Cambridge Univ. Press, Cambridge, UK 1994)
9. L. Bartelt-Berger: Lasersystem aus kohärent gekoppelten Grundmode-Dioden-Lasern. Dissertation, Universität Stuttgart, Stuttgart (1998)
10. R. Beach, W. J. Benett, B. L. Freitas, D. Mundinger, B. J. Comaskey, R. W. Solarz, M. A. Emanuel: Modular microchannel-cooled heat sink for high average power laser-diode array, IEEE J. QE **28**, 966–976 (1992)
11. P. Loosen, T. Ebert, J. Jandeleit. H.-G. Treusch: Packing high-power diode-laser bars using micro-channel copper mounts, Proc. LEOS '97, Orlando, FL, IEEE CH36243, 434–435 (1999)
12. ISO/DIS-Standard 11 146, International Organization for Standardization, Geneva, Switzerland (http://www.iso.ch/) (1999)
13. A. E. Siegman: New Developments in Laser Resonators, SPIE Proc. **1224** (1990), 2
14. M. Born, E. Wolf: *Principles of Optics* (Pergamon, New York 1980) p. 189
15. P. Albers, H. J. Heimbeck, E. Langenbach: IEEE J. QE **28**, 1088–1100 (1992)
16. J. Endriz: Brightness Conserving Optical System for Modifying Beam Symmetry, United States Patent 5,168,401 (1992)
17. B. Ehlers, K. Du, M. Baumann, H.-G. Treusch, P. Loosen, R. Poprawe: SPIE Proc. **3097**, 639–644 (1997)
18. Product information, Rofin-Sinar, Hamburg/Dilas, Mainz (http://www.rofin-sinar.com) (1999)
19. Product information, Laserline, Koblenz, Germany (http://www.laserline.de) (1999)
20. S. Heineman, L. Leiniger: SPIE Proc. **3267**, 116–124 (1998)
21. B. De Odorico, C. Hewing: Feinwerktech. Messtech. **105/106**, 451 (1997) (in German)
22. V. Sturm, H.-G. Treusch, P. Loosen: SPIE Proc. **3097**, 717–726 (1997)
23. P. Loosen, G. Treusch, C. R. Haas, U. Gardenier: SPIE Proc. **2382**, 78–88 (1995)
24. J. Biesenbach, P. Loosen, H. G. Treusch, V. Krause, A. Kösters, S. Zamel, W. Hilgers: SPIE Proc. **2263**, 152–163 (1994)
25. Product information, Limo Corp./Dortmund, Germany (http://www.limo.de) (1999)
26. D. Hänsch, H. Pütz, H. G. Treusch, A. Gillner, R. Poprawe: welding of plastics with diode lasers, in Proc. Int. Conf. ICALEO '98, Orlano, FL 1998 (Laser Institute of America, Orlando, FL) in press

27. C. Dawes: *Laser Welding* (Woodhead, Abington, Cambridge, UK 1992); Fraunhofer LT: Deep Penetration Welding with High-Power Diode-Lasers, Laser Magazin **4**, 31 (1999), in German

28. L. J. Mawst, J. L. Bhattacharya, D. Botez: 8 W continuous wave front-facet power from broad waveguide Al-free 980 nm diode lasers, Appl. Phys. Lett. **69**, 1532 (1996)

29. J. W. Tomm, A. Baerwolff, R. Puchert, A. Jaeger, T. Elsaesser: Optical probes as tools for the investigation of aging properties of high-power laser diode arrays, SPIE Proc. **3244**, 576–586 (1997)

30. S. Ramanujan, H. G.Winful: Spontaneous emission induced filamentation in flared amplifers, IEEE J. QE **32**, 784 (1996)

31. O. Hess, T. Kuhn: Spatio-temporal dynamics of semiconductor lasers: theory, modeling and analysis, Prog. Quant. Electron. **20**, 85 (1996)

32. D. J. Bossert, G. C. Dente, M. L. Tilton: Beam quality in multiwatt semiconductor amplifiers, SPIE Proc. **3284**, 63–71 (1998)

33. M. Pessa, J. Nappi, P. Savolainen, A. Ovtchinnikov, M. Toivonen, R. F. Murison, H. M. Asonen: State-of-the-art aluminum-free 980-nm laser diodes, SPIE Proc. **2682**, 161–168 (1996)

34. V. V. Wong, S. D. DeMars, A. Schoenfelder, R. J. Lang: Angled grating distributed feedback laser with 1.2 W cw single-mode, diffraction-limited output at 1.06 μm, OSA Tech. Dig. Ser. **12** (Optical Society of America, Washington, DC 1998) paper CMJ1

35. D. L. Cunningham, , R. D. Jacobs: Commercial applications of high-powered laser diodes. SPIE Proc. **2382**, 72–77 (1995)

36. A. Rabl: Comparison of solar concentrators. Solar Energy **18,** 93 (1976)

37. J. R. Leger: Microoptical components applied to incoherent and coherent laser arrays. in D. Botez, D. R. Scifres (Eds.): *Diode Laser Arrays*, (Cambridge Univ. Press 1994) pp. 123–179

38. W. Lauterborn, T. Kurz, M. Wiesenfeldt: *Kohärente Optik* (Springer, Berlin, Heidelberg 1993)

39. A. E. Siegman: *Lasers* (University Science Books, Mill Valley, CA 1986)

40. W. Koechner: *Solid-State Laser Engineering*, Springer Ser. Opt. Sci. **1**, 5th edn. (Springer, Berlin, Heidelberg 1999)

41. J. Buus: *Single Frequency Semiconductor Lasers* Tutorial Texts Opt. Engin. **5** (SPIE Optical Engineering, Bellingham, WA 1991)

42. M. Ohtsu: *Highly Coherent Semiconductor Lasers* (Artech House, Norwood, MA 1992)

43. A. Yariv: *Optical Electronics,* 4th edn. (Harcourt Brace Jovanovich, Philadelphia 1991)

44. L. B. Jeunhomme: *Single-Mode Fiber Optics,* 2nd edn. Opt. Engin. **4** (Dekker, New York 1990)

45. C. K. Kao: *Optical Fiber* (Pelegrinus, London 1988)

46. S. S. Orlov, A. Yariv, S. Van Essen: Coupled-mode analysis of fiber-optic add-drop filters for dense wavelength-division multiplexing, Opt. Lett. **22**, 688 (1997)

47. J. Franz, V. K. Jain: *Optical Communication Systems* (Wiley Narosa, New Delhi 1996)

48. N. W. Carlson: *Monolithic Diode-Laser Arrays.* Springer Ser. Electron. Photon. **33** (Springer, Berlin, Heidelberg 1994)

49. L. J. Mawst, D. Botez: High-power coherent sources based on antiguided structures, SPIE Proc. **2397**, 526–533 (1995)

50. D. Botez: High-power monolithic phase-locked arrays of antiguided semiconductor diode lasers, IEEE Proc. J. **139**, 14–23 (1992)

51. D. Botez, T. Holcomb: Bloch-function analysis of resonant arrays of antiguided diode lasers, Appl. Phys. Lett. **60**, 539–541 (1992)

52. J. P. Hohimer, G. R. Hadley, D. C. Craft, T. H. Shiau, S. Sun: Stable-mode operation of leaky-mode diode laser arrays at high pulsed and CW currents. App. Phys. Lett. **58**, 452–454 (1991)

53. D. Mehuys, W. G. Streifer, R. G. Waarts, D. F. Welch: Modal analysis of linear Talbot-cavity semiconductor lasers. Opt. Lett. **16**, 823 (1991)

54. R. H. Rediker, K. A. Rauschenbach, R. P. Schloss: Operation of a coherent ensemble of five diode lasers in an external cavity, IEEE J. QE **27**, 1582–1593 (1991)

55. J. S. Osinski, D. Mehuys, D. F. Welch, R. G. Waarts, J. S. Major Jr., K. M. Dzurko, R. J. Lang: Phased array of high-power coherent, monolithic flared amplifier master oscillator power amplifiers, Appl. Phys. Lett. **66**, 556 (1995)

56. J. Levy, K. Roh: Coherent array of 900 semiconductor laser amplifiers, SPIE Proc. **2382** (1995)

57. R. Lang: Injection locking properties of a semiconductor laser, IEEE J. QE **18**, 976 (1982)

58. I. Petitbon, P. Gallion, G. Debarge, C. Chabran: Locking bandwidth and relaxation oscillations of an injection-locked semiconductor laser, IEEE J. QE **24**, 148 (1988)

59. M. Tempus, W. Lüthy, H. P. Weber: Coherent recombination of laser beams with interferometrical phase control, Appl. Phys. B **56**, 79 (1993)

60. W. M. Neubert, K. H. Kudielka, W. R. Leeb, A. L. Scholz: Experimental demonstration of an optical phased array antenna for laser space communications, Appl. Opt. **33**, 3820 (1994)

61. L. Berger, U. Brauch, A. Giesen, H. Hügel, H. Opower, M. Schubert, K. Wittig: Coherent fiber coupling of laser diodes, SPIE Proc. **2682**, 39 (1996)

62. B. R. Frieden: Lossless conversion of a plane laser wave to a plane wave of uniform irradiance, Appl. Opt. **4**, 1400 (1965)

63. E. Hecht, A. Zajac: *Optics* (Addison-Wesley, Reading 1974)

64. S. Erhard: Institut für Strahlwerkzeuge, Universität Stuttgart; private communication

65. J. R. Leger, G. J. Swanson, M. Holz: Efficient side lobe suppression of laser diode arrays, Appl. Phys. Lett. **50**, 1044 (1987)

66. P. Ehbets, H. P. Herzig, R. Dändliker, P. Regnault, I. Kjelberg: Beam shaping of high-power laser diode arrays by continuous surface-relief elements, J. Mod. Opt. **40**, 637 (1993)

67. G. L. Schuster, J. R. Andrews: Coherent beam combining: optical loss effects on power scaling, Appl. Opt. **34**, 6801 (1995)

68. B. Lücke, G. Hergenhan, U. Brauch, M. Scholl, A. Giesen, H. Opower, H. Hügel: Autostable Injection-Locking of a 4 × 4 VCSEL-Array with on Chip Master Laser, SPIE **3946**, (2000)

New Concepts for Diode-Pumped
Solid-State Lasers

Andreas Tünnermann[1], Holger Zellmer[1], Wolfram Schöne[2], Adolf Giesen[3],
and Karsten Contag[4]

[1] Friedrich-Schiller-Universität Jena, Institut für Angewandte Physik,
Max-Wien-Platz 1, D-07743 Jena, Germany
tuennermann@iap.uni-jena.de
[2] LASOS Laser-Fertigung GmbH,
Carl-Zeiss-Promenade 10, D-07745 Jena, Germany
lasos@zeiss.de
[3] Universität Stuttgart, Insitut für Strahlwerkzeuge,
Pfaffenwaldring 43, D-70569 Stuttgart, Germany
giesen@ifsw.uni-stuttgart.de
[4] Forschungsgesellschaft für Strahlwerkzeuge mbH,
Nobelstrasse 15, D-70569 Stuttgart, Germany
contag@fgsw.uni-stuttgart.de

Abstract. Solid-state lasers are attractive sources of coherent radiation for various scientific and technological applications. But the different fields of applications increasingly demand more powerful, efficient, and rugged lasers with higher beam quality. Hence, at present a new generation of laser systems, based on diode lasers, has begun to dominate. Through replacement of the discharge lamps formerly used as pump sources by diode lasers, the inherent advantages of the latter regarding compactness, efficiency, and reliability are directly transferred to the laser system as a whole. Moreover, the diode lasers allow for the realization of completely new laser concepts, and as a consequence, for the demonstration of laser output parameters that were not accessible so far. After a short introduction to the field of solid-state lasers, we will focus on two very promising examples of these new concepts, namely the fiber laser and the thin-disk laser, which have attracted much attention recently.

1 Fundamental Concepts of Diode-Pumped Solid-State Lasers

The most widely used solid-state laser material today is Neodymium-doped Yttrium Aluminum Garnet (Nd:YAG), operated at a laser wavelength of $1.06\,\mu m$. In addition to this, a number of different laser materials have been developed during the last few years in order to serve various scientific and technological applications, especially in medicine and metrology. The emission wavelengths of these materials are located in the near infrared (Table 1). In the case that visible radiation is required, the infrared laser emission has to be converted using nonlinear techniques like Second-Harmonic Generation (SHG). Although some ions offer possible laser transitions in the visible, the

R. Diehl (Ed.): High-Power Diode Lasers, Topics Appl. Phys. **78**, 369–408 (2000)
© Springer-Verlag Berlin Heidelberg 2000

operation of such lasers has been hampered on the one hand by unfavorable laser parameters and on the other by the lack of suitable pump sources, as in the usual laser scheme the pump wavelength has to be shorter than the laser wavelength. Only recently have new laser concepts made possible the realization and exploitation of more-complex laser schemes, opening the route to direct visible laser emission (Sect. 2.5).

Table 1. Selected solid-state laser materials with corresponding laser and pump wavelengths

Solid-state laser material	Laser wavelength (μm)	Pump wavelength (μm)
Yb:YAG[a]	1.03	0.94
Nd:YAG[a]	1.06, 1.32, 1.44	0.81
Tm:YAG[a], Tm:YLF[b]	2.1	0.78, 0.80
Er:YAG[a], Er:YLF[b], Er:BYF[c]	3.0	0.79, 0.97

[a] YAG: Yttrium-Aluminum-Garnet, [b] YLF: Yttrium-Lithium-Fluoride, [c] BYF: Barium-Yttrium-Fluoride

The typical solid-state laser comprises a laser-active medium with the shape of a rod. This geometry is most easily manufactured and gives the output beam a generic rotational symmetry, which is often favorable. Other geometries like rectangular slabs are used if special laser parameters have to be realized or special characteristics of the gain medium are required. In general, the majority of solid-state lasers for industrial applications are optically excited by noble-gas arc lamps. Mounting the laser rod and the arc lamp together in a reflecting chamber gives the simple and rugged setup shown in Fig. 1. Unfortunately, the broadband spectral emission of arc lamps only poorly matches the narrow absorption lines typical for the active ions in solid-state laser media. Therefore, only a small fraction of the absorbed pump power is transferred to the upper laser level. In addition to efficiency problems, a strong heating of the laser material results, which decreases beam quality and attainable output power. Typically, less than 3% of the electrical input power to these laser systems is converted into laser radiation [1].

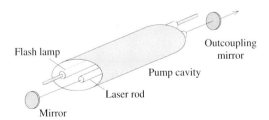

Fig. 1. Conventional lamp-pumped solid-state laser

Compared to arc lamps, diode lasers with their efficiencies of more than 50% are ideal pump sources for solid-state lasers. Becoming available in quantities during the last decade, powerful and reliable diode lasers triggered the development of a wide spectrum of new concepts and geometries for solid-state lasers. The narrow spectral emission of diode lasers can be tuned in order to match exactly the pump transition of the laser ion, thus minimizing the heat production inside the solid-state medium. This leads to higher efficiencies and beam qualities. Moreover, the lifetime of diode lasers exceeds 10 000 h – to be compared to less than 1000 h for the lifetime of an arc lamp. This enables a reliable and maintenance-free long-term operation of the overall laser system.

Another characteristic feature of diode lasers is the directionality of their emission. In contrast to lamps the diode-laser radiation can be focused rather easily to small spots. This gives access to very high pump intensities, which in turn allows us to operate new laser transitions in practical, efficient devices. Moreover it is possible to selectively excite the region of the laser-active medium which is filled by the laser mode. This good overlap between the pump volume and the resonator's mode volume results in very high efficiencies close to the theoretical limit established by the quantum defect, which is the energetic difference between pump and laser photons. In addition, a carefully designed system operates in the lowest-order transverse resonator mode, which gives the optimum beam quality. A typical example for this type of laser geometry is shown in Fig. 2a. Numerous investigations have been carried out in order to optimize the transfer of pump power into the laser mode volume, taking into account the characteristics of the laser material and of the pump diodes [2,3,4,5].

Fig. 2b shows a side-pumping setup, which is favorable in some cases, especially in high-average-power systems, where the necessary amount of pump power cannot be delivered through the small end faces of the laser crystal. Within the limits discussed in the following section, side pumping permits easy scaling of the output power by increasing the length of the laser medium.

As a consequence of their combination of characteristics, diode lasers are highly attractive pump sources for solid-state lasers. In addition to improving existing setups, they allow for the realization of completely new laser concepts. Examples are fiber lasers and thin-disk lasers, which will be discussed in detail later on in this chapter.

1.1 Thermal Considerations

The optical excitation process in solid-state lasers inevitably produces heat inside the gain material due to non-radiative transitions. Depending on the given experimental conditions, lamp pumping generates up to four times more heat than diode pumping at the same level of power delivered to the upper laser level [6], but even in the case of excitation by diode lasers the heat load is not zero. Usually this is described by the fractional thermal load, which is

(a)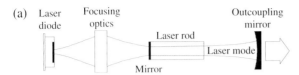

(b)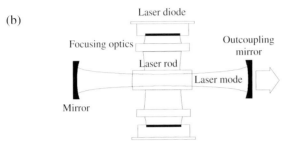

Fig. 2. Diode-pumped solid-state lasers with (**a**) longitudinal pumping and (**b**) side pumping

the ratio of generated heat power to the absorbed optical pump power. For example, if the widely used laser material Nd:YAG is pumped by diode lasers at 808 nm, this quantity has a value of about 0.4 as long as no laser power is extracted. Under conditions of laser extraction its value drops to about 0.3 [7,8].

It is exactly this heat load which usually limits the performance of solid-state lasers in terms of output power and beam quality. Heat generation takes place in the volume of the gain material, while cooling is performed at the edges, either by a flow of coolant or by conduction to a heat sink. This situation leads to a non uniform temperature distribution inside the pumped laser medium and consequently to mechanical stress. Hence the ultimate limit for the amount of pump power that can be applied to a piece of laser material and therefore for the attainable output power is given by stress fracture.

In the conventional end-pumping geometry (Fig. 2a), the highest tensile stress occurs at the pumped end face. For Nd:YAG, the obtainable output power from a rod end-pumped through both end faces has been estimated to be about 60 W [9]. This limit can be pushed to higher values by using rods with undoped end caps, thus removing the heat from the physical end faces [10,11]. Higher output powers can be realized by incorporating more than one rod into the resonator, at the expense of increased complexity, or by using long rods with low absorption, which prohibits the selective excitation of the transversal fundamental mode in most cases.

Applying more pump power is easier in side-pumping setups, just by increasing the rod length while keeping the pump-power density constant. Nd:YAG rods can tolerate pump powers on the order of 300 W/cm of length in safe operation [12,13]; larger values probably lead to crystal damage by tensile stress at the rod circumference. With these pump powers output powers exceeding 1 kW out of a single laser crystal are possible.

But mechanical fracture is not the only thermally induced effect to be considered in solid-state laser design. In general, the refractive index of the laser medium is dependent on temperature. Thus the temperature profile leads to a corresponding refractive-index profile inside the gain medium, which therefore has a lensing effect on the laser radiation. Especially in end-pumped systems the deformation of the end faces also contributes to this thermally induced lens. Under the assumption of temperature-independent heat conductivity, a homogeneously heated long rod cooled at its circumference has a parabolic profile of its refractive index [1]. The resulting lensing effect can in principle easily be compensated for with standard optical components inside the resonator, but only for a fixed pump-power, because thermal lensing of the rod is pump-power dependent. In real laser devices the variation of heat conductivity with temperature and a spatially non-uniform distribution of absorbed pump power lead to aberrations of the produced lens, which severely affect extraction efficiency and output beam quality [9,14]. The strong absorption of diode-laser pump radiation can result in very inhomogeneous pump-radiation distributions. Therefore, in order to fully exploit the advantage of generating less heat compared to lamp-pumping, attention has to be paid to the distribution of absorbed pump power [12].

Another thermal effect degrading the laser performance is stress-induced birefringence, which results in depolarization and bifocusing [1]. Depolarization causes losses in the presence of polarizing elements within the resonator. The bifocusing effect has been shown to limit the achievable output power in transversal fundamental-mode operation from rod lasers [15]. Up to 80 W of CW output power in fundamental-mode operation from a single Nd:YAG laser rod has been demonstrated experimentally [16]. A possibility to avoid the limitation due to bifocusing may be the use of laser materials in which the thermally induced birefringence is only a small perturbation to their natural birefringence [15]. Unfortunately, these materials tend to have other drawbacks like low thermal conductivity or low stress-fracture resistance which limit their use in high-power applications.

The thermal effects discussed so far make thermal management an essential part of solid-state laser design. In order to effectively remove heat from the gain material, a high ratio of cooled surface to pumped volume is favorable [34]. Starting from a conventional rod cooled at the circumference, this can be achieved by making it very long and thin, resulting in a fiber geometry. It can also be realized by making it very short and cooling the end faces, which leads to a thin-disk design. These are two promising new laser concepts which recently attracted much attention. They will be treated in the remainder of this review.

2   Fiber Lasers

Fiber lasers offer the possibility to overcome the limits in scaling the output power of solid-state lasers while maintaining the beam quality [17,18]. The beam quality of the resulting laser beam is determined by the refractive-index profile of the fiber which in turn is defined by the geometrical dimensions and numerical aperture of the active-waveguide structure. Independent of external influences, the laser oscillates in the transverse fundamental mode. This means that compared to conventional (even diode-pumped) solid-state lasers, fiber lasers are completely immune to thermo-optical effects. Effects such as thermally induced lens formation and stress-induced birefringence in the active region resulting in a degradation of the beam quality as a function of the pumping power and reduced efficiency are not observed even at high output powers. Furthermore, for fiber lasers the thermal load caused by the pumping process is spread over a longer region as well, as due to the greater surface-to-volume ratio, heat is more efficiently removed, and thus the observed temperature increase in the laser core in comparison to solid-state diode-pumped lasers is small [19]. Therefore the reduction of the quantum efficiency of the active medium with increasing temperature during laser operation plays a subordinate role in fiber-laser geometries.

Since the length of the active medium needed is determined by the pump absorption requirements when designing an end-pumped bulk laser device, the pump spot size and therefore the pump intensity is limited by the confocal parameter of the pump radiation. This situation is somewhat different in a waveguide device. Pump radiation which is launched within the fiber core continues to propagate down the fiber until it is completely absorbed while maintaining the spot size diameter, resulting in large values of the pump intensity times interaction-length product. Due to the high intensity in the active core and the large interaction length of pump radiation and laser radiation, multiple laser transitions are observed in active fibers. Weak absorption bands and weak or three-level transitions can be exploited in CW laser operation, which in general cannot easily be achieved in solid-state lasers. Even visible and ultraviolet laser transitions have been operated by producing an efficient population and inversion of energetic highly excited states in active ions through the absorption of two or more low-energy photons. Near-infrared pumping radiation can therefore directly be converted into visible laser radiation (up-conversion laser process, [20]). Table 2 shows a selection of the most-important laser transitions in the mid-infrared to ultraviolet spectral range [21,22,23].

Gain values which can be achieved by fiber lasers for single transitions in the active region are in the 30 dB range. So far the maximum achievable output power in laser operation is limited by the available pump power. For diode-pumped monomode fiber lasers the beam quality and output power of the diode laser limits the power scalability to a few hundred milliwatts.

Table 2. The emission spectrum of fiber lasers ranges from the mid infrared to the ultraviolet

Wavelength (μm)	Element	Wavelength (μm)	Element	Wavelength (μm)	Element
3.4	Er	1.31	Pr	0.715	Pr
2.9	Ho	1.2	Ho	0.695	Pr
2.75	Er	1.08	Pr	0.651	Sm
2.3	Tm	1.06	Nd	0.635	Pr
2.04	Ho	1.04	Yb	0.61	Pr
1.9	Tm	0.98	Er	0.55	Ho
1.72	Er	0.975	Yb	0.546	Er
1.66	Er	0.94	Nd	0.52	Pr
1.55	Er	0.91	Pr	0.491	Pr
1.47	Tm	0.88	Pr	0.48	Tm
1.38	Ho	0.85	Er	0.455	Tm
1.34	Nd	0.8	Tm	0.38	Nd

2.1 Laser-Active Waveguides

In contrast to conventional solid-state lasers in which at least a free optical beam path is formed in the laser resonator, the beam formation and guiding in a fiber-laser device is realized in optical waveguides. In general, these waveguides are based on rare-earth-doped optical dielectric materials. Materials used such as silica, phosphate glass and fluoride glass display attenuation of about 10 dB/km, many orders of magnitude less than in solid-state laser crystals. Both the absorption bands as well as the emission bands of the rare-earth ions display a spectral broadening compared to crystalline solid-state laser materials due to the interaction with the glass matrix. This reduces the frequency stability and bandwidth requirements of the pump-light source. Thus, wavelength selection of suitable diode-laser pump sources, and even temperature stabilization of the pumping source is not required in general.

Figure 3 shows the structure of an active fiber. The fiber consists of a rare-earth-doped active core with a refractive index n_1, generally surrounded by a pure silica-glass cladding of refractive index $n_2 < n_1$ so that, based on total internal reflection on the interface surface between the core and the cladding, waveguiding takes place in the core. The core of the fiber laser is both the active medium and the waveguide for pump radiation and generated laser radiation. The complete fiber is protected against external influences by a polymer coating.

The beam quality of the fiber laser is determined only by the given optical characteristics of the refractive-index profile of the waveguide. If the fiber core satisfies the conditions for the dimensionless fiber parameter V as follows:

$$V = \frac{2\pi a}{\lambda}\sqrt{n_1^2 - n_2^2} = \frac{2\pi a}{\lambda}\text{NA} < 2.40, \tag{1}$$

where a is the radius of the core and λ the wavelength of the laser radiation, only the transversal fundamental mode can propagate through the fiber. For

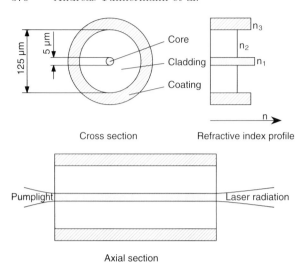

Cross section   Refractive index profile

Axial section

Fig. 3. Geometry of a single-clad laser fiber. The rare-earth-doped active core is surrounded by a cladding of undoped fused silica. A polymer coating is used to protect the fiber from environmental influences

these fibers the term monomodal or single-mode fiber is used. The diameter of the core is typically 3 to 8 μm. For larger core diameters higher-order transversal modes can also occur in which case the term multimodal fiber is used. The numerical aperture NA defines the sine value of the angle to the fiber axis up to which radiation is being coupled to the fiber. The number of modes Z propagating in the fiber is approximated for large values of the fiber parameter V according to $Z = V^2/2$ [24]. To reduce optical spreading of the modes in the coating of the fiber, the coating has to have a higher refractive index $n_3 > n_2$.

Commonly, in order to optically excite the fiber laser, the pump radiation is coupled through the fiber facet into the laser core. However, in the case of longitudinal pumping the pump radiation has to be coupled to the waveguide which has dimensions of a few microns. Therefore highly brilliant pump radiation sources are required to excite monomodal fibers, which are so far limited to output powers of about 1 W. To scale up the pump power one needs a larger fiber diameter that is adapted to the beam-parameter product of a high-power diode laser-array. However, the enlarged active core of the fiber allows higher transversal mode oscillations, resulting in reduced beam quality. An alternative is to use separate cores for pumping and lasing, the so-called double-clad design described below [25]. With this concept the operation of a fiber laser with an output power of several tens of watts was possible for the first time [26,27].

2.2   Double-Clad Fiber Lasers

In the case of double-clad fibers the pump radiation is not directly launched into the active core but into a surrounding multimode core known as the pump core. In order to realize the pump core which also acts as cladding for the active core with optical-waveguide characteristics, the surrounding coating must have a smaller refractive index. Normally a fluorine-doped silica glass or a highly transparent polymer with a low refractive index such as silicone or Teflon is used. The diameter of the pump core is typically a few hundred microns and its numerical aperture $NA \cong 0.3$ to 0.7 (Fig. 4).

The pump radiation which is launched into the pump core is coupled into the laser core over the entire fiber length and is absorbed there by the rare-earth ions, so the upper laser levels are excited. Using this technique the multimodal pumping radiation from the high-power diode lasers can be effectively converted into laser radiation with excellent beam quality.

The efficiency of a fiber laser is strongly dependent on the coupling of the pump radiation from the pump core to the laser core. In conventional double-clad fibers with centered laser cores and circular pump cores the pump radiation is only partially absorbed, and a large portion is transmitted independent of the fiber length and doping concentration. The proportion of the unused pumping power is determined only by the geometry of the double-core fiber. In general, for fibers with $100\,\mu m$ pump core diameter and $5\,\mu m$ laser core diameter, only about 30% of the pump radiation can be absorbed.

In order to improve the pump-radiation absorption and thus achieve a more-effective laser operation, the propagation of the pump radiation in the

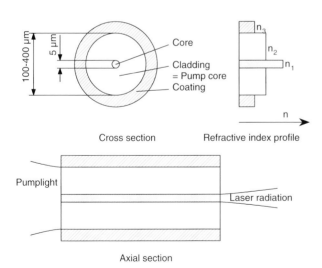

Fig. 4. Double-clad fiber laser. The active single-mode core is surrounded by a pump core with dimensions of a few hundred microns and a numerical aperture in the range 0.3 to 0.7

pump core must be carefully considered. Thus modeling and optimization of the absorption process has been a main area of interest in high-power fiber laser studies in recent years. In the following section absorption characteristics will be discussed briefly.

2.3   Pump-Radiation Absorption in Double-Clad Fibers

To describe the pump-radiation propagation and absorption in double-clad fiber lasers, two different methods can be applied. The *wave-optical approach* is based on solutions of the wave equation in the form of Bessel functions for cylindrically symmetric waveguides. The *ray-propagation method* is based on geometrical optics and uses ray-tracing algorithms; within this model pump-radiation absorption in fibers can be determined regardless of the waveguide geometry.

In the wave optical-picture, the field distribution can be determined in cylindrical optical waveguides by solving the wave equation. For a specific mode LP_{lm} with mode numbers l and m using cylindrical coordinates r and φ the radial distribution of the electrical-field amplitude E_{lm} of the fiber mode is described by Bessel functions (J_l, K_l) and the azimuthal distribution by sine functions with phase δ:

$$E_{1,m}^{core} = a_1 \sin(l\varphi + \delta) J_l(\gamma r) \ ,$$
$$E_{1,m}^{cladding} = a_2 \sin(l\varphi + \delta) K_l(\gamma r) \ , \tag{2}$$

Using an approximation (similar to the Wentzel–Kramers–Brillouin approximation known in quantum mechanics) it is possible to obtain the missing constants a_1 and a_2 as well as the spreading constant $\gamma = \gamma(m, l, V)$ for a defined fiber. The intensity of the individual modes is obtained by the square of the field strength and is depicted in Fig. 5 for some low-order LP modes. It can easily be seen that fiber modes with azimuthal mode number $l > 0$ (by far the majority of fiber modes) show no intensity on the fiber axis (cf. helical radiation). Double-clad fibers with centrally arranged laser cores therefore show reduced optical efficiencies.

In order to perform a more quantitative analysis of the pump-radiation absorption, the overlap between the intensity of all modes induced in the fiber and the active core of the fiber must be determined. The absorption coefficient for each individual LP mode is obtained from the overlap of its intensity and the rare-earth distribution in the fiber. A disadvantage of this method to determine the pump-radiation absorption is, however, that it is only applicable to rectilinear fibers with circular pumping cores. This description is no longer applicable to the generally used coiled fibers.

An essentially more versatile method used to describe the pump-radiation absorption in double-clad fibers is obtained by applying ray-tracing methods to optical waveguides, which is valid as long as the dimensions of the optical waveguide are greater than the wavelength of the radiation.

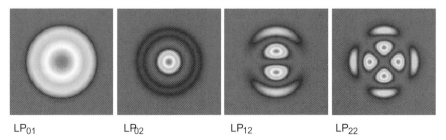

LP$_{01}$ LP$_{02}$ LP$_{12}$ LP$_{22}$

Fig. 5. Intensity distribution of low-order LP$_{lm}$ fiber modes. The azimuthal mode number l and the radial mode number m correspond to the number of minima in the intensity distribution in azimuthal and radial directions, respectively. With increasing radial mode number the intensity in the center of the fiber decreases. *Left*: the fundamental mode LP$_{01}$

From the ray-tracing point of view the pump radiation launched within a circular core can be divided into two categories: meridional rays and helical rays (Fig. 6). In principle only meridional rays passing through the center of the pumping core and thus through the laser core can be absorbed. Helical rays cannot fall below a certain inner radius r_i and thus cannot be absorbed. The limitations to fibers with circular pump-core cross sections do not apply to ray-tracing methods; all geometries can be simulated.

To improve the pump-radiation absorption and accordingly to increase the optical efficiency of double-clad fibers, two basically different concepts may be followed. Firstly, fiber designs can be developed which prevent the propagation of helical radiation. Helical radiation resulting from mode mixing is utilized in the second method.

A distinct improvement in the pump-radiation absorption especially for short fibers has been made possible by the development of new pump-core geometries which utilize a disturbance of the rotational symmetry of the fiber along the longitudinal axis (z axis) to hamper the propagation of helical rays. Rectangular and D-shaped pumping-core cross sections enforce a

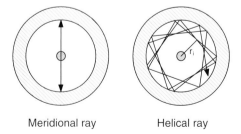

Meridional ray   Helical ray

Fig. 6. Meridional and helical rays in the pump cladding of a double-clad fiber. Helical rays cannot fall short of a certain inner radius r_i, and thus cannot be absorbed in the active core

chaotic spreading of the pump radiation and ensure that the pump radiation coupled into the pump core crosses the active core during propagation. Thus complete pump-radiation absorption occurs even within a short fiber length. The calculated pump-radiation absorption in such fibers with novel geometries is shown in Fig. 7. Fibers with similar cross-sectional areas were simulated in order to compare the different designs.

Absorption of helical radiation can also be achieved by geometries in which the active core is positioned decentered in the pump core rather than at its center. Such off-center cores then lie in a region crossed by helical radiation which can therefore be absorbed. The pump-radiation absorption of a conventional centralized double-clad fiber is also shown. The above-mentioned absorption saturation occurs within a short fiber length.

As an alternative to the costly manufacturing of fibers with non-round pumping cores a concept was developed by which a suitable mode mixing improves the pump-radiation absorption in circular fibers with centered pump cores. The curvature in the course of the fiber causes pump radiation from the outer regions of the pumping core to be added to that of the center core. A form shown to be especially favorable for pump-radiation absorption arises when the fiber is coiled in a kidney shape (Fig. 8). The successive fiber parts with different radii and directions of curvature give rise to an excellent pump-radiation absorption. Figure 9 shows the pump-radiation absorption for a kidney-shaped coiled fiber. Noticeable is the step-shaped trend of the absorption especially for small radii. This effect, which has also been experimentally observed, can be explained by the increased pump-radiation absorption directly following a change in the direction of curvature.

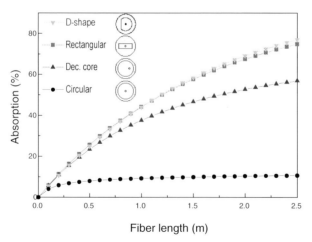

Fig. 7. Normalized pump-radiation absorption in double-clad fibers with novel geometries. The highest pump-radiation absorption for a given fiber length is observed in D-shaped or rectangular fibers. In double-clad fibers with centered active cores, only a small fraction of the pump radiation can be absorbed

Fig. 8. Double-clad fiber coiled in kidney shape in order to increase the pump-radiation absorption. The high intensity inside the core causes visible up-conversion fluorescence

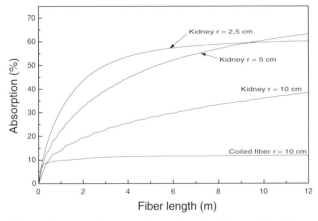

Fig. 9. Pump-radiation absorption in fibers with kidney-shaped curvature. The parameter r denotes the largest bending radius of the kidney

A disadvantage of the kidney-shaped coil geometry is that generally very long fibers ($80\,\text{m} - 100\,\text{m}$) are required for total pump-radiation absorption. Especially for three-level laser systems in which case reabsorption of the laser radiation occurs, long fibers are not acceptable, and so these geometries are reserved for low-loss fibers with four-level laser transitions. However, the fibers are easy to manufacture and show a compatibility with standard fiber-optical components.

2.4   High-Power Laser Operation

Fibers used in high-power experiments have silica pumping cores of diameter $400\,\mu\text{m}$ and numerical aperture $\text{NA} \cong 0.38$, so that a high-power diode-laser array with a beam-parameter product up to $80\,\text{mm}$ mrad can be coupled into the core. The laser core of the fiber has a diameter of 10 to 12 µm at a numerical aperture of NA ≅ 0.17. Therefore these fibers are slightly multimodal at an emission wavelength of about 1 µm. The fiber core is doped with approximately 1300 ppm (mol) Nd^{3+} ions as laser-active material. In order to increase the absorbed pump power, an 80 m-long central symmetrical fiber was coiled in kidney shape. A diode-laser module with 100 W output power at 810 nm wavelength was used as a pump source. Using this diode-laser module a pump power of 72 W was coupled into the pump core of the double-clad fiber. Figure 10 shows the experimental arrangement of the high-power fiber laser. The pump radiation of the diode laser is coupled into the core of the active fiber through the resonator mirror by means of coupling optics. On the output side of the fiber a highly reflective mirror is present to reflect non-absorbed pump radiation back into the fiber. For optimum laser output the transmission of the mirror at the laser wavelength is about 90%. (Due to the high amplification in rare-earth-doped fibers the optimum outcoupling is clearly higher than that of conventional solid-state lasers, and laser operation is even possible without feedback.)

From a double-clad fiber laser with optimized pump-radiation absorption in kidney-shaped coiled geometry a CW output power of 32.5 W for a launched power of 72 W was achieved. The optical efficiency as a function of the pumping power is about 46%. For the D-shaped fiber with the same pump-core diameter an output power of about 30 W at a significantly shorter fiber length is observed. Computer simulations indicate that comparable output powers can be expected from ytterbium-doped fibers with slightly increased overall efficiency. However, suitable commercial high-power pump-laser sources operating around 915 nm have become available only recently.

The beam quality is given by an $M^2 = 1.3$, corresponding to monomodal laser emission. Higher-order modes which are guided in the fiber due to its core geometry are effectively suppressed in laser operation due to higher losses compared to the fundamental mode. The linear power characteristic (Fig. 11) indicates that although the power density is well above 3 MW/cm^2, no saturation effects owing to nonlinearities or thermally induced effects occur. Hence for the fiber lasers used we can obtain linear power scaling to even higher output powers by increasing the pump power.

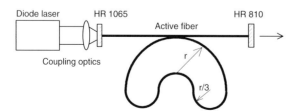

Fig. 10. Experimental setup of a high-power fiber laser

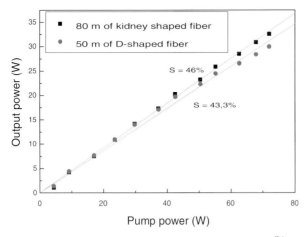

Fig. 11. Input–output curve of a high-power Nd^{3+}-doped fiber laser. The slope efficiency is 46%. The fiber length is 80 m

The spectral properties of fiber lasers are fundamentally different from those of conventional solid-state lasers. As a result of spectral broadening of the laser transition of the neodymium ions in silica host material the emission linewidth is typically 15 nm (Fig. 12). Fiber lasers are therefore attractive for novel applications in measuring techniques such as coherence radar. In addition, the spectral width prevents the onset of Stimulated Brillouin Scattering (SBS) which causes a limitation of the transferable power through monomodal fibers in narrow-band lasers, depending on the fiber length, to a few watts.

The limits of power scalability are defined by the damage threshold of the fiber facets and laser mirrors, as well as by Stimulated Raman Scattering

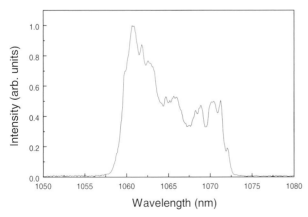

Fig. 12. Emission spectrum of a fiber laser operating at 20 W output power

(SRS). The damage threshold is approached at an output power of about 200 W for a single-modal core at 1 μm; the threshold for SRS is in the range between 200 W and 300 W, depending on the fiber length so that a scaling in the 100 W region for diffraction-limited output from a fiber device should be feasible.

2.5   Fiber-Laser Emission in the Visible Spectral Region

The high interaction length-intensity product, only attainable in fiber lasers, permits the realization of novel up-conversion-based laser schemes. By means of a stepwise absorption of the pump radiation, highly excited states can be occupied, so that laser emission occurs at short wavelengths, without the need to apply parametric optical methods [28]. Laser oscillation is achieved at room temperature for various rare-earth ions in a fluoride-glass matrix in the visible and ultraviolet spectral regions whereby in ZBLAN glass (ZrF_4-BaF_2-LaF_3-AlF_3-NaF) with a phonon energy of 580 cm^{-1}, in general the lowest lasing threshold, is observed. Common silica and phosphate glasses with phonon energies higher than 1000 cm^{-1} are not suitable as host materials for the manufacture of up-conversion lasers, as non-radiative relaxation processes limit the transition lifetimes of the rare-earth doping ions, and thus prevent an efficient multi-photon excitation.

   Using fluoride-glass fibers, all-solid-state laser systems with output powers up to some 100 mW in the visible spectral range have been demonstrated. Laser schemes in praseodymium, erbium and thulium are shown in Fig. 13. In comparison to erbium and thulium ions, whereby a multi-photon excitation of the upper laser level occurs for a pump radiation at 970 nm and 1122 nm, respectively [21,29,30], the excitation of praseodymium ions is realized by

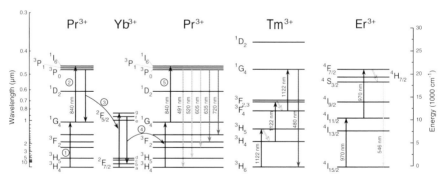

Fig. 13. Laser schemes for up-conversion lasers based on thulium, erbium and praseodymium/yttrium doping with emission wavelengths in the visible spectral region. Inversion results from multi-photon long-wavelength pump-radiation absorption. Laser emission occurs at 480 nm (Tm), 546 nm (Er) and 635 nm (Pr). *Double arrows* indicate non radiative relaxation processes

an energy transfer from the sensitizer material ytterbium to the laser-active praseodymium at a pump wavelength of 835 nm [31,32].

An optimum conversion efficiency is observed for ZBLAN optical waveguides with typical doping levels of 1000 ppm whereby the core diameter of the active waveguide is 3.5 µm and the numerical aperture NA ≅ 0.2. The cut-off wavelength for these geometries is near 0.9 µm. Nevertheless it is possible to produce diffraction-limited visible radiation in these waveguides as higher transversal modes suffer greater losses than transversal fundamental modes, and the fundamental modes can reduce the hole inversion in the active core of the fiber.

The dependence of the laser-output power on the pump wavelength for erbium-doped fibers is shown in Fig. 14. Over a tuning range of $\Delta_\lambda = 5$ nm a reduction in laser-output power is practically not observable, as it is possible to optimize the interaction length and therefore realize an almost-complete absorption of the pump radiation even beyond the absorption maximum. A laser spectrum of a praseodymium fiber laser is shown in Fig. 15. The spectral bandwidth of the system for a central wavelength of 635 nm is $\Delta_\lambda = 2$ nm. Optical-to-optical efficiencies up to 30% have been obtained for pumping powers of 300 mW. Recently, we achieved a power scaling of this visible-spectrum system by a first ever transfer of the double-clad concept to up-conversion fiber lasers [33].

The fiber-laser concept thus offers the possibility to obtain highly efficient and compact sources of coherent radiation in the visible and near-infrared spectral ranges. As a result of their structure fiber lasers show an immunity to thermo-optical effects while at the same time allow a simple miniaturization. This, combined with diode-laser pumping sources, provides an almost perfect

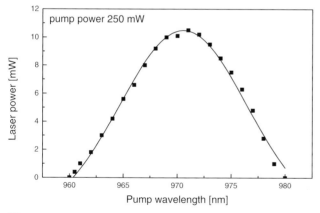

Fig. 14. Laser output power of an erbium-doped fiber laser in the visible spectral region as a function of the pump-laser wavelength. The upper laser level is excited by two-photon absorption of near-infrared diode-laser radiation

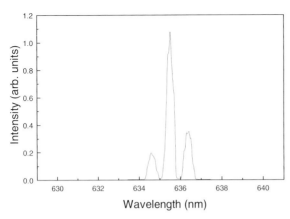

Fig. 15. Emission spectrum of a praseodymium/ytterbium fiber laser in the red spectral region excited by a diode laser with an emission wavelength of 840 nm. The integrated output power amounts to 50 mW

source of coherent radiation for a multitude of applications in science and technology.

3   Thin-Disk Laser

In this section the new thin-disk laser design for diode-pumped solid-state lasers will be introduced and discussed in detail. It will be demonstrated that, realizing this concept, high-power diode-pumped solid-state lasers become feasible which simultaneously yield high efficiency and good beam quality. In addition, this design also shows excellent results with quasi-three-level laser-active materials like Yb:YAG, which cannot be operated with good results in the classical scheme using thick rods or slabs.

As discussed in Sect. 1.1 thermal management of the waste heat generated in the laser-active medium is an essential part of each solid-state laser design. To remove the waste heat effectively out of the material, we use a very thin disk (thickness about 200 µm, diameter up to 12 mm) with one face mounted on a heat sink. So the heat is effectively conducted into the heat sink. This is the basic concept of the thin-disk laser. Additionally, to reduce the waste heat generated in the laser crystal it is necessary to use laser-active materials with low energy difference between pump and laser photons resulting in high efficiency and a low fraction of waste heat.

In the following the design considerations of the thin-disk laser are explained in detail and also the results of some numerical simulations are shown demonstrating the potential of this new concept. In the last part experimental results are discussed for two different pump designs. Special attention is paid to demonstrate the scalability laws to high output power with good beam quality and high efficiency.

3.1   Design Considerations

The basic principle of the thin-disk laser design is the use of a very thin laser crystal disk with one face mounted on a heat sink [35]. Figure 16 shows the fundamental scheme of this laser design. The thickness of the crystal disk is small compared to the diameter of the pumped area. The disk itself is anti-reflection coated for both pump and laser wavelengths at the front side and high-reflection coated for both wavelengths at the rear side which is mounted on the cooling device. The pump radiation is imaged onto the disk from the front side under oblique angles passing the disk at least twice. (The unabsorbed portion of the pump radiation is reflected at the rear side of the crystal.) The laser resonator is built by the disk itself and an outcoupling mirror in front of the thin disk as shown in Fig. 16. Alternatively, the disk may act as a folding mirror in a V-shaped resonator.

For keeping the disk thin and for achieving sufficient ($> 80\%$) absorption, the pump radiation passes the disk several times, thus exploiting the non-absorbed fraction. Pump designs with 8 and 16 passes of the pump radiation through the disk have been developed and will be discussed in detail in Sect. 3.4.

This thin-disk design reduces the volume-to-cooling-surface ratio of the crystal drastically, as discussed in Sect. 1.1 and also in [34]. Therefore, high pump-power densities can be applied without a high-temperature rise within the crystal. Furthermore, together with a flat-top pump-beam profile, this disk geometry leads to an almost homogeneous and one-dimensional heat flux perpendicular to the surface of the disk. Thermal distortions are therefore strongly reduced compared to conventional cooling schemes. To a first approximation no thermal lens of the disk occurs and also no thermally induced birefringence can be observed. As a result, the times-diffraction-limit-factor M^2 of such a thin-disk laser can be close to 1 even at high output power and also with high efficiency [36,37,38].

The output power of such a thin-disk laser can be scaled easily by increasing the diameter of the pumped area keeping the pump-power density con-

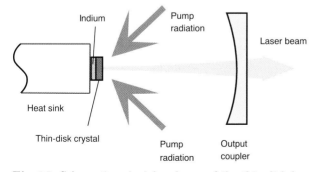

Fig. 16. Schematic principle scheme of the thin-disk laser design

stant. Because of the nearly one-dimensional heat flux no additional thermal distortions will occur, so that the beam-quality aspects remain unaffected. Another approach to scale the laser-output power into even the kilowatt range is the use of several disks in one resonator, where each disk is used as a folding mirror for the laser radiation inside the resonator [39].

Due to the low temperature rise inside the laser-active material, this thin-disk laser design is particularly suited for quasi-three-level laser systems. In this case a high pump-power density is necessary to reach laser threshold but without heating the crystal too much, since the threshold power itself is strongly dependent on the crystal temperature [27,40].

Especially the quasi-three-level system Yb:YAG is a very promising laser-active material since the quantum defect for Yb, which is the energy difference between pump and laser photons, is less than 10% [7] and because there are no losses due to excited-state absorption, up-conversion or cross-relaxation in Yb-doped materials. This material seems to be most appropriate for high-power laser systems if the temperature can be kept low at high pump-power densities that are necessary for exciting this quasi-three-level system. The thin-disk laser design may therefore play an important role for the employment of Yb-doped laser-active materials with superior properties, as will be shown later.

3.2   Numerical Simulation of the Thin-Disk Laser

In order to be able to predict and to explain experimental results more accurately as compared to simple analytical descriptions, a comprehensive numerical model has been developed for CW operation of the thin-disk laser system [41].

In this model CW operation of the thin-disk laser is numerically simulated in three distinct steps which are also based upon the experimental setups which we have realized and which are described in detail in Sect. 3.4.

1. In the first step the distribution of the absorbed pump power within the crystal is calculated using Monte-Carlo ray tracing taking into account variations of pump-power absorption with crystal temperature and population inversion.
2. In the second step the temperature distribution within the crystal disk is calculated. The sources of heat considered for this calculation are the conversion loss between pump and laser photon energy as well as the absorption losses of the pump and laser radiation within the coatings.
3. The population inversion and the laser-output power are calculated in the third step assuming an ideal plane-parallel and infinitesimally short resonator.

Steps 1 to 3 are iterated until a steady state is achieved. These steps are discussed in greater detail in the following sections. For a detailed presentation see also [41].

3.2.1 Absorption Profile of the Pump Radiation

Calculating the distribution of the absorbed pump power within the crystal through Monte-Carlo ray tracing means that the optical paths of randomly generated pump photons are followed from the source through the total optical system, regarding absorption as a statistical process.

The starting conditions for the pump radiation are chosen such that the wavelength distribution, the angle distribution and the power-density distribution match the experimentally determined parameters of the pump sources which are used for the experiments. With this procedure any pump source like the end of a single fiber, the end of a fiber bundle or a complete stacked laser-diode array imaged through an aperture (which can also be regarded as the source) can be used as a source for pumping the thin-disk laser. Every photon starts with a random (lateral) position at the source, with a random angle to the axis which is perpendicular to the source, and with a random wavelength.

The paths of the photons are then followed through the system along the various mirrors and through the crystal until they are absorbed, miss a mirror or the disk or leave the system again after the designated number of passes through the crystal. Photons absorbed in the crystal disk are counted in elements of a mesh, into which the crystal is divided. Such a mesh typically consists of 20 layers perpendicular to the crystal axis (z axis), 60 rings in the radial direction and 40 angular segments (Fig. 17).

In these calculations, when following the paths of the photons through the crystal, the variation of the absorption coefficient with temperature and the bleaching of absorption due to population-inversion are taken into account.

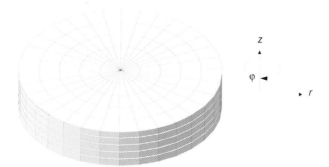

Fig. 17. Schematic drawing of the crystal mesh for the calculation of absorption

3.2.2 Temperature Profile within the Crystal

Since all pump designs for the thin-disk lasers discussed in this review result in pump profiles within the disk which are nearly symmetric around the disk axis (z axis), the temperature of the crystal is calculated for pump profiles of rotational symmetry. Also the output power is calculated for rotationally symmetric beams.

The generated heat, which results from the energy difference between pump photons and laser photons (8.6% for Yb:YAG) is taken from the calculated absorption distribution by angular averaging and is applied to elements of a two-dimensional mesh of a finite-difference program.

To calculate the temperature rise inside the crystal the heat conductivity and its dependence on temperature and doping concentration have also to be taken into account. Since all these interrelations are poorly described in the literature for the range of operation in a disk, the heat conductivity for Yb:YAG as a function of the doping concentration (5–25 at. % doping concentration) and the temperature (330 K to 670 K) has been measured [42]. Figure 18 shows the results for various doping concentrations of Yb:YAG as well as for undoped YAG. It is remarkable that the heat conductivity at room temperature for Yb-doped YAG is strongly reduced compared to undoped YAG. On the other hand, its dependence on the doping concentration and the temperature is relatively small whereas the temperature dependence for undoped YAG is significantly higher. For the calculations the following relation between heat conductivity λ_{th}, temperature T and atomic doping concentration c_{dop} is used

$$\lambda_{\mathrm{th}}(T, c_{\mathrm{dop}}) = (7.28 - 7.3c_{\mathrm{dop}}) \left(\frac{300\,\mathrm{K} - 96\,\mathrm{K}}{T - 96\,\mathrm{K}} \right)^{(0.48 - 0.46c_{\mathrm{dop}})} \frac{\mathrm{W}}{\mathrm{m\,K}} \,. \qquad (3)$$

For undoped YAG the relation

$$\lambda_{\mathrm{th}}(T) = 10.41 \left(\frac{300\,\mathrm{K} - 96\,\mathrm{K}}{T - 96\,\mathrm{K}} \right)^{0.63} \frac{\mathrm{W}}{\mathrm{m\,K}} \qquad (4)$$

is used. These relations are expressed by the fitted curves to the experimental data in Fig. 18.

The effect of the heat-sink geometry and the highly reflective coating on the thermal characteristics of the disk are determined by the values of the heat conductivity of the materials and the heat resistance of the coating, respectively. Especially the heat resistance of the coating leads to an additional temperature increase in the disk. For high pump-power densities, when the waste-heat flux through the coating is also high (100–$1000\,\mathrm{W/cm^2}$) this temperature increase may exceed 50–$100°\,\mathrm{C}$. Therefore special coatings have been developed with high reflectivity and relatively low heat resistance. These coatings consist of a metal layer and a stack of a small number of dielectric coatings between crystal and metal for increasing the reflectivity to about 99.9%. The crystal itself is mounted on a copper plate (thickness 0.5 to 1 mm) which is cooled from the back side using water or ethanol (for temperatures below $0°\,\mathrm{C}$). To achieve a low heat resistance between the copper plate and the cooling fluid an impingement cooling system is used where multiple

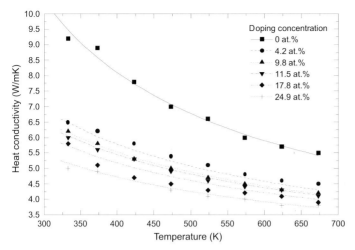

Fig. 18. Heat conductivity of Yb:YAG and undoped YAG in the temperature range from 330 K to 670 K

water jets impinge on the rear side of the copper plate. Indium is used for fixing the crystal to the cooling device and for reducing the stress between cooling plate and crystal. Finite-Element Models (FEMs) have been used for calculating the temperature distribution inside the complete composite disk system (copper–indium–coating–disk), taking into account the heat sources inside the disk as described above and the absorption of pump and laser radiation inside the coating which also contributes to the total temperature rise.

3.2.3   Output Power and Population Inversion

Knowing the temperature distribution in the disk and the absorbed pump-power density in each mesh element of the crystal disk, it is possible to calculate the laser power which can be extracted from an infinitesimally short resonator. In the following the equations for calculating the output power are given for a quasi-three-level system like Yb:YAG.

The behavior of the laser can be calculated using the rate equation

$$\frac{dN_1}{dt} = W_P - \frac{N_1}{\tau} - \sigma_e(\omega, T)\frac{I_{eff}}{\hbar\omega_{las}}$$
$$\times \left[N_1 - (N - N_1)\exp\left(\frac{\hbar(\omega_{las} - \mu)}{kT}\right)\right] \tag{5}$$

with the number of excited ions per time and volume W_P, the fluorescence lifetime of the upper-state manifold τ, the energy of a laser photon $\hbar\omega_{las}$, the density of excited ions N_1, the total density of ions N, the emission cross section $\sigma_e(\omega, T)$ and the temperature T (for the excitation potential

cross section $\sigma_e(\omega, T)$ and the temperature T (for the excitation potential $\hbar\mu$ see (11)). I_{eff} is the effective resonator power density, which results from multiple passes through the crystal and varies with the axial position z inside the crystal ($z = 0$ is located at the back side of the crystal):

$$I_{eff}(z) = I_{res} \sum_{i=1}^{m/2} \left\{ \exp\left[gl\left(2i - 1 - \frac{z}{l}\right)\right] + \exp\left[gl\left(2i - 1 + \frac{z}{l}\right)\right] \right\} . \quad (6)$$

Here m is the number of passes through the crystal, g is the gain coefficient, l is the crystal thickness and I_{res} the resonator-power density after reflection at the outcoupling mirror.

The second equation used for calculating the resonator behavior is the condition that gain and loss are balanced in steady-state operation:

$$m \int_0^l \sigma_e(z) \left\{ N_1(z) - [N - N_1(z)] \exp\left(\frac{\hbar(\omega - \mu)}{kT} \right) \right\} z\,dz$$
$$+ \ln R_{OC} + \ln(1 - L) = 0 \quad (7)$$

with R_{OC} denoting the reflectivity of the output coupler and L the resonator internal losses.

Using (5) and the population of the upper-state manifold $N_1(z)$ and the resonator-power density I_{res} are adjusted iteratively for every radial position of the mesh. The total output power is then given by the radial integral

$$P_{out} = \frac{1 - R_{OC}}{R_{OC}} 2\pi \int_r I_{res}(r) r\,dr . \quad (8)$$

For solving this equation system it is necessary to determine the gain g. For a quasi-three-level system the expression for the gain coefficient $g(\omega, T)$ that includes stimulated emission and absorption is used

$$g(\omega, T) = N_1 \sigma_e(\omega, T) - N_0 \sigma_a(\omega, T) . \quad (9)$$

According to [50] the absorption and stimulated emission cross sections $\sigma_a(\omega, T)$ and $\sigma_e(\omega, T)$ are connected at a given frequency ω and a given temperature T by a detailed balance relation

$$\sigma_a(\omega, T) = \sigma_e(\omega, T) \exp[\hbar(\omega - \mu)/kT] . \quad (10)$$

Here $\hbar\mu$ is a temperature-dependent excitation potential, given by the relation

$$\exp(-\hbar\mu/kT) = \frac{N_1^0}{N_0^0} , \quad (11)$$

where N_1^0 and N_0^0 are the population densities of the upper and lower manifolds at thermal equilibrium of the unpumped material.

With this relation, (9) can be rewritten as

$$g(\omega, T) = \sigma_e(\omega, T) \left\{ N_1 - N_0 \exp\left[\hbar(\omega - \mu)/kT\right] \right\} . \quad (12)$$

Since all the parameters (absorption, temperature, inversion, and output power) are interdependent and influence each other it is necessary to solve the complete system iteratively. So all the calculations have to be repeated until a steady-state solution has been found.

3.3 Results and Discussion

In this section the results of various experiments are presented and discussed. Also results from the numerical calculations are demonstrated and compared with the measurements. For these measurements two different pump designs have been used which are therefore explained in detail, followed by a section reporting the most important results from the calculations before the measurements themselves will be depicted and discussed.

All results reported in the following sections but the last have been achieved using Yb:YAG as the laser-active medium. Due to its low quantum defect, the low heat generation in this material and the broad absorption spectrum, Yb:YAG is a nearly ideal material for diode-pumped solid-state lasers [7,43,44]. The quasi-three-level nature of this material and the resulting strong thermal problems with the classical laser designs can be overcome by using this material in the thin-disk geometry.

3.3.1 Pump Design for Eight and Sixteen Passes of Pump Radiation

As explained in Sect. 3.2 it is important to keep the disk thin if all the advantages of this design should be exploited. On the other hand, for achieving a high optical efficiency, nearly all the pump power must be absorbed within the disk. Therefore, crystal thickness, doping concentration and number of absorbing passes for the pump radiation have to be carefully balanced.

In the first experiments eight passes of the pump radiation through the disk could be realized. Figure 19 shows a three-dimensional view of the implemented setup. In these experiments the pump radiation was delivered to the laser using a fiber bundle with a numerical aperture of 0.37. Four spherical mirrors and one flat mirror beside the disk are used to image the radiation from the fiber bundle onto the disk four times.

First, the pump radiation is imaged directly from the end of the fiber bundle, which is located next to the disk, onto the disk by one of the spherical mirrors yielding two passes of the pump radiation through the disk. The non-absorbed portion of the pump radiation is then re-imaged onto the surface of the flat mirror beside the disk. This mirror is tilted so that the reflected radiation is imaged again onto the crystal using the third spherical mirror, resulting in passes numbers three and four. The so-far non-absorbed pump radiation is then re-imaged by the last spherical mirror onto the crystal again so that the pump radiation follows the total path back to the fiber bundle resulting in the passes numbers five to eight through the disk. In a 400 μm

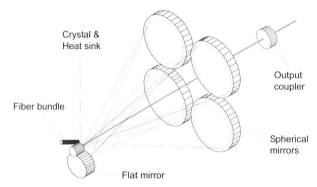

Fig. 19. Pump design for eight passes of the pump radiation

thick Yb:YAG disk doped with 10 at. % more than 80% of the pump radiation is absorbed.

The results obtained with this design, which will be discussed later, show a significant temperature dependence of the optical efficiency, especially for the quasi-three-level system Yb:YAG. To reduce this effect and especially to increase the optical efficiency, a new pump scheme with more passes of the pump radiation has been developed [45].

Additionally, the complexity of the pump design with several imaging mirrors has been reduced using only one parabolic mirror for imaging all the pump beams onto the thin disk. This new pump scheme has been realized for sixteen passes of the pump radiation as shown in Fig. 20. The pump radiation from the source (that may be the end of a single fiber, the end of a fiber bundle or the radiation from a high-power stacked diode-laser array imaged through an aperture) is first collimated by a lens to a beam parallel to the optical axis of the parabolic mirror which focuses the off-axis incident pump radiation to its focal plane where the thin-disk crystal is located. The size of the image is given by the ratio of the focal lengths of the collimating lens and the parabolic mirror times the size of the source.

The non-absorbed pump radiation hits the parabolic mirror a second time after two passes through the crystal where it is collimated again. After redirecting the pump radiation onto another site on the parabolic mirror using a single roof prism the pump radiation is once more focused onto the crystal

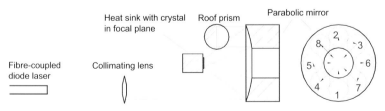

Fig. 20. Pump design for sixteen passes of the pump radiation

(this gives absorption passes numbers three and four) and the transmitted part is collimated again. Two other prisms are used to repeat this procedure until the parabolic mirror is used completely. Eight absorption passes are realized in this way. In Fig. 20 the sequence of the segments on the parabolic mirror used as the pump beam traverses through the setup is indicated by the numbers 1 to 8. In order to get another eight absorption passes the pump radiation which is not absorbed so far is reflected back by a plane mirror so that the pumping optics is passed in the opposite direction.

If it is possible to use even more segments on the parabolic mirror the number of the absorption passes can be further increased by insertion of additional roof prisms. The higher the brightness of the pump source and the lower the power-density requirements for pumping the disk the more passes of the pump radiation can be realized.

3.3.2   Results from the Numerical Calculations

The numerical simulations are performed according to the model described in Sect. 3.3 and using the real geometries of the two designs described above. Also the real cooling-finger design has been used to calculate the temperature distribution inside the disk. Some additional parameters used in the model are given in Table 3.

Table 3. Parameters used for modeling the thin-disk laser

Parameter	Value
P_{pump}	500 W
$\emptyset_{\mathrm{pump}}$	3.23 mm
$\mathrm{NA}_{\mathrm{pump}}$	0.093
λ_{pump}	940.3 nm
$\Delta\lambda_{\mathrm{pump}}$ (FWHM)	4 nm
f_{parabol}	60 mm
$\emptyset_{\mathrm{parabol}}$	150 mm
R_{parabol}	99.2%
T_{prism}	99.6%
c_{dop}	10 at. %
N	$13.8 \times 10^{20}\,\mathrm{cm}^{-3}$
τ	951 μs
$\sigma_{\mathrm{e}}(T)$	$[0.942 + 35.78\exp(-0.01152\,T/\,\mathrm{K})] \times 10^{-20}\,\mathrm{cm}^2$
L	0.1%
m	2
T_{c}	15° C
R_{th}	0.121 K cm^2/ W
$\lambda_{\mathrm{pump}}/\lambda_{\mathrm{las}}$	0.913
f_{heat}	0.06
$E_{\mathrm{i}}(\,\mathrm{cm}^{-1})$	0, 565, 612, 785, 10327, 10624, 10930

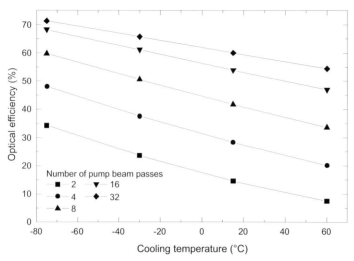

Fig. 21. Calculated optical efficiency as function of the temperature of the cooling fluid for different numbers of passes of the pump radiation through the disk

Figure 21 shows the calculated optical efficiency (laser power to incident pump power) as a function of the temperature of the cooling fluid for different numbers of passes of the pump radiation through the crystal. For each data point the thickness of the crystal has been optimized in the simulations. The doping concentration for these calculations was 10 at. % throughout.

As can be seen from Fig. 21 the efficiency for Yb:YAG strongly depends on the number of pump passes, especially for high temperatures. This is due to three effects which are related to the reduced crystal thickness with higher number of pump passes: first, the optimum total absorbed power in the disk increases for maximum efficiency with an increasing number of passes; second, the maximum crystal temperature decreases because of the reduced crystal thickness and third, since the number of Yb^{3+} ions which must be pumped to transparency also scales linearly with the crystal thickness, the threshold pump-power density of a quasi-three-level system is lower for thinner crystals. It is deduced from these results that increasing the number of pump passes is a very effective way to enhance the optical efficiency.

The temperature distribution inside the disk is shown in Fig. 22 for sixteen absorption passes. These calculations were performed for the optimal crystal thickness and for 500 W of pump power at an incident power density of 5 kW/cm^2. The temperature of the cooling fluid was taken to be 15° C. Figure 23 shows the temperature on the optical axis as a function of the z position. All the parameters for the simulation are listed in Table 3.

Figure 24 shows the optical efficiency as a function of the pump power for eight and sixteen passes of the pump radiation for the pump designs used, operated at a cooling-fluid temperature of 15° C for a pump power

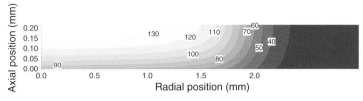

Fig. 22. Temperature distribution inside the disk, mounted on a cooling plate, for the case of sixteen absorption passes

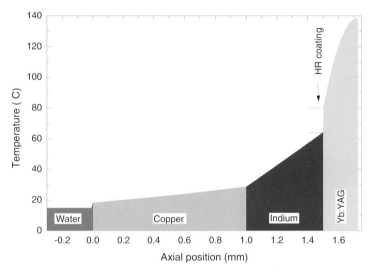

Fig. 23. Temperature of the cooling system and of the disk along the laser axis

of 60 W. Additionally, measured values are also shown for sixteen passes, indicating a very good agreement between simulations and measurements. An absolute increase in optical efficiency of more than 10% at room temperature is predicted for the new pump design (sixteen passes) compared to the old one (eight passes).

3.3.3   Experimental Results

Comparable results for both pump designs are presented in Fig. 25 for eight passes and in Fig. 26 for sixteen passes of pump radiation. The results are quite similar for multimode and for TEM_{00}-mode operation. For the multimode operation a resonator length of 6.5 cm and an outcoupler with a curvature $r = 0.5$ m have been used. This results in an output power of more than 35 W with optical efficiencies of 55.2% (eight passes) and 57.7% (sixteen passes). Also the times-diffraction-limit-factor M^2 is roughly the same (6.5–7.1).

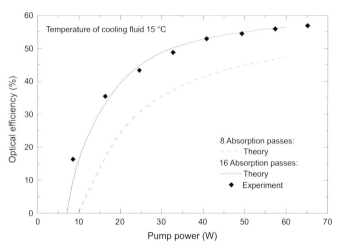

Fig. 24. Optical efficiency as function of the incident pump power for eight and for sixteen passes

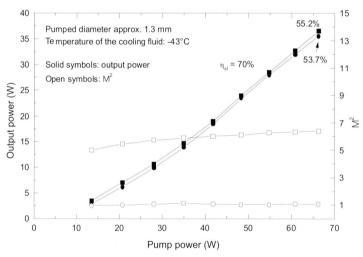

Fig. 25. Laser output power and times-diffraction-limit-factor for the pump design with eight absorption passes

Increasing the length of the resonator between 30 and 50 cm results in a TEM_{00}-mode operation with $M^2 = 1.06$ for eight passes and $M^2 = 1.01$ for sixteen passes with an efficiency of more than 50% for both pump designs. The only difference is the temperature of the cooling fluid which is $-43°$ C for eight passes and $+15°$ C for sixteen passes of the pump radiation.

Besides these results, further experiments were carried out to measure the maximum output power and the times-diffraction-limit-factor at various resonator configurations. Figure 27 demonstrates that the optical efficiency

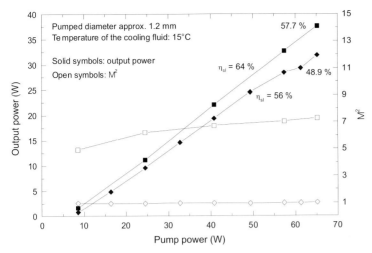

Fig. 26. Laser output power and times-diffraction-limit-factor for the pump design with sixteen absorption passes

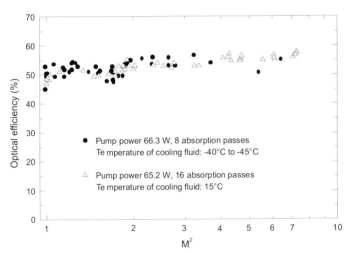

Fig. 27. Optical efficiency versus times-diffraction-limit-factor for eight and sixteen absorption passes of the pump radiation

is the same for both pump designs and nearly independent of the measured beam quality.

Comparing the results for eight and sixteen absorption passes it is rather obvious that, for achieving the same optical efficiency, the temperature of the cooling fluid can be $40°$ C to $60°$ C higher for sixteen absorption passes than for eight absorption passes. Figure 28 compares several measurements with eight absorption passes to one measurement with sixteen absorption passes, demonstrating the advantage of the higher number of pump passes at

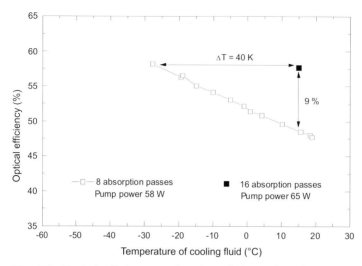

Fig. 28. Optical efficiency as function of the cooling-fluid temperature for eight and sixteen passes of the pump radiation

the cooling-fluid temperature of 15° C. Also the excellent agreement to the simulated efficiencies in Fig. 21 is remarkable.

3.3.4 Power Scalability

As already mentioned in Sect. 3.1 the power of the thin-disk laser can be increased by increasing the pump spot diameter at constant pump-power density. It is mandatory to keep the pump-power density constant especially for quasi-three-level systems like Yb:YAG since the threshold-power density is high.

Figure 29 shows the results for a disk pumped with more than 1 kW of optical pump power using sixteen passes for absorbing the pump radiation. The maximum output power was nearly 500 W with an optical efficiency of 48%. For this laser the pump radiation from two stacked diode-laser arrays was focused into a rod (made from quartz) to homogenize the pump-beam profile. The end of the rod was then imaged onto the disk using the design shown in Fig. 20. The diameter of the pumped area was measured to be 6 mm, and the maximum incident power density was 4 kW/cm^2.

Figure 30 shows the efficiency as a function of the cooling-fluid temperature for different high-power thin-disk lasers pumped with eight and sixteen pump passes, respectively. Comparing this efficiency for a high pump power (Fig. 30) with that for a lower pump power (Fig. 28) clearly shows the influence of some three-dimensional (waste) heat flux inside the disk for smaller pump spot diameters, reducing the temperature of the disk. Nevertheless,

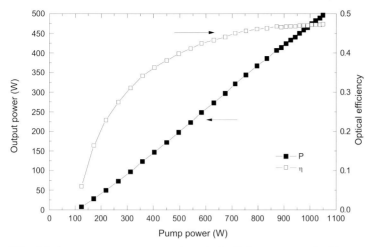

Fig. 29. Laser output power and optical efficiency versus incident pump power

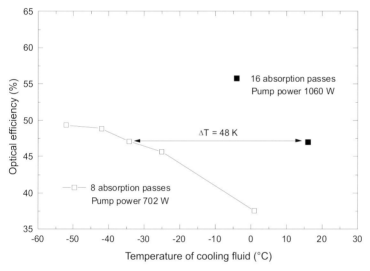

Fig. 30. Optical efficiency as function of the cooling-fluid temperature for eight and sixteen absorption passes of the pump radiation

50% optical efficiency is achievable for high pump and laser power setups to be operated at a cooling-fluid temperature of 15° C.

Another way for increasing the output power is the use of several disks in one resonator. In this case each of the disks acts as a folding mirror in the cavity resulting in a zig-zag configuration of the laser resonator. Figure 31 shows the design of the resonator used for these scaling experiments. In this case the pump radiation is split for pumping both crystals.

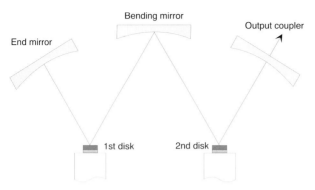

Fig. 31. Laser resonator design with two disks in series

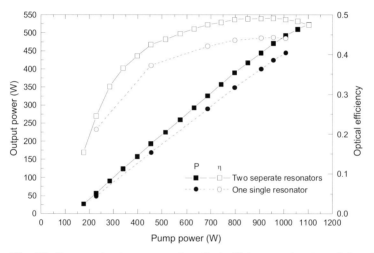

Fig. 32. Laser output power and optical efficiency using two disks with two independent resonators (*squares*) and combined within one resonator (*triangles*)

Figure 32 demonstrates the power scalability by using two disks according to the setup shown in Fig. 31. The efficiency is nearly the same compared to the results with one disk (the small reduction is mainly due to the additional losses from the additional mirrors inside the resonator). Applying this concept, the power of the thin-disk laser can be scaled into the kilowatt range.

3.3.5   Other Materials than Yb:YAG

The thin-disk laser design can be used with the same advantages as for Yb:YAG for a wide range of other laser materials. So far, this design has been operated with the following materials

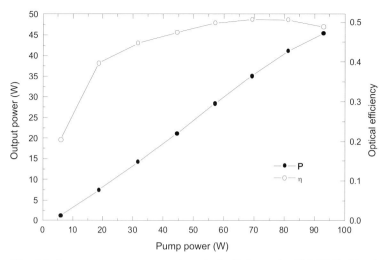

Fig. 33. Laser output power and optical efficiency for Nd:YAG thin-disk laser

- Nd:YAG (1.064 µm, 1.32 µm, and 946 nm) [46]
- Nd:YVO$_4$ [47,48]
- Tm:YAG [49].

Figure 33 demonstrates laser results using Nd:YAG as laser-active material. 45 W CW output power at 50% optical efficiency has been demonstrated. It is expected that many more laser-active materials can be operated with the thin-disk laser design also leading to a strong reduction of thermal effects. Therefore, high output power, high efficiency and good focusability are no longer contradictory for solid-state lasers.

3.4   Conclusion

The thin-disk laser design is a novel concept that allows us to build diode-pumped solid-state lasers with high output power (up to the kilowatt range), with high efficiency ($> 50\%$ optical–optical) and excellent beam quality ($M^2 \approx 1$ up to several hundred watts, $M^2 \ll 10$ for lasers in the kilowatt range), simultaneously. Additionally, with this design quasi-three-level laser materials can be used that yield the highest efficiency values for high-power lasers. Among these, Yb-doped materials will play an important role in industrial high-power laser systems of the future.

List of Symbols

a	Radius of fiber core
a_1, a_2	Constants in the radial field distribution of fiber modes
c_{dop}	Atomic doping concentration
E_{i}	Energy levels of laser ion
E_{lm}	Electrical-field amplitude of the fiber mode with mode numbers l, m
f_{heat}	Factor of heat generation in HR layer
f_{parabol}	Focal length of the parabolic mirror
g	Gain coefficient
$\hbar$	Planck constant divided by 2π
I_{res}	Resonator internal intensity
I_{eff}	Effective resonator internal intensity
J_l, K_l	Bessel functions of order l
k	Boltzmann constant
l	Crystal thickness
L	Resonator internal loss
LP_{lm}	Laguerre Polynome with mode numbers l, m
m	Number of laser-beam passes through the crystal per resonator roundtrip
M^2	Times-diffraction-limit factor
N_0^0, N_1^0	Density of laser ions in lower and upper multiplet at thermal equilibrium
P_{out}	Output power
t	Time
$\Delta \lambda_{\mathrm{pump}}$	Spectral width of the pump source
l, m	Mode numbers of fiber modes
N	Density of laser ions
N_0, N_1	Density of laser ions in lower and upper multiplett
n_1	Refractive index of fiber core
n_2	Refractive index of fiber cladding
n_3	Refractive index of fiber coating
NA	Numerical aperture of a fiber
$\mathrm{NA}_{\mathrm{pump}}$	Numerical aperture of the pump source
P_{pump}	Pump power
R_{OC}	Reflectivity of output coupler
R_{parabol}	Reflectivity of the parabolic mirror
R_{th}	Thermal resistance
r, φ	Cylindrical coordinates
T	Temperature
T_{c}	Cooling temperature
T_{prism}	Transmission of a prism
V	Fiber parameter (dimensionless)

W_P	Number of excited ions per time and volume
γ	Spreading constant of a fiber
Z	Number of modes propagating in a fiber
λ	Laser wavelength
$\Delta\lambda$	Spectral width
λ_{th}	Thermal conductivity
λ_{las}	Laser wavelength
λ_{pump}	Peak pump wavelength
μ	Chemical potential
σ_a, σ_e	Cross section of absorption/emission
τ	Fluorescence lifetime
ω	Angular frequency
ω_{las}	Angular frequency of the laser radiation
δ	Phase
$\emptyset_{parabol}$	Diameter of the parabolic mirror
$\emptyset_{pump}$	Diameter of the pump source

References

1. W. Koechner: *Solid-State Laser Engineering*, 5th edn., Springer Ser. Opt. Sci. **1** (Springer, Berlin, Heidelberg 1999)
2. T. W. Fan, A. Sanchez: Pump Source Requirements for End-Pumped Lasers, IEEE J. QE **26**, 311 (1990)
3. F. Salin, J. Squier: Geometrical Optimizaton of Longitudinally Pumped Solid-State Lasers, Opt. Commun. **86**, 397 (1991)
4. P. Laporta, M. Brussard: Design Criteria for Mode Size Optimization in Diode-Pumped Solid-State Lasers, IEEE J. QE **27**, 2319 (1991)
5. Y. F. Chen, C. F. Kao, T. M. Huang, C. L. Wang, S. C. Wang: Influence of Thermal Effect on Output Power Optimization in Fiber-Coupled Laser-Diode End-Pumped Lasers, IEEE J. Sel. Top. Quant. Electron. **3**, 29 (1997)
6. U. Brauch, M. Schubert: Comparison of Lamp and Diode Pumped CW Nd:YAG Slab Lasers, Opt. Commun. **117**, 116 (1995)
7. T. Y. Fan: Heat Generation in Nd:YAG and Yb:YAG, IEEE J. QE **29**, 1457 (1993)
8. B. Comaskey, B. D. Moran, G. F. Albrecht, R. J. Beach: Characterization of the Heat Loading of Nd-Doped YAG, YOS, YLF and GGG Excited at Diode Pumpimg Wavelengths, IEEE J. QE **31**, 1261 (1995)
9. S. C. Tidwell, J. F. Seamans, M. S. Bowers, A. K. Cousins: Scaling CW Diode-End-Pumped Nd:YAG Lasers to High Average Powers, IEEE J. QE **28** 997 (1992)
10. M. Tsunekane, N. Taguchi, H. Inaba: Improvement of Thermal Effects in a Diode-End-Pumped, Composite Tm:YAG Rod with Undoped Ends, Appl. Opt. **37**, 3290 (1998)
11. R. Weber, B. Neuenschwander, M. MacDonald, M. B. Roos, H. P. Weber: Cooling Schemes for Longitudinally Diode Laser-Pumped Nd:YAG Rods, IEEE J. QE **34**, 1046 (1998)

12. D. Golla, M. Bode, S. Knoke, W. Schöne, F. von Alvensleben, A. Tünnermann: High Power Operation of Nd:YAG Rod Lasers Pumped by Fiber-Coupled Diode Lasers, *Advanced Solid-State Lasers 1996,* OSA Trends Opt. Photon. Ser. **1**, 198 (1996)

13. A. Takada, Y. Akiyama, T. Takase, H. Yuasa, A. Ono: Diode Laser-Pumped CW Nd:YAG Lasers with More than 1 kW Output Power, *Advanced Solid-State Lasers 1999,* OSA Tech. Dig. Ser. (1999) paper MB 18, p. 69

14. N. Hodgson, H. Weber: Influence of Spherical Aberration of the Active Medium on the Performance of Nd:YAG Lasers, IEEE J. QE **29**, 2497 (1993)

15. V. Magni: Resonators for Solid-State Lasers with Large-Volume Fundamental Mode and High Alignment Stability, Appl. Opt. **25**, 107 (1986)

16. W. Schöne, S. Knoke, S. Schirmer, A. Tünnermann: Diode-Pumped CW Nd:YAG Lasers with Output Powers Up to 750 W, *Advanced Solid-State Lasers 1997,* OSA Trends Opt. Photon. Ser. **10**, 292 (1997)

17. P. W. France: *Optical Fiber Lasers and Amplifiers* (Blackie, Glasgow 1991)

18. D. Richardson, J. Minelly, D. Hanna: Fiber Laser Systems Shine Brightly, Laser Focus World **33**, 87, Sept. (1997)

19. L. Zenteno: High Power Double-Clad Fiber Lasers, J. Lightwave Technol. **11**, 1435 (1993)

20. D. S. Funk, J. G. Eden: Glass-Fiber Lasers in the Ultraviolet and Visible, IEEE J. Sel. Top. Quant. Electron. **1**, 784 (1995)

21. M. C. Brierley, J. F. Massicott, T. J. Whitley, C. A. Millar, R. Wyatt, S. T. Davey, D. Szebesta: Visible Fibre Lasers, BT Technol. J. **11**, 128 (1993)

22. P. Urquhart: Rewiew of Rare Earth Doped Fibre Lasers and Amplifiers, IEE Proc. J. **135**, 385 (1988)

23. A. Shikida, H. Yanagita, H. Toratani: Al-Zr Fluoride Glass for Ho3+-Yb3+ Green Upconversion, J. Opt. Soc. Am. B **11**, 928 (1994)

24. S. Geckeler: *Lichtwellenleiter für die optische Nachrichtenübertragung,* 3rd edn. (Springer, Berlin, Heidelberg 1990)

25. E. Snitzer, H. Po, F. Hakimi, R. Tumminelli, B. C. McCollum: Double Clad Offset Core Nd Fibre Laser, Optical Fiber Sensors Conf. (1988) postdeadline paper PD5

26. H. Po, J. D. Cao, B. M. Laliberte, R. A. Minns, R. F. Robinson, B. H. Rockney, R. R. Tricca, Y. H. Zhang: High Power Neodymium-Doped Single Transverse Mode Fiber Laser, Electron. Lett. **29**, 1500 (1993)

27. H. Zellmer, U. Willamowski, A. Tünnermann, H. Welling, S. Unger, V. Reichel, H.-R. Müller, J. Kirchhof, P. Albers: High Power CW Nd-Doped Fibre Laser Operating at 9.2 W with High Beam Quality, Opt. Lett. **20**, 578 (1995)

28. J. Y. Allain, M. Monerie, H. Poignant: Blue Upconversion Fluorozirconate Fibre Laser, Electron. Lett. **26**, 166 (1990)

29. I. J. Booth, C. J. Mackechine, B. F. Ventrudo: Operation of Diode Laser Pumped Tm3+ ZBLAN Upconversion Fiber Laser at 482 nm, IEEE J. QE **32**, 118 (1996)

30. R. Paschotta, P. R. Barber, A. C. Tropper, D. C. Hanna: Characterization and Modeling of Thulium:ZBLAN Blue Upconversion Fiber Lasers, J. Opt. Soc. Am. B **14**, 1213 (1997)

31. P. Xie, T. R. Gosnell: Room-Temperature Upconversion Fiber Laser Tunable in The Red, Orange, Green and Blue Spectral Regions, Opt. Lett. **20**, 1014 (1995)

32. T. R. Gosnell: Avalanche Assisted Upconversion in Pr3+/Yb-Doped ZBLAN Glass, Electron. Lett. **33**, 411 (1997)
33. H. Zellmer, K. Plamann, G. Huber, H. Scheife, A. Tünnermann: Visible Double-Clad Upconversion Fibre Laser, Electron. Lett. **34**, 565 (1998)
34. K. Ueda, N. Uehara: Laser Diode Pumped Solid State Lasers for Gravitational Wave Antenna, SPIE Proc. **1837**, 336 (1993)
35. A. Giesen, H. Hügel, A. Voss, K. Wittig, U. Brauch, H. Opower: Scalable Concept for Diode-Pumped High-Power Solid-State Lasers, Appl. Phys. B **58**, 365 (1994)
36. A. Giesen, U. Brauch, I. Johannsen, M. Karszewski, C. Stewen, A. Voss: High-Power Near Diffration-Limited and Single-Frequency Operation of Yb:YAG Thin Disc Laser, *Advanced Solid-State Lasers 1996,* OSA Trends Opt. Photon. Ser. **1**, 11 (1996)
37. A. Giesen, U. Brauch, I. Johannsen, M. Karszewski, U. Schiegg, C. Stewen, A. Voss: Advanced Tunability and High-Power TEM$_{00}$-Operation of the Yb:YAG Thin Disc Laser, *Advanced Solid-State Lasers 1997,* OSA Trends Opt. Photon. Ser. **10**, 280 (1997)
38. M. Karszewski, U. Brauch, K. Contag, S. Erhard, A. Giesen, I. Johannsen, C. Stewen, A. Voss: 100 W TEM$_{00}$ Operation of Yb:YAG Thin Disc Laser with High Efficiency, *Advanced Solid-State Lasers 1998,* OSA Trends Opt. Photon. Ser. **19**, 296 (1998)
39. K. Contag, U. Brauch, A. Giesen, I. Johannsen, M. Karszewski, U. Schiegg, C. Stewen, A. Voss: Multi-Hundred Watt Diode Pumped Yb:YAG Thin Disc Laser, *Solid State Lasers VI*, SPIE Proc. **2986**, 2 (1997)
40. A. Giesen, M. Karszewski, C. Stewen, A. Voss, L. Berger, U. Brauch: Recent Results of the Scalable Diode-Pumped Yb:YAG Thin Disk Laser, *Advanced Solid-State Lasers,* OSA Proc. **24**, 330 (1995)
41. K. Contag, U. Brauch, S. Erhard, A. Giesen, I. Johannsen, M. Karszewski, C. Stewen, A. Voss: Simulations of the Lasing Properties of a Thin Disk Laser Combining High Output Powers with Good Beam Quality, SPIE Proc. **2989**, 23 (1997)
42. K. Lisak, W. Hohenauer: Thermophysikalische Charakterisierung von YAG Kristallen, Ergebnisbericht zu Auftrag 86, 87-98 (Austrian Research Centers, Seibersdorf 1998)
43. P. Lacovara, H. K. Choi, C. A. Wang, R. L. Aggarwal, T. Y. Fan: Room-Temperature Diode-Pumped Yb:YAG Laser, Opt. Lett. **16**, 1089 (1991)
44. H. W. Bruesselbach, D. S. Sumida, R. A. Reeder, R. W. Byren: Low-Heat High-Power Scaling Using InGaAs-Diode-Pumped Yb:YAG Lasers, IEEE J. Sel. Topics Quant. Electron. **3**, 105 (1997)
45. S. Erhard, A. Giesen, M. Karszewski, T. Rupp, C. Stewen, I. Johannsen, K. Contag: Novel Pump Design of Yb:YAG Thin Disc Laser for Operation at Room Temperature with Improved Efficiency, *Advanced Solid-State Lasers 1999,* OSA Trends Opt. Photon. Ser. **26**, 38 (1999)
46. A. Giesen, G. Hollemann, I. Johannsen: Diode-Pumped Nd:YAG Thin Disc Laser, *CLEO 99*, OSA Tech. Dig. Ser. (1999) p. 29
47. R. Koch, G. Hollemann, R. Clemens, H. Voelckel, A. Giesen, A. Voss, M. Karszewski, C. Stewen: Effective Near Diffration Limited Diode Pumped Thin Disk Nd:YVO$_4$ Laser, OSA Tech. Dig. Ser. **11**, 480 (1997)
48. G. Hollemann, H. Zimer, A. Hirt: Pulsed-Mode Operation of a Diode-Pumped Nd:YVO$_4$ Thin Disc Laser, OSA Tech. Dig. Ser. **16**, 543 (1998)

49. A. Diening, B. M. Dicks, E. Heumann, G. Huber, A. Voss, M. Karszewski, A. Giesen: High Power Tm:YAG Thin.Disc Laser, OSA Tech. Dig. Ser. **16** (1998) paper CWF46
50. D. E. McCumber: Theory of Phonon-Terminated Optical Masers, Phys. Rev. A **134**, 299 (1964)

Index

Topics in Applied Physics